Tenth International Workshop on Health Text Mining and Information Analysis (LOUHI 2019)

Hong Kong, China
3 November 2019

ISBN: 978-1-7138-0083-5

EMNLP-IJCNLP 2019

**Tenth International Workshop
on Health Text Mining
and Information Analysis
LOUHI 2019**

Proceedings of the Workshop

November 3, 2019
Hong Kong

Introduction (TBD)

The International Workshop on Health Text Mining and Information Analysis (LOUHI) provides an interdisciplinary forum for researchers interested in automated processing of health documents. Health documents encompass electronic health records, clinical guidelines, spontaneous reports for pharmacovigilance, biomedical literature, health forums/blogs or any other type of health-related documents. The LOUHI workshop series fosters interactions between the Computational Linguistics, Medical Informatics and Artificial Intelligence communities. The eight previous editions of the workshop were co-located with SMBM 2008 in Turku, Finland, with NAACL 2010 in Los Angeles, California, with Artificial Intelligence in Medicine (AIME 2011) in Bled, Slovenia, during NICTA Techfest 2013 in Sydney, Australia, co-located with EACL 2014 in Gothenburg, Sweden, with EMNLP 2015 in Lisbon, Portugal, with EMNLP 2016 in Austin, Texas; in 2017 was held in Sydney, Australia; and in 2018 was co-located with EMNLP 2018 in Brussels, Belgium. This year the workshop is co-located with EMNLP 2019 in Hong Kong.

The aim of the LOUHI 2019 workshop is to bring together research work on topics related to health documents, particularly emphasizing multidisciplinary aspects of health documentation and the interplay between nursing and medical sciences, information systems, computational linguistics and computer science. The topics include, but are not limited to, the following Natural Language Processing techniques and related areas:

- Techniques supporting information extraction, e.g. named entity recognition, negation and uncertainty detection

- Classification and text mining applications (e.g. diagnostic classifications such as ICD-10 and nursing intensity scores) and problems (e.g. handling of unbalanced data sets)

- Text representation, including dealing with data sparsity and dimensionality issues

- Domain adaptation, e.g. adaptation of standard NLP tools (incl. tokenizers, PoS-taggers, etc) to the medical domain

- Information fusion, i.e. integrating data from various sources, e.g. structured and narrative documentation

- Unsupervised methods, including distributional semantics

- Evaluation, gold/reference standard construction and annotation

- Syntactic, semantic and pragmatic analysis of health documents

- Anonymization/de-identification of health records and ethics

- Supporting the development of medical terminologies and ontologies

- Individualization of content, consumer health vocabularies, summarization and simplification of text

- NLP for supporting documentation and decision making practices

- Predictive modeling of adverse events, e.g. adverse drug events and hospital acquired infections

- Terminology and information model standards (SNOMED CT, FHIR) for health text mining

- Bridging gaps between formal ontology and biomedical NLP

The call for papers encouraged authors to submit papers describing substantial and completed work but also focus on a contribution, a negative result, a software package or work in progress. We also encouraged to report work on low-resourced languages, addressing the challenges of data sparsity and language characteristic diversity.

This year we received a high number of submissions (50), therefore the selection process was very competitive. Due to time and space limitations, we could only choose a small number of the submitted papers to appear in the program.

Each submission went through a double-blind review process which involved three program committee members. Based on comments and rankings supplied by the reviewers, we accepted 23 papers. Although the selection was entirely based on the scores provided by the reviewers, we regretfully had to set a relatively high threshold for acceptance. The overall acceptance rate is 46%. After the decision about acceptance, 2 papers were withdrawn by the authors. During the workshop, 11 papers will be presented orally, and 10 papers will be presented as posters.

Finally, we would like to thank the members of the program committee for providing balanced reviews in a very short period of time, and the authors for their submissions and the quality of their work.

Sanna Salanterä, University of Turku, Finland
Stefan Schulz, Graz General Hospital and University Clinics, Austria
Maria Skeppstedt, Institute for Language and Folklore, Sweden
Amber Stubbs, Simmons College, USA
Hanna Suominen, CSIRO, Australia
Suzanne Tamang, Stanford University School of Medicine, USA
Sumithra Velupillai, KTH, Royal Institute of Technology, Sweden, and King's College London, UK
Özlem Uzuner, MIT, USA
Pierre Zweigenbaum, LIMSI, CNRS, Université Paris-Saclay, Orsay, France

Additional Reviewers:

Andre Lamurias, Universidade de Lisboa, Portugal
Bridget McInnes, Virginia Commonwealth University, USA
Meliha Yetisgen, University of Washington, USA

Table of Contents

Conference Program

November 3, 2019

9:00–10:30   Session 1

9:00   *Introduction*

9:05   *Cross-document coreference: An approach to capturing coreference without context*
Kristin Wright-Bettner, Martha Palmer, Guergana Savova, Piet de Groen and Timothy Miller

09:30   *Poster booster*

09:45   *Poster session*

*Comparing the Intrinsic Performance of Clinical Concept Embeddings by Their
Field of Medicine*
John-Jose Nunez and Giuseppe Carenini

*On the Effectiveness of the Pooling Methods for Biomedical Relation Extraction
with Deep Learning*
Tuan Ngo Nguyen, Franck Dernoncourt and Thien Huu Nguyen

*Syntax-aware Multi-task Graph Convolutional Networks for Biomedical Relation
Extraction*
Diya Li and Heng Ji

BioReddit: Word Embeddings for User-Generated Biomedical NLP
Marco Basaldella and Nigel Collier

*Leveraging Hierarchical Category Knowledge for Data-Imbalanced Multi-Label
Diagnostic Text Understanding*
Shang-Chi Tsai, Ting-Yun Chang and Yun-Nung Chen

Experiments with ad hoc ambiguous abbreviation expansion
Agnieszka Mykowiecka and Malgorzata Marciniak

*Multi-Task, Multi-Channel, Multi-Input Learning for Mental Illness Detection using
Social Media Text*
Prasadith Kirinde Gamaarachchige and Diana Inkpen

Extracting relevant information from physician-patient dialogues for automated clinical note taking
Serena Jeblee, Faiza Khan Khattak, Noah Crampton, Muhammad Mamdani and Frank Rudzicz

Biomedical Relation Classification by single and multiple source domain adaptation
Sinchani Chakraborty, Sudeshna Sarkar, Pawan Goyal and Mahanandeeshwar Gattu

Assessing the Efficacy of Clinical Sentiment Analysis and Topic Extraction in Psychiatric Readmission Risk Prediction
Elena Alvarez-Mellado, Eben Holderness, Nicholas Miller, Fyonn Dhang, Philip Cawkwell, Kirsten Bolton, James Pustejovsky and Mei-Hua Hall

10:30–11:00   Break

11:00–12:30   Session 2

11:00   *What does the language of foods say about us?*
Hoang Van, Ahmad Musa, Hang Chen, Stephen Kobourov and Mihai Surdeanu

11:25   *Dreaddit: A Reddit Dataset for Stress Analysis in Social Media*
Elsbeth Turcan and Kathy McKeown

11:50   *Towards Understanding of Medical Randomized Controlled Trials by Conclusion Generation*
Alexander Te-Wei Shieh, Yung-Sung Chuang, Shang-Yu Su and Yun-Nung Chen

12:15   *Building a De-identification System for Real Swedish Clinical Text Using Pseudonymised Clinical Text*
Hanna Berg, Taridzo Chomutare and Hercules Dalianis

12:40–14:00   Lunch

14:00–15:30   Session 3

14:00   *Invited Talk*
              TBA

14:40   *Automatic rubric-based content grading for clinical notes*
              Wen-wai Yim, Ashley Mills, Harold Chun, Teresa Hashiguchi, Justin Yew and
              Bryan Lu

15:05   *Dilated LSTM with attention for Classification of Suicide Notes*
              Annika M Schoene, George Lacey, Alexander P Turner and Nina Dethlefs

15:30–16:00   Break

16:00–17:40   Session 4

16:00   *Writing habits and telltale neighbors: analyzing clinical concept usage patterns
              with sublanguage embeddings*
              Denis Newman-Griffis and Eric Fosler-Lussier

16:25   *Recognizing UMLS Semantic Types with Deep Learning*
              Isar Nejadgholi, Kathleen C. Fraser, Berry De Bruijn, Muqun Li, Astha LaPlante
              and Khaldoun Zine El Abidine

16:50   *Ontological attention ensembles for capturing semantic concepts in ICD code pre-
              diction from clinical text*
              Matus Falis, Maciej Pajak, Aneta Lisowska, Patrick Schrempf, Lucas Deckers, Sha-
              dia Mikhael, Sotirios Tsaftaris and Alison O'Neil

17:15   *Neural Token Representations and Negation and Speculation Scope Detection in
              Biomedical and General Domain Text*
              Elena Sergeeva, Henghui Zhu, Amir Tahmasebi and Peter Szolovits

Cross-document coreference: An approach to capturing coreference without context

Kristin Wright-Bettner[1], **Martha Palmer**[1], **Guergana Savova**[2],
Piet de Groen[3], **Timothy Miller**[2]
[1]Department of Linguistics, University of Colorado, Boulder, CO 80309
[2]Boston Children's Hospital and Harvard Medical School, Boston, MA 02115
[3]Department of Medicine, University of Minnesota, Minneapolis, MN 55455
kristin.wrightbettner@colorado.edu
martha.palmer@colorado.edu
[2]{first.last}@childrens.harvard.edu
degroen@umn.edu

Abstract

In this paper, we discuss a cross-document coreference annotation schema that we developed to further automatic extraction of timelines in the clinical domain. Lexical senses and coreference choices are determined largely by context, but cross-document work requires reasoning across contexts that are not necessarily coherent. We found that an annotation approach that relies less on context-guided annotator intuitions and more on schematic rules was most effective in creating meaningful and consistent cross-document relations.

1 Introduction

The ability to learn cross-document coreference and temporal relationships in clinical text is crucial for the automatic extraction of comprehensive patient timelines of events (Raghavan et al., 2014). To that end, we present a gold corpus of 198 clinical-narrative document sets, where each set consists of three notes for a given patient (594 individual notes total). Each file is annotated with intra-document temporal, coreference, and bridging relations (SET-SUBSET, WHOLE-PART, CONTAINS-SUBEVENT), and each set is annotated with cross-document coreference and bridging relations.

The goal of the current project was to leverage the inherited, intra-document annotations from two prior projects (discussed in Section 2) to capture longer, more developed timelines of patient information. We did this by creating human-annotated cross-document coreference and bridging links and then using inference to combine this information with the knowledge already gained from the intra-document temporal and coreference/bridging links.

In this paper, we discuss the impacts of cross-document-specific phenomena on human annotation and machine learning, most notably the effect of disjunct narratives on cross-document coreference judgments. Cohesive discourse is a crucial linguistic tool for determining coreference, yet the cross-document relations annotation task fundamentally takes place across discontinuous narratives. We found an approach that is governed more by annotation rules than annotator intuition to be most effective, producing an inter-annotator agreement score of 93.77% for identical relations. While an approach that moves away from linguistically-intuitive judgments may seem surprising at first, it is in fact quite fitting for a task that is inherently void of the discourse-level linguistic cues that humans employ to make those intuitive associations.

We also discuss other cross-document phenomena, inter-annotator agreement, and, briefly, areas for future work. Related work is discussed throughout.

2 The THYME colon cancer corpus

This annotation effort merged and expanded on document-level annotations created by two prior projects – a temporal relations project[1] (Styler et al., 2014), and a coreference and bridging relations project.[2] These two projects will be referred to as THYME 1 (Temporal History of Your Medical Events) and Clinical Coreference.

---

[1] Corpus publicly available from TempEval. Guidelines available at
http://clear.colorado.edu
/compsem/documents/THYME_guidelines.pdf.
[2] Clinical Coreference Annotation Guidelines available at
http://clear.colorado.edu/compsem/documents/c
oreference_guidelines.pdf.

Proceedings of the 10th International Workshop on Health Text Mining and Information Analysis (LOUHI 2019), pages 1–10
Hong Kong, November 3, 2019. ©2019 Association for Computational Linguistics
https://doi.org/10.18653/v1/D19-62

Relation Type	Description	Link
IDENTICAL (IDENT)	M1 refers to the same event/entity as M2	*[M1] IDENT [M2]*
SET-SUBSET (S-SS)	M2 refers to one or more members of a larger group, represented by M1.	*[M1]$_{SET}$ - [M2]$_{SUBSET}$*
CONTAINS-SUBEVENT (CON-SUB)	M1 temporally contains M2, and M2 is inherently part of the structure of M1.	*[M1] CON-SUB [M2]*
WHOLE-PART (W-P)	M2 is compositionally part of a larger entity, represented by M1.	*[M1]$_{WHOLE}$ - [M2]$_{PART}$*

Table 1: Gold-annotated cross-document relation types in the THYME colon cancer corpus. M1 refers to Markable 1 and M2 refers to Markable 2. Markables include events, entities, or temporal expressions. W-P was used only for entities; CON-SUB only for events. All four relation types are coreference or bridging links rather than temporal links, except for CON-SUB, which conveys both temporal and structural information and is represented as a temporal link (TLINK) in our annotation tool. This TLINK type is discussed in Section 3.1.

The corpus [3] consists of de-identified physicians' notes on colon cancer patients. The examples used throughout this paper are artificially created; however, we have done our best to replicate the relevant linguistic contexts. Each set of three notes consists of a clinical report, a pathology report, and a second clinical report, in that chronological order and spanning a period of weeks or months. Capturing such temporally-extensive information gives us the ability to track the status of the disease over time and responses (or not) to treatment.

The THYME colon cancer corpus now includes: a) intra-document gold annotations for all markables (events, entities, and temporal expressions) and several types of temporal, coreference, and bridging relations; and b) cross-document gold annotations for four coreference and bridging relation types, which represent a subset of the intra-document types [4] and are described in Table 1.

---

[3]This corpus has also been annotated according to the Penn Treebank, PropBank, and Unified Medical Language System (UMLS) Semantic Network schemas (Albright et al., 2013), though these data did not influence the current project.

[4] The intra-document relations additionally include the following types: *CONTAINS, BEFORE, OVERLAP, BEGINS-ON, ENDS-ON, NOTED-ON,* and *APPOSITIVE.* These are all temporal relations, except APPOSITIVE, which is a coreference relation. All were used by either Clinical Coreference or THYME 1 (Styler et al., 2014), except for *CONTAINS-SUBEVENT* and *NOTED-ON,* which are new to the current project. All are discussed in detail in our guidelines: https://www.colorado.edu/lab/clear/projects/computational-semantics/annotation.

Many prior studies have noted the intractability of creating cross-document gold annotations on large corpora (Day et al., 2000, for example). Each cross-document effort has therefore restricted the scope of their annotations in some way (e.g., Song et al., 2018; Cybulska and Vossen, 2014) and/or developed machine-produced annotations for cross-document relations, rather than human-produced (Raghavan et al., 2014; Dutta and Weikum, 2015; Baron and Freedman, 2008; Gooi and Allen 2004; etc.). We likewise restricted our approach by limiting the cross-document relations to the groups of three files which represent each patient, and by limiting the number of annotated relation types. However, the THYME corpus is the largest dataset of gold-annotated clinical narratives to-date that we are aware of, in terms of types of markables and relations annotated.

We are indebted to the contributions of the projects that preceded us. Much of the technical and conceptual groundwork had already been laid for our task. In particular, the notion of narrative containers (Styler et al., 2014; Pustejovsky and Stubbs, 2011) informed our addition of the CONTAINS-SUBEVENT temporal link and our cross-document annotation process.

However, we found that the segregation of tasks during the creation of the single-file gold annotations caused a variety of technical and conceptual conflicts once their outputs were merged. Furthermore, aspects of the temporal task suffered from focusing only on local-context relations; a global grasp of the text, which coreference annotation facilitates, reveals temporally significant information that may be otherwise missed or misinterpreted.

Early experiments showed that conflicts in the merged annotations rendered meaningful cross-document annotation untenable. To reconcile these conflicts, we therefore introduced an intra-document corrections-style manual annotation pass prior to cross-document double-annotation and adjudication.

3 Cross-document annotation: Process, assumptions, phenomena

It has been well-attested that determining cross-document relations poses a unique set of challenges for both systems and annotators. Song et al. (2018) discuss the cognitive strain on annotators, and others have observed the decrease in linguistic cues that occurs cross-document (Raghavan et al., 2014; Hong et al., 2016). In this paper, we are most interested in the latter, particularly the impacts of cross-document mode on identical relations.

Many coreference annotation guidelines, including ours, use a straightforward definition of coreference, which may be summarized as two different mentions in a text having the same real- or hypothetical-world referent (e.g., Cybulska and Vossen, 2014; Richer Event Description Annotation Guidelines, 2016[5]; Cohen et al., 2017). This definition leads to a binary approach to identical judgments – two mentions either refer to the same thing or they do not. Annotators are forced to make a polar choice about representations of meaning, when those representations in fact exist on a spectrum. This is not a new discovery: "There are cases where variant readings of a single lexical form would seem to be more appropriately visualized as points on a continuum – a single fabric of meaning with no clear boundaries" (Cruse, 1986). However, the natural language processing community is still learning how to deal with this.

Others have identified the problems that this oversimplified definition creates for annotation: "Degrees of referentiality as well as relations that do not fall neatly into either coreference or non-coreference—or that accept both interpretations—are a major reason for the lack of inter-coder agreement in coreference annotation" (Recasens, 2010). Hovy et al. (2013) also recognized the need for a more nuanced approach and introduced membership and subevent relations as a result.

Furthermore, both Recasens and Hovy discuss the role that pragmatics plays in determining coreference:

- "Two mentions fully corefer if their activity/event/state DE [discourse element] is identical in all respects, *as far as one can tell from their occurrence in the text*" (Hovy et al., 2013, emphasis added).
- "We redefine coreference as a scalar relation between two (or more) linguistic expressions that refer to discourse entities *considered to be at the same granularity level relevant to the linguistic and pragmatic context*" (Recasens et al., 2011, emphasis added).

Context, therefore, contributes in a crucial way to determining sense for a given lexical unit – and therefore also to determining coreference relations for that unit. We agree with Cruse, Recasens, and Hovy and observe the unique challenge this poses for cross-document annotation, since distinct narratives do not share a coherent discourse context. Recasens et al. (2011) propose that categorization and meaning are constructed in a *temporary*, active process; in cross-document work, we are attempting to create meaningful relations between temporally disconnected discourses. Put differently, the coherence of context is decreased while the number of contexts for lexical senses is increased.

While not surprising, this phenomenon does have interesting consequences for annotation. In fact, by the definitions given above, "doing" coreference between disjunct linguistic and pragmatic contexts could be viewed, on some level, as impossible.

But not all hope is lost. Particularly for our corpus, texts are very closely related and it is possible to create meaningful relations. However, the phenomenon just described requires an approach to cross-document coreference annotation that is unique from within-document. We dealt with this primarily by adding a subevent relation that was governed more by annotation rules and less by annotators' intuitions. We present the reasons for and outcome of this approach in the next section, followed by discussion of other cross-document phenomena and our technical cross-document linking process.

---

[5] https://github.com/timjogorman/RicherEv entDescription/blob/master/guidelines.md

3.1 An approach to coreference across disjunct contexts

Consider the following single-file example:

(1) *October 15th, 2015 – Dr. Wu performed* **resection** *of the primary tumor. Ms. Smith's recovery from* **surgery** *has been without complication.*

The choice here about whether to link *resection* and *surgery* as coreferential is likely to produce a disagreement. Annotator A may decide they are IDENTICAL (IDENT) since they clearly refer on some level to the same cancer treatment procedure; a significant semantic relationship would be lost if we did not link them. Annotator B, however, may decide *resection* refers only to the literal act of removing the tumor, while *surgery* points to the entire procedure. Essentially, the annotators disagree about whether the two terms are "close enough" on the meaning spectrum to warrant an IDENT link. More precisely, the disagreement stems from different interpretations of semantic granularity – Annotator A's identity "lens" is more coarse-grained, while Annotator B's is more fine-grained.

Consider a second example:

(2) *PLAN:* **Resection** *of primary tumor and gallbladder removal. Patient is scheduled for* **surgery** *on October 15th, 2015.*

Here, the finer-grained approach to *Resection* and *surgery* is supported – required, in fact – by the context. No coreference relationship is possible since the surgery clearly consists of two subprocedures, the tumor resection and the gallbladder removal.

Now consider the two examples together, where (2) is from the chronologically earlier note and (1) is from the later note in a single set:

(3) **Note A**: *PLAN:* **Resection** *of primary tumor and gallbladder removal. Patient is scheduled for* **surgery** *on October 15th, 2015.*
- No coreference link
- *surgery* CONTAINS *Resection*
- *surgery* CONTAINS *removal*

Note B: *October 15th, 2015 – Dr. Wu performed* **resection** *of the primary tumor. Ms. Smith's recovery from* **surgery** *has been without complication.*
- *resection* IDENTICAL *surgery*

The IDENT link shown for Note B represents the original gold annotation in our data, i.e., the more coarse-grained approach to identity described above. This is arguably the better perspective here, based on Recasens' definition of coreference above ("discourse entities considered to be at the same granularity level relevant to the linguistic and pragmatic context"); Note B's narrative is quite broad-brushed and supports what Hovy terms a "wide reading" of *resection* (Hovy et al., 2013). Pragmatically, *resection* and *surgery* are the same in Note B; pragmatically, they are not the same in Note A.

The predicament for cross-document linking is obvious. If we link *resection*$_A$ to *resection*$_B$, this entails that *resection*$_A$ is IDENT to *surgery*$_B$; if we then link *surgery*$_A$ to *surgery*$_B$, this now entails that the procedure temporally contains itself. If we leave *resection*$_A$ unlinked to *resection*$_B$ to avoid this conflict, problematically, we miss the relation between identical strings that refer to the same event (*?resection of primary tumor IDENT resection of primary tumor*), not to mention that leaving these unlinked would be extremely counterintuitive for annotators.

This type of situation is common in cross-document work. Since identity judgments are based on granularity levels that are in turn determined by the pragmatics of the narrative, and since the pragmatic contexts of two or more disjunct narratives are not necessarily coherent, cross-document mode frequently forces annotators to choose between: (a) not linking two mentions that are obviously and significantly semantically related, or (b) linking these mentions and thereby forcing logically-conflicting information as in (3), which in turn renders the existing temporal links much less meaningful.

To account for this variation in context-determined granularity, we introduced the CONTAINS-SUBEVENT (CON-SUB) link, which says that EVENT B is both temporally contained by EVENT A and it composes part of EVENT A's essential structure (modeled after the subevent relation in O'Gorman et al., 2016). We added this new relation intra-document in the corrections pass, as well as in the later cross-document pass. For examples like (3), this meant the Note B IDENT relation was re-interpreted as a subevent relation: *surgery* CON-SUB *resection*. This allowed us to preserve the close semantic connection between the two EVENTs in both narratives, while avoiding the logical conflicts that would have rendered our output much less meaningful and informative. We can also assume the inter-annotator agreement achieved (discussed in section 4) is much higher than it would have been had we left annotators in the predicament shown in (3).

The consistency noted above was achieved by an approach that relied less on discourse cues and more on general semantic distinctions. Instead of allowing annotators to intuitively judge between wide and narrow readings (borrowing Hovy's terms again) of lexical items based on the context, we required IDENT and CON-SUB relations to be based more on the dictionary definitions of the terms. This is because we could not predict the granularity distinctions that cross-document information would expose, as shown in (3). For example, annotators were required to differentiate between "general" surgery terms (e.g., *surgery, procedure, operation*, etc.) and "specific" surgery terms (*colectomy, resection, excision*, etc.), such that the general term nearly always contained the specific term as a subevent. This compensated for the majority of granularity distinctions in the THYME corpus (though not all, since there can always be more fine-grained levels of nuance). This framework therefore facilitated more straightforward cross-document linking, though it did also force annotators to make some counterintuitive within-document choices since senses *are* influenced by the context.

Song et al. (2018) took an opposite approach to cross-document coreference linking through their use of event hoppers, which permit "coreference of two events that are *intuitively* the same although certain features may differ" (emphasis added). We found this approach did not suit our needs since the ultimate goal was to capture a coherent timeline of clinical events, and intuitive coreference linking produced temporal conflicts, as shown above.

While coreference linking is not possible on the cross-document level in the same nuanced and intuitive way that it is within-document, there is still a great deal of important information we can capture. The texts in our corpus are topically very similar and there are typically a lot of corroborating details, such as dates and locations (again, these have been de-identified, but in a consistent fashion). Additionally, the clinically-delineated sections and the note types and structure provide clues about how to interpret the events; for example, due to the date and descriptive details, we can know which procedure in the clinical note the pathology note refers to, even if the overall procedure is not explicitly mentioned in the pathology note.

Time constraints prevented us from adding CON-SUB for all event types. We annotated it for four event categories, chosen based on clinical significance and demonstrated need due to cross-document conflicts like the one in (3): (a) patient treatment events, including surgical procedures and chemotherapy/radiation treatments; (b) cancer events (*cancer, adenocarcinoma, tumor*, etc.); (c) medications; and (d) chronic disease events.

Due to other conflicts arising from the disconnected contexts that the subevent relation was not able to reconcile, we permitted the cross-document adjudicators (but not annotators) to make within-document annotation changes when absolutely necessary.

In summary, we found that it is possible to capture meaningful cross-document coreference relations, but the approach must differ from intra-document annotation because pragmatically-directed within-document intuitions may conflict in unpredictable ways on the cross-document level.

3.2 Other cross-document phenomena

We have discussed in depth the way identical judgments are affected by disjunct contexts. We discuss two more cross-document phenomena here: (1) the use of inference in linking stative events; and (2) how cross-document work exposes typos and misinformation. Notably, this could be leveraged to identify mistakes in the text, which may contribute to current efforts to reduce medical errors in patient treatment.

Inference and stative events

Cross-document coreference is typically easier for punctual events (such as tests and procedures) and harder for durative events that can change in value over time (for example, a mass that is initially benign but becomes malignant). As with many other cross-document challenges, this issue is also present within-document, but is exacerbated in the cross-document setting because context is reduced. Consider the following example:

(4) Note A (March 24 2012 SECTIONTIME):
*Pulse Rate: **Regular***
Note B (March 26 2012 SECTIONTIME):
*Heart: **Regular** rate*

Here we have two clinically-relevant states associated with two different times: the regularity of the patient's heart rate on March 24, 2012, and the regularity of the patient's heart rate on March 26, 2012. The question for a cross-doc annotator is whether these two EVENTs are IDENTICAL.

For the current example, it is likely the regular condition has continued, but the fact is we do not know, especially since the patient may have a medical condition that causes sporadic irregularity. Furthermore, we might be initially inclined to infer sameness due to the close temporal proximity of the two measurements (two days apart), but that thought trajectory quickly leads to problems: When are two continuous events not temporally near enough to infer sameness? A week? A month? How do we decide?

Song et al. (2018) discuss a similar example across four notes, in which they corefer the first three events because they occur in "about the same time period and same place" (occurring over the timespan of a month), but they do not corefer the fourth event "as it happened at a different time" (about four months after the most recent other mention). However, it is not clear how they determined that a month is a reasonably close enough timespan to infer sameness, while four months is not.

Our approach, therefore, was that when condition or attributive EVENTs – events that vary in value – are measured or identified at two different times, they should not be linked, unless there is explicit linguistic evidence (e.g., use of the present perfect tense) they are the same event. Essentially, we decided that temporal proximity alone was not enough to infer an identical relation for two condition/value events.

Of course, inference is a source of inter-annotator disagreement for other cross-document choices as well. A comprehensive analysis is outside the scope of this paper, but the topic is discussed further in the following point.

How cross-document annotation exposes mistakes in the data

We discuss this in detail not only because it has implications for discovering misinformation in the text, but also because it demonstrates two more significant challenges to cross-document clinical annotation: the heavy cognitive burden on annotators, and the need for clinical knowledge. Consider the following example:

(5) **Note A** (DOCTIME: August 21, 2012):
*We have ordered a **CT** abdomen and pelvis to rule out liver metastases prior to surgery. Mr. Olson will also need an EKG and bloodwork. Testing was negative.*
- *CT* assigned DocTimeRel of *AFTER*, i.e., it occurs after DOCTIME (Aug 21, 2012).

Note B (DOCTIME: September 30, 2012):

*CT abdomen and pelvis was compared to the prior **study** of August 20, 2012, Mr. Olson had low-anterior resection.*
- *August 20, 2012* CONTAINS *study*

CT_A and $study_B$ are in fact IDENTICAL. Combined with the temporal information noted above, this entails that the same event both occurs after Aug. 21, 2012, and is temporally contained by Aug. 20, 2012 – a logical impossibility.

We know they are the same event based primarily on real-world knowledge of the standard order of medical procedures, as follows: It is clear in Note B that there are two different CT scans. The question facing a cross-document annotator is which one, if either, is IDENT to CT_A? We know explicitly from the text that CT_A occurred prior to the patient's surgery. CT_B occurred after the patient's surgery, since, however cryptically, it references observation of the surgery ("Mr. Olson had low-anterior resection"). Therefore, CT_A and CT_B are not referring to the same scan.

Now the question is whether $study_B$ is IDENT to CT_A. The initial evidence is to the contrary – $study_B$ is explicitly said to occur on Aug. 20, while *CT* is inferably after (or later in the day on) Aug. 21. However: (a) it is unusual to have two CT scans back-to-back, without further discussion; (b) an Aug. 20 CT is not discussed in the Aug. 21 note; (c) in Note A, immediately after noting that several tests have been ordered, the text says, "Testing was negative." Based on the verb tenses in the paragraph, the assumption would likely be that *Testing* here refers to other tests, not the ones just ordered. However, the flow of discourse suggests otherwise, along with the fact that no other prior testing is referred to in the same section. With the additional information we have from Note B, a more reasonable interpretation presents itself: $study_B$ is IDENT to CT_A, and Aug. 20 is the correct date of the scan. The note was likely originally written on Aug. 20, prior to the scan that was done later that day, and was later updated with the test results but without any indication of the update being written at a later time. This analysis was confirmed by review of all notes by our medical expert consultant.

There are several noteworthy observations about this: First, there is quite a bit of oncological knowledge required to notice the conflict above. Furthermore, the non-standard syntax in Note B would make it easy for an annotator to miss the fact that CT_B is after the resection.

Example	Text	Within-doc links	Cross-doc links
File Set 1	Note A: *...screening tests...* Note B: *...screening tests...MRI*	$tests_B$ SET-SUBSET MRI_B	$tests_A$ IDENT $tests_B$
File Set 2	Note A: *...screening tests...* Note B: *...MRI...*	None	$tests_A$ SET-SUBSET MRI_B
File Set 3	Note A: *...screening tests...MRI* Note B: *...screening tests...MRI*	$tests_A$ SET-SUBSET MRI_A $tests_B$ SET-SUBSET MRI_B	$tests_A$ IDENT $tests_B$ MRI_A IDENT MRI_B

Table 2: For File Set 1, there is no cross-doc S-SS link between $tests_A$ and MRI_B because this can be inferred from the cross-doc IDENT link and the within-doc S-SS link shown. For File Set 3, the fact that MRI_A has the same referent as MRI_B is not inferable from the intra-document structural links; hence, we create a cross-doc IDENT link. (Crucially, all examples assume that context allows us to know that these mentions do in fact refer to the same testing events.)

Second, even armed with the necessary clinical knowledge, there is still a fair amount of inference involved in making the above choice. However, note that *all* of the annotation options here, including the option to not link at all, require a lot of inference (as is the nature of many cross-document analyses). There are different types of inference based on different kinds of information. While we decided that temporal closeness is not enough by itself to infer a relation for condition/value events, we decided here that medical knowledge of standard processes is enough to infer a relation.

Third, assuming the above observations were made, an impossible annotation choice presents itself: Do we make the coreference link even though it forces a temporal conflict, or do we keep the timeline clean and lose the coreference relation? We decided on the former, and kept track of the noted temporal conflicts in order to inform systems training.

Finally, note the time, attention, and careful thought process required for determining this single cross-document link. While certainly not all decisions are this demanding, the amount of time necessary to produce high-caliber annotations should be apparent. It took highly-experienced annotators about 1.5 hours on average to complete one document set, or an estimated 891 hours total for two annotators and one adjudicator to produce 198 gold sets with a total of 10,560 cross-document links. This does not include time spent on initial annotation experiments, process and guidelines development, annotator training, and post-processing steps.

3.3 Cross-document annotation process

To manage the potentially vast number of cross-document links, we established a set of assumptions about inferable relations that guided the following process and are further discussed in Table 2 (note: "structural links" refers to links that have a hierarchical rather than identical relationship: *CON-SUB, S-SS, W-P*):

(a) Link topmost mention to topmost mention. We assume the other relations can be inferred from within-document chains.

(b) If there is a within-document structural link between two markables, do not create that same link cross-document for the same two events/entities. Put differently, create cross-document structural links only when *both* components of the relation do not have a cross-document IDENT link. Again, we assume that other relations can be inferred.

(c) Always create IDENT links whenever appropriate.

4 Inter-annotator agreement

We scored inter-annotator agreement (IAA) only for annotation categories that were new to the current project, i.e., intra-document CON-SUB links and all cross-document links. Furthermore, we only scored annotator-annotator agreement (not annotator-gold), since adjudicators were permitted to change single-file annotations while annotators were not. The total number of gold markables and relations are shown in Table 3; IAA results are shown in Table 4 and are averaged over all the documents (both tables shown on following page).

The IDENT score is much higher than the structural linking scores because the structural links were only created in cases where neither component of the link had a cross-document IDENT relation (see Section 3.3). These relations were therefore brand-new and had to be identified

Markables (594 documents)	143,147 total	Relations, within-doc and cross-doc (594 documents)	70,572 total	Cross-doc relations (198 documents)	10,762 total
TIMEX3s	7,796	Temporal links	35,428 total	IDENTICAL	9,102
Entities*	47,355	CONTAINS	14,037	SET-SUBSET	405
EVENTs	86,172	CON-SUB**	4,718	WHOLE-PART	13
SECTIONTIME	1,230	BEFORE	4,217	CON-SUB	1,242
DOCTIME	594	OVERLAP	5,091		
		BEGINS-ON	1,200		
		ENDS-ON	557		
*Entities are referred to as MARKABLEs in our guidelines, due to the naming practice of the prior Clinical Coreference project.		NOTED-ON	5,608	**CON-SUB is listed twice under the second column since it's both a temporal link and a bridging link.	
		Aspectual links	873 total		
		INITIATES	259		
		CONTINUES	302		
		TERMINATES	278		
		REINITIATES	34		
		Coreference and bridging links	38,337 total		
		IDENTICAL	23,827		
		SET-SUBSET	5,907		
		WHOLE-PART	3,885		
		CON-SUB**	4,718		

Table 3: Total gold markables and relations for the THYME colon cancer corpus.

without the benefit of a single coherent discourse, as discussed in depth above. On the other hand, annotators were able to draw on the information conveyed in intra-document relations when determining cross-document IDENT relations.

The WHOLE-PART (W-P) IAA score is zero because there were very few cross-document W-P relations in the corpus, under our guidelines. W-P is used only for entities, and we did not do W-P cross-document linking for anatomical entities (due to the massive amount of mentions, the spider-webbed relations, and the number of vague terms – *tissue portions*, etc. – we only created IDENT anatomy relations at the cross-document level). Therefore, the only cross-narrative W-P relations were between organizations/departments and members of those entities, which were only rarely knowable from the text.

The CONTAINS-SUBEVENT (CON-SUB) agreement score is likely higher than the SET-SUBSET (S-SS) score because we applied it to four specific event categories (see Section 3.1) that consist of oft-repeated terms. S-SS, on the other hand, had no such constraints, making this relation much more challenging to identify over the scope of three often-lengthy documents. Furthermore, while some set-member relations are obvious, others are not. For example:

(6) **Note A**: *Pt denies alcohol or tobacco use*.
   Note B: *He denies drinking*.
   • *use*$_{NEG}$ S-SS *drinking*$_{NEG}$

Intra-document IAA	Cross-document IAA
CON-SUB: 34.14%	IDENTICAL: 93.77%
	CON-SUB: 36.43%
	SET-SUBSET: 6.88%
	WHOLE-PART: 0.00%

Table 4: Intra-document and cross-document inter-annotator agreement scores in terms of percentage agreement.

One of our annotators identified the S-SS link shown, while the other did not. In the future, more examples and/or constraints of fringe S-SS relations in the annotation guidelines could be developed to improve S-SS agreement.

5 Conclusion

As demonstrated, developing an extensive timeline of patient events that occur over multiple weeks and months is an extremely complicated process. Understanding the breadth of complexity and the heavy demands on annotators is necessary for projecting annotation budgets and timelines, and for understanding the nature and quality of the resulting data for predicting machine learning performance. Two of the most pressing areas for future research include: (a) further development and testing of our approach to cross-document linking presented in section 3.1; and (b) development of a comprehensive methodology for incorporating medical expertise, as alluded to in section 3.2 (building on but extending beyond the light-annotation tasks methodology proposed by Stubbs, 2013). It is critical that wherever possible the annotation process is based on clear rules rather than annotator intuition as the former lends itself to automation whereas the latter at best results in a non-scalable solution with a narrow field of implementation. Developing these rules requires medical domain expertise.

Our results for cross-document coreference annotation leave ample room for improvement. Yet we believe that the approaches discussed here will serve as another significant step in the development of automatic extraction of event timelines in medical data.

Acknowledgments

The work was supported by funding R01LM010090 from the National Library Of Medicine. The content is solely the responsibility of the authors and does not necessarily represent the official views of the National Library Of Medicine or the National Institute of Health.

We would like to thank: Dana Green especially for annotation and insightful annotation input; Ahmed Elsayed and Dave Harris for medical annotation and advice; James Martin for schema development advice; Wei-Te Chen and Skatje Myers for technical support; Michael Regan, Matthew Oh, Hayley Coniglio, Samuel Beer, and Jameson Ducey for annotation; and Adam Wiemerslage for IAA and post-processing scripts.

References

Daniel Albright, Arrick Lanfranchi, Anwen Fredriksen, William F Styler IV, Colin Warner, Jena D Hwang, Jinho D Choi, Dmitriy Dligach, Rodney D Nielsen, James Martin, Wayne Ward, Martha Palmer, Guergana K Savova. 2013. Towards comprehensive syntactic and semantic annotations of the clinical narrative. *Journal of the American Medical Informatics Association*, 20(5): 922–930. https://doi.org/10.1136/amiajnl-2012-001317.

Alex Baron and Marjorie Freedman. 2008. Who is Who and What is What: Experiments in Cross-Document Co-Reference. In *Proceedings of the 2008 Conference on Empirical Methods in Natural Language Processing*, Association for Computational Linguistics, pages 274–283. https://www.aclweb.org/anthology/D08-1029.

K. Bretonnel Cohen, Arrick Lanfranchi, Miji Joo-young Choi, Michael Bada, William A. Baumgartner Jr., Natalya Panteleyeva, Karin Verspoor, Martha Palmer, Lawrence E. Hunter. 2017. Coreference annotation and resolution in the Colorado Richly Annotated Full Text (CRAFT) corpus of biomedical journal articles. *BMC Bioinformatics*, 18:372.

D. A. Cruse. 1986. *Lexical Semantics*. Cambridge University Press, Cambridge, UK.

Agata Cybulska and Piek Vossen. 2014. Using a sledgehammer to crack a nut? Lexical diversity and event coreference resolution. In *Proceedings of the Ninth International Conference on Language Resources and Evaluation (LREC-2014)*, European Language Resources Association, pages 4545-4552. http://www.lrec-conf.org/proceedings/lrec2014/pdf/840_Paper.pdf.

David Day, Alan Goldschen, John Henderson. 2000. A Framework for Cross-Document Annotation. In *Proceedings of the Second International Conference on Language Resources and Evaluation (LREC'00)*, European Language Resources Association. http://www.lrec-conf.org/proceedings/lrec2000/pdf/201.pdf.

Sourav Dutta and Gerhard Weikum. 2015. Cross-Document Co-Reference Resolution using Sample-Based Clustering with Knowledge Enrichment. *Transactions of the Association for Computational Linguistics*, 3, pages 15–28. https://doi.org/10.1162/tacl_a_00119.

Chung Heong Gooi and James Allan. 2004. Cross-Document Coreference on a Large Scale Corpus. In *Proceedings of the Human Language Technology Conference of the North American Chapter of the Association for Computational Linguistics: HLT-NAACL 2004*, Association for Computational Linguistics, pages 9-16. https://www.aclweb.org/anthology/N04-1002.

Yu Hong, Tongtao Zhang, Tim O'Gorman, Sharone Horowit-Hendler, Heng Ji, Martha Palmer. 2016. Building a Cross-document Event-Event Relation Corpus. In *Proceedings of the 10th Linguistic Annotation Workshop held in conjunction with ACL 2016 (LAW-X 2016)*, Association for Computational Linguistics, pages 1–6. https://aclweb.org/anthology/W16-1701.

Edward Hovy, Teruko Mitamura, Felisa Verdejo, Jun Araki, Andrew Philpot. 2013. Events are Not Simple: Identity, Non-Identity, and Quasi-Identity. In *Workshop on Events: Definition, Detection, Coreference, and Representation*, Association for Computational Linguistics, pages 21–28. https://www.aclweb.org/anthology/W13-1203.

Tim O'Gorman, Kristin Wright-Bettner, Martha Palmer. 2016. Richer Event Description: Integrating event coreference with temporal, causal and bridging annotation. In *Proceedings of the 2nd Workshop on Computing News Storylines (CNS 2016)*, Association for Computational

Linguistics, pages 47–56. https://aclweb.org/anthology/W16-5706.

James Pustejovsky and Amber Stubbs. 2011. Increasing informativeness in temporal annotation. In *Proceedings of the 5th Linguistic Annotation Workshop*, Association for Computational Linguistics, pages 152–160. https://www.aclweb.org/anthology/W11-0419.

Preethi Raghavan, Eric Fosler-Lussier, Noémie Elhadad and Albert M. Lai. 2014. Cross-narrative temporal ordering of medical events. In *Proceedings of the 52nd Annual Meeting of the Association for Computational Linguistics (Volume 1: Long Papers)*, Association for Computational Linguistics, pages 998–1008. https://aclweb.org/anthology/P14-1094.

Marta Recasens. 2010. "Coreference: Theory, Annotation, Resolution and Evaluation." PhD Thesis. University of Barcelona. http://stel.ub.edu/cba2010/phd/phd.pdf.

Marta Recasens, Eduard Hovy, M. Antònia Martí. 2011. Identity, Non-identity, and Near-identity: Addressing the complexity of coreference. *Lingua, 121*(6): 1138-1152. https://doi.org/10.1016/j.lingua.2011.02.004.

Marta Recasens, M. Antònia Martí, Constantin Orasan. 2012. Annotating Near-Identity from Coreference Disagreements. In *Proceedings of the Eighth International Conference on Language Resources and Evaluation (LREC-2012)*, European Languages Resources Association, pages 165-172. http://www.lrec-conf.org/proceedings/lrec2012/pdf/674_Paper.pdf.

Zhiyi Song, Ann Bies, Justin Mott, Xuansong Li, Stephanie Strassel, Christopher Caruso. 2018. Cross-Document, Cross-Language Event Coreference Annotation Using Event Hoppers. In *Proceedings of the Eleventh International Conference on Language Resources and Evaluation (LREC-2018)*, European Languages Resources Association, pages 3535-3540. https://www.aclweb.org/anthology/L18-1558.

Amber Stubbs. 2013. "A Methodology for Using Professional Knowledge in Corpus Annotation." PhD Thesis. Brandeis University. http://amberstubbs.net/docs/AmberStubbs_dissertation.pdf.

William F. Styler IV, Steven Bethard, Sean Finan, Martha Palmer, Sameer Pradhan, Piet C de Groen, Brad Erickson, Timothy Miller, Chen Lin, Guergana Savova and James Pustejovsky. 2014. Temporal Annotation in the Clinical Domain. *Transactions of the Association for Computational Linguistics, 2*, pages 143–154. https://doi.org/10.1162/tacl_a_00172.

Comparing the Intrinsic Performance of Clinical Concept Embeddings by Their Field of Medicine

John-Jose Nunez[1,2] and Giuseppe Carenini[1]
[1] Department of Computer Science, [2] Department of Psychiatry
The University of British Columbia, Canada
{jjnunez, carenini}@cs.ubc.ca

Abstract

Pre-trained word embeddings are becoming increasingly popular for natural language-processing tasks. This includes medical applications, where embeddings are trained for clinical concepts using specific medical data. Recent work continues to improve on these embeddings. However, no one has yet sought to determine whether these embeddings work as well for one field of medicine as they do in others. In this work, we use intrinsic methods to evaluate embeddings from the various fields of medicine as defined by their ICD-9 systems. We find significant differences between fields, and motivate future work to investigate whether extrinsic tasks will follow a similar pattern.

1 Introduction

The application of natural language processing (NLP) and machine learning to medicine presents an exciting opportunity for tasks requiring prediction and classification. Examples so far include predicting the risk of suicide or accidental death after a patient is discharged from general hospitals (McCoy et al., 2016) or classifying which patients have peripheral vascular disease (Afzal et al., 2017). A common resource across NLP for such tasks is to use high-dimensional vector word representations. These word embedding include the popular *word2vec* system (Mikolov et al., 2013) which was initially trained on general English text, using a skip-gram model on a Google News corpus.

Due to considerable differences between the language of medical text and general English writing, prior work has trained medical embeddings using specific medical sources. Generally, these approaches have trained embeddings to represent medical concepts according to their 'clinical unique identifiers' (CUIs) in the Unified Library Management System (ULMS) (Bodenreider, 2004). Words in a text can then be mapped to these CUIs (Yu and Cai, 2013). Various sources have been used, such as medical journal articles, clinical patient records, and insurance claims (De Vine et al., 2014), (Minarro-Giménez et al., 2014), (Choi et al., 2016).

Prior authors have sought to improve the quality of these embeddings, such as using different training techniques or more training data (Beam et al., 2018). In order to judge the quality of these embeddings, they have primarily used evaluation methods quantifying intrinsic qualities, such as their ability to predict drug-disease relations noted in the National Drug File - Reference Terminology (NDF-RT) ontology (Minarro-Giménez et al., 2014), or whether similar types of clinical concepts had cosine similiar vectors (Choi et al., 2016).

To date these embeddings have been both trained and evaluated on general medical data. That is, no fields of medicine were specified or excluded; data could be from an obstetrician delivering a baby, a cardiologist placing a stent, or a dermatologist suggesting acne treatment. It is unclear how well such embeddings perform for a specific field of medicine. For example, we can consider psychiatry, the field of medicine concerned with mental illnesses such as depression or schizophrenia. Prior work has shown that psychiatric symptoms are often described in a long, varied, and subjective manner (Forbush et al., 2013) which may present a particular challenge for training these embeddings and NLP tasks generally.

As these pre-trained embeddings may increasingly be used for down-stream NLP tasks in spe-

Proceedings of the 10th International Workshop on Health Text Mining and Information Analysis (LOUHI 2019), pages 11–17
Hong Kong, November 3, 2019. ©2019 Association for Computational Linguistics
https://doi.org/10.18653/v1/D19-62

cific fields of medicine, we seek to determine whether embeddings from one field perform relatively well or poorly relative to others. Specifically, we aim to follow prior work using intrinsic evaluation methods, comparing the geometric properties of embedding vectors against others given known relationships. This will offer a foundation for future work that may compare the performance on extrinsic NLP tasks in different medical fields. Finding relative differences may support that certain medical fields would benefit from embeddings trained on data specific to their field, or using domain adaptation techniques as sometimes used in the past (Yu et al., 2017).

2 Methods

2.1 Sets of Embeddings

We sought to compare a variety of clinical concept embeddings trained on medical data. Table 1 contains details of the sets compared in this project, all of which are based on *word2vec*. We obtained DeVine200 (De Vine et al., 2014), ChoiClaims300, and ChoiClinical300 (Choi et al., 2016) all from the latter's Github. We downloaded BeamCui2Vec500 (Beam et al., 2018) from this site. Unfortunately, we were unable to obtain other sets of embeddings mentioned in the literature (Minarro-Giménez et al., 2014), (Zhang et al., 2018) (Xiang et al., 2019).

2.2 Determining a Field of Medicine's Clinical Concepts

A clinical concept's corresponding field of medicine is not necessarily obvious. In order to have an objective and unambiguous classification, we utilized the ninth revision of the International Statistical Classification of Diseases and Related Health Problems (ICD-9) (Slee, 1978). This is a widely used system of classifying medical diseases and disorders, dividing them into seventeen chapters representing medical systems/categories such as mental disorders, or disease of the respiratory system. While the 10th version is available, we chose this version based on prior work using it, and the pending release of the 11th version. We will use these ICD9 systems to define the different medical fields.

We determined a CUI's field of medicine according to a CUI-to-ICD9 dictionary available from the UMLS (Bodenreider, 2004). We consider pharmacological substance related to a field of medicine system if it treats or prevents a disease with an ICD9 code within a particular ICD9 system. We determine this by using the NDF-RT dictionary, which maps CUIs of substances to the CUIs of conditions they treat or prevent, and then convert these CUIs to the ICD9 systems as before. As such, A CUI representing a drug may have multiple ICD9 systems and therefore medical fields.

2.3 Evaluation Methods

We sought to compare multiple methods for evaluating the quality of a medical field's embeddings based on prior work. We were unable to use Yu et al's (2017) method, based on comparing the correlation of vector cosine similarity against human judgements from the UMNSRS-Similarity dataset (Pakhomov, 2018) due to there being too few examples across many medical fields. The code for all implemented methods will be publicly available upon publication of this work from the first author's GitHub.

Medical Relatedness Measure (MRM) This method from Choi et al (2016) is based on quantifying whether concepts with known relations are neighbours of each other. They use known relationships between drugs and the diseases they treat or prevent, and also the relations between diseases that are grouped together in the Clinical Classifications Software (CCS) hierarchical groupings, a classification from the Agency for Healthcare Research and Quality (Cli). The scoring utilizes Discounted Cumulative Gain, which attributes a diminishing score the further away a known relationship is found if within k neighbours.

In our implementation, we calculate the Medical Relatedness Measure (MRM) based on the 'coarse' groupings from the CCS hierarchies. Scores are calculated for CUIs that represent diseases with a known ICD9 code. The mean MRM is then calculated for all CUIs within a given ICD9 system. The implementation was adapted from Python 2.7 code available from the original author's Github. We calculate MRM as:

$$\text{MRM}(V, F, k) = \frac{1}{|V(F)|} \sum_{v \in V(F)} \frac{1}{|V(G)|} \sum_{i=1}^{k} \frac{1_G(v(i))}{log_2(i+1)}$$

Where V are medical conditions, F a field of medicine, $V(F)$ the medical conditions within an ICD-9 system/field of medicine, G the CCS group that medical condition $v \in V(F)$ is part of, and $V(G)$ the subset of medical conditions found in

Name	Dimension	Number	Number of Training Data	Type of Training Data
DeVine200	200	52,102	17k + 348k	clinical narratives journal abstracts
ChoiClaims300	300	14,852	4m	health insurance claims
ChoiClinical300	300	22,705	20m	clinical narratives
BeamCui2Vec500	500	109,053	60m + 20m + 1.7m	health insurance claims clinical narratives full journal texts

Table 1: Characteristics of the embeddings compared, including the name referred, the embedding dimensions, the number of embeddings in the dataset, and the type of data used to train them.

Drug	Actual Medical Field	Predicted Medical Field	Correct?
Fluoxetine	Mental Disorders	Mental Disorders	Yes
Sertraline	Mental Disorders	Neoplasms	No
Risperidone	Mental Disorders	Mental Disorders	Yes
Olanzapine	Mental Disorders	Mental Disorders	Yes
Valproic Acid	Mental Disorders Diseases of the Nervous System	Mental Disorders Congenital Abnormalities	Yes
Lamotragine	Mental Disorders Diseases of the Nervous System	Diseases of the Skin Diseases of the Nervous System	No
		Mental Disorders SysVec Score:	**4/6 = 0.67**

Table 2: Illustrative example showing how System Vector Accuracy (SysVec) would be calculated for the medical field "Mental Disorders" if it contained only six drugs. Predicted medical field is the medical field/ICD9 system vector closest to the drug, or n closest fields if a drug treats conditions in n multiple fields. System vectors are the normalized mean vector of that system's medical conditions.

this group. 1_G is 0 or 1 depending on whether $v(i)$, the ith closest neighbour to a condition v, is in the same group. k neighbours are considered.

To illustrate this, consider calculating the MRM for F "Diseases of the Musculoskeletal System". It involves summing the scores for its conditions, such as *rheumatoid arthritis ($v \in V(F)$)*. This condition is part of the CCS-coarse grouping (G), "Rheumatoid arthritis and related disease". This group contains twelve conditions, such as *Felty's syndrome* and *Rheumatoid lung*. With Choi et al's choice of $k = 40$, the score for *rheumatoid arthritis* would depend on how many of the eleven other conditions in this group are within the 40 nearest neighbours ($v(i)$) to *rheumatoid arthritis*, and would give a higher score the nearer they are, the highest being if they are the eleven nearest neighbours.

Medical Conceptual Similarity Measure (MCSM) The other method used by Choi et al's work evaluates whether embeddings known to be of a particular set are clustered together. They use conceptual sets from the UMLS such as 'pharmacologic substance' or 'disease or syndrome'. Discounted Cumulative Gain is again used, based on whether a CUI has other CUIs of its set within k neighbours.

We reimplement this method, but instead of using the UMLS conceptual sets, we create sets from the ICD9 systems, again giving a score to neighbours that are diseases or drugs from the same field of medicine/ICD9 system. Again, this was adapted from code from Choi et al's Github. The Medical Conceptual Similarity Measure (MCSM) can be represented as:

$$\text{MCSM}(V, F, k) = \frac{1}{|V(F)|} \sum_{v \in V(F)} \sum_{i=1}^{k} \frac{1_F(v(i))}{log_2(i+1)}$$

Similar to MRM, F is a medical field/ICD9 system, $V(F)$ the medical conditions within a system, and 1_F 0 or 1 depending on whether neighbour $v(i)$ is also in this medical field.

For illustration, consider an example calculating the MCSM for the medical field/system (F) "Infectious and Parasitic Diseases". This involves calculating the score for the medical condition (v) *primary tuberculous infection*. If *rifampin*, an antibiotic, was found to be nearby, it would contribute to the MCSM, as it treats conditions in "Infectious and Parasitic Diseases" and so would be classified as being part of this system. On the other hand, if the respiratory illness *asthma* was one of the k nearest neighbours, it would add nothing to the MCSM score, as it is a disease in a different system, "Diseases of the Respiratory System".

Significance against Bootstrap Distribution (Bootstrap) Beam et al (2018) also evaluate how well known relationships between concepts are represented by embedding vector similarity. For a given known relation, they generate a bootstrap distribution by randomly calculating cosine similarities between embedding vectors of the same class (eg. a random drug and disease when evaluating drug-disease relations). For a given known relation, they consider that the embeddings produced an accurate prediction if their cosine similarity is within the top 5%, the equivalent of $p < 0.05$ for a one-sided t-test.

Our implementation considers the may-treat or may-prevent known relationships from the NDF-RT dataset. We calculate the percentage of known relations for drug-disease pair within each medical field. Beam et al have not yet made their code publicly available, so we reimplemented this technique in Python.

System Vector Accuracy (SysVec) We implement a new, simple method to evaluate a medical field's embeddings. A representative vector is calculated for each medical field/ICD9 system by taking the mean of the normalized embedding vectors of a field's diseases. We then consider all of the drugs known to treat or prevent a disease of a given medical field. A field's *System Vector Accuracy* is then the percentage of these drugs whose most similar (by cosine similarity) system vector is this field's. A higher score indicates better performance. We implemented this method in Python.

For example, a system vector for "Mental Disorders" would be calculated from the embeddings for diseases such as *schizophrenia* and *major depressive disorder*. "Mental Disorders'" *System Vector Accuracy* is the percentage of its medications (e.g. *fluoxetine, risperidone, paroxetine*) whose embedding vectors are more similar to the "Mental Disorders" system vector than all others. *Fluoxetine* is an anti-depressant medication solely used to treat "Mental Disorders", so we would expect its vector to be more similar to this system vector than, say, the system vector representing "Diseases of the Skin and Subcutaneous Tissue".

Some drugs treat or prevent diseases in n multiple medical field. For a field, such a drug is classified as being accurately predicted if the field's system vector is amongst the n most similar system vectors. For instance, *valproic acid* is an anticonvulsant used to treat both mental disorders and those of the nervous system. "Mental Disorders'" *System Vector Accuracy* would take into account whether its system vector was one of the $n=2$ most similar system vectors. For further illustration, Table 2 shows an example SysVec calculation.

2.4 Comparing Scores

Comparing Sets of Embeddings We calculated the mean scores for an embedding set, only including embeddings with corresponding ICD9 values and present in all of the compared sets. For the MCSM and MRM scores, we conducted two-way paired t-tests between the scores from each embedding set, adjusted with the Bonferroni correction. For the binary Bootstrap and SysVec scores, we judged statistical significance by calculating z-scores and their corresponding Bonferroni corrected p-values.

A negative control set of embeddings was constructed by taking the embeddings from Beam et al (2018) and randomly arranging which clinical concepts an embedding corresponds to.

Comparing Fields of Medicine As the embeddings from Beam et al (2018) are most recent, trained on the most data, and have significantly higher scores than the other embeddings compared, we used these embeddings to compare scores from the different fields of medicine. This set also contained the most embeddings, allowing more embeddings from each field to be compared.

We sought to determine whether a field of medicine's embeddings were significantly worse or better than the average. As such, for each field of medicine we calculated the mean score from each evaluation method. We then used statistical tests to compare a field's scores from a given evaluation method with the same scores from all other fields. For MCSM and MRM scores we used two-tailed t-tests, and for Bootstrap and SysVec, z-scores, all corrected with the Bonferroni correction.

To aggregate a medical field's results, we calculated a 'Net Significance' metric by taking how many of the four method's scores were significantly above the mean, minus how many were significantly below. We found this more interpretable than other methods such as aggregating normalized scores.

3 Results

3.1 Differences Between Sets of Embeddings

Comparing the sets of embeddings (Table 3) shows consistent differences. BeamCui2Vec500's scores are the highest across all methods, and this difference is very significant, with p-value $\ll 10^{-5}$ after Bonferonni correction. The ChoiClaims300 embeddings seem next best, and the remaining sets still have much higher scores than those of the negative control.

3.2 Differences Between Medical Systems

Differences are also observed between embeddings from the various fields of medicine as represented by the ICD-9 systems (Table 4). For instance, embeddings related to the medical field Mental Disorders have scores significantly above the mean score across all systems for two evaluation methods, while those of the field Symptoms, Signs, and Ill-defined Conditions are significantly below for three. Due to a smaller number of documented drug-disease relationships across two medical fields, scores were not calculated with those methods using these relationships.

4 Discussion and Future Direction

To our knowledge, this is the first investigation into whether clinical concept embeddings from a given field of medicine perform relatively well or poorly compared to others. We conducted this investigation comparing available sets of such embeddings, using a variety of previously described intrinsic evaluation methods in addition to a new one. Given that one set of embeddings performed better than others, we used this set to compare the different fields of medicine, and found significant results between various fields.

The superior performance of one set of embeddings - those from Beam et al (2018) - are consistent with the depth and breadth of data used to train these embeddings. Training used three different types of data, including that from health insurance claims, clinical narratives, and full texts from medical journals. The size of the dataset was also much larger than that of the others. Our work validates their findings that their embeddings offer the best performance. However, it would be interesting to also consider the recent clinical concept embeddings developed by (Xiang et al., 2019). They use a similar amount of data (50 million) as Beam et al, using a large dataset from electronic health records, and apply a novel method to incorporate time-sensitive information. At the time of submission, we were unable to obtain their embeddings, and so leave this comparison to future work.

Examining the differences between fields of medicine, we note that the poor performance of embeddings from the system "Symptoms, Signs, and Ill-defined Conditions" may support validity of the results. This collection of miscellaneous medical conditions would not be expected to have the intrinsic vector similarity and cohesion evaluated by our evaluation methods.

Further work may explore why the other systems have varied performance. We wonder if the observed results correlate with possible distinctiveness of the various medical fields. For example, one of the best performing systems was "Neoplasms". The conditions in this field are often unambiguous - a cancer like *non-small cell lung cancer* has little other meaning - and the drugs used for these diseases tend to be similarly specific. On the other hand, poorly performing systems such as "Diseases of the Skin and Subcutaneous Tissue" and "Diseases of the Musculoskeletal System and Connective Tissue" often utilize immunosuppressant medications that are used across many fields of medicine. Future work could investigate this conjecture by comparing scores when restricting what clinical concepts are compared, such as only common or distinct medications.

This work evaluated embeddings using intrinsic measures of embedding quality. This presents some advantages, but also the most obvious limitation and direction for future work. These intrinsic methods allowed a consistent evaluation to be carried out between medical fields, and allowed a wide variety of embedding sets to be compared. The methods all evaluate qualities that well-trained embeddings should have, though still represent artificial use-cases. Evaluating these embeddings on extrinsic, down-stream tasks may provide more practically relevant comparisons. However, these tasks will need to be comparable

Embedding Set	MRM	MCSM	Bootsrap	SysVec
Negative Control	0.02	1.24	0.05	0.35
DeVine200	0.24	5.14	0.27	0.79
ChoiClaims300	0.43	5.34	0.42	0.80
ChoiClinical300	0.33	4.49	0.42	0.74
BeamCui2Vec500	0.52	6.39	0.67	0.90

Table 3: Mean scores for embedding sets for each evaluation method. See Methods section for abbreviations

ICD-9 Systen	MRM	MCSM	Bootstrap	SysVec	Net Significance
All Systems (Negative Control)	*0*	*1.08*	*0.04*	*0.25*	-
All Systems	0.55	8.07	0.89	0.63	-
Infectious and Parasitic Diseases	*0.45*	*7.72*	**0.93**	**0.92**	0
Neoplasms	**0.62**	**9**	0.94	0.55	+2
Endocrine, Nutritional and Metabolic Diseases, and Immunity Disorders	*0.44*	*5.64*	0.89	0.53	-2
Diseases of the Blood and Blood-forming Organs	*0.31*	*4.36*	0.82	0.79	-2
Mental Disorders	0.53	**9.34**	0.96	**0.83**	+2
Diseases of the Nervous System and Sense Organs	**0.76**	**8.44**	0.87	*0.33*	+1
Diseases of the Circulatory System	0.59	8.12	**0.96**	**0.72**	+2
Diseases of the Respiratory System	*0.36*	*5.85*	0.94	**0.82**	+1
Diseases of the Digestive System	**0.61**	7.93	*0.77*	0.62	0
Diseases of the Genitourinary System	**0.61**	*6.82*	0.86	0.58	0
Complications of Pregnancy, Childbirth, and the Puerperium	*0.51*	**10.27**	-	-	0
Diseases of the Skin and Subcutanous Tissue	*0.37*	*5.1*	0.81	0.58	-2
Diseases of the Musculoskeletal System and Connective Tissue	*0.47*	8.22	0.88	*0.29*	-2
Congenital Anomalies	0.5	*6.24*	0.73	0.73	-1
Certain Conditions Originating in the Perinatal Period	*0.48*	**9.84**	-	-	0
Symptoms, Signs, and Ill-defined Conditions	*0.26*	*2.68*	*0.77*	0.56	-3
Injury and Poisoning	**0.59**	**9.09**	*0.75*	*0*	0

Table 4: Comparison of mean scores using different evaluation methods for the fields of medicine as represented by their ICD-9 system. The row All Systems shows the mean score for each method across embeddings from all systems. A **bold** score indicates that a system's score was significantly above the All Systems score, while an *italic* score indicates it was below. Significance is judged by having a p-value <0.05 after Bonferroni correction. Net Significance is the number of these significant differences above the All Systems score minus the number below. A system's score is not calculated if there are fewer than ten examples for a method. See Methods section for evaluation method abbreviations. All scores in this table are calculated using the embeddings from Beam et al.

and available for multiple medical fields. For instance, the recent work by Xiang et al (2019) compared embeddings trained by different methodologies on a task predicting the onset of heart failure (Rasmy et al., 2018). This would be an appropriate task to judge embeddings from "Diseases of the Circulatory System"; others would be needed for other systems. We also plan to investigate the validity of these intrinsic evaluation methods by comparing them to extrinsic results.

Another future direction could be to investigate what could be done to improve performance in the fields with lower scores. For instance, Zhang et al (2018) used domain adaptation techniques for psychiatric embeddings, and this could also be carried out for those systems we identified as doing poorly. Alternatively, one could train embeddings solely on data from one field of medicine and investigate how this affects performance.

References

Clinical Classifications Software (CCS), 2003. page 54.

Naveed Afzal, Sunghwan Sohn, Sara Abram, Christopher G. Scott, Rajeev Chaudhry, Hongfang Liu, Iftikhar J. Kullo, and Adelaide M. Arruda-Olson. 2017. Mining peripheral arterial disease cases from narrative clinical notes using natural language processing. *Journal of Vascular Surgery*, 65(6):1753–1761.

Andrew L. Beam, Benjamin Kompa, Inbar Fried, Nathan P. Palmer, Xu Shi, Tianxi Cai, and Isaac S. Kohane. 2018. Clinical Concept Embeddings Learned from Massive Sources of Multimodal Medical Data. *arXiv:1804.01486 [cs, stat]*.

Olivier Bodenreider. 2004. The Unified Medical Language System (UMLS): Integrating biomedical terminology. *Nucleic Acids Research*, 32(Database issue):D267–D270.

Youngduck Choi, Chill Yi-I Chiu, and David Sontag. 2016. Learning Low-Dimensional Representations of Medical Concepts. *AMIA Summits on Translational Science Proceedings*, 2016:41–50.

Lance De Vine, Guido Zuccon, Bevan Koopman, Laurianne Sitbon, and Peter Bruza. 2014. Medical Semantic Similarity with a Neural Language Model. In *Proceedings of the 23rd ACM International Conference on Conference on Information and Knowledge Management*, CIKM '14, pages 1819–1822, New York, NY, USA. ACM.

Tyler B. Forbush, Adi V. Gundlapalli, Miland N. Palmer, Shuying Shen, Brett R. South, Guy Divita, Marjorie Carter, Andrew Redd, Jorie M. Butler, and Matthew Samore. 2013. "Sitting on Pins and Needles": Characterization of Symptom Descriptions in Clinical Notes". *AMIA Summits on Translational Science Proceedings*, 2013:67–71.

Thomas H. McCoy, Victor M. Castro, Ashlee M. Roberson, Leslie A. Snapper, and Roy H. Perlis. 2016. Improving Prediction of Suicide and Accidental Death After Discharge From General Hospitals With Natural Language Processing. *JAMA psychiatry*, 73(10):1064–1071.

Tomas Mikolov, Kai Chen, Greg Corrado, and Jeffrey Dean. 2013. Efficient Estimation of Word Representations in Vector Space. *arXiv:1301.3781 [cs]*.

José Antonio Minarro-Giménez, Oscar Marín-Alonso, and Matthias Samwald. 2014. Exploring the application of deep learning techniques on medical text corpora. *Studies In Health Technology And Informatics*, 205:584–588.

Serguei Pakhomov. 2018. Semantic Relatedness and Similarity Reference Standards for Medical Terms.

Laila Rasmy, Yonghui Wu, Ningtao Wang, Xin Geng, W. Jim Zheng, Fei Wang, Hulin Wu, Hua Xu, and Degui Zhi. 2018. A study of generalizability of recurrent neural network-based predictive models for heart failure onset risk using a large and heterogeneous EHR data set. *Journal of Biomedical Informatics*, 84:11–16.

Vergil N. Slee. 1978. The International Classification of Diseases: Ninth Revision (ICD-9). *Annals of Internal Medicine*, 88(3):424.

Yang Xiang, Jun Xu, Yuqi Si, Zhiheng Li, Laila Rasmy, Yujia Zhou, Firat Tiryaki, Fang Li, Yaoyun Zhang, Yonghui Wu, Xiaoqian Jiang, Wenjin Jim Zheng, Degui Zhi, Cui Tao, and Hua Xu. 2019. Time-sensitive clinical concept embeddings learned from large electronic health records. *BMC Medical Informatics and Decision Making*, 19(2):58.

Sheng Yu and Tianxi Cai. 2013. A Short Introduction to NILE. *arXiv:1311.6063 [cs]*.

Zhiguo Yu, Byron C. Wallace, Todd Johnson, and Trevor Cohen. 2017. Retrofitting Concept Vector Representations of Medical Concepts to Improve Estimates of Semantic Similarity and Relatedness. *arXiv:1709.07357 [cs]*.

Yaoyun Zhang, Hee-Jin Li, Jingqi Wang, Trevor Cohen, Kirk Roberts, and Hua Xu. 2018. Adapting Word Embeddings from Multiple Domains to Symptom Recognition from Psychiatric Notes. *AMIA Summits on Translational Science Proceedings*, 2017:281–289.

On the Effectiveness of the Pooling Methods
for Biomedical Relation Extraction with Deep Learning

Tuan Ngo Nguyen[†], Franck Dernoncourt[‡] and Thien Huu Nguyen[†]
[†] Department of Computer and Information Science, University of Oregon
[‡] Adobe Research
{tnguyen, thien}@cs.uoregon.edu, dernonco@adobe.com

Abstract

Deep learning models have achieved state-of-the-art performances on many relation extraction datasets. A common element in these deep learning models involves the pooling mechanisms where a sequence of hidden vectors is aggregated to generate a single representation vector, serving as the features to perform prediction for RE. Unfortunately, the models in the literature tend to employ different strategies to perform pooling for RE, leading to the challenge to determine the best pooling mechanism for this problem, especially in the biomedical domain. In order to answer this question, in this work, we conduct a comprehensive study to evaluate the effectiveness of different pooling mechanisms for the deep learning models in biomedical RE. The experimental results suggest that dependency-based pooling is the best pooling strategy for RE in the biomedical domain, yielding the state-of-the-art performance on two benchmark datasets for this problem.

1 Introduction

In order to analyze the entities in text, it is crucial to understand how the entities are related to each other in the documents. In the literature, this problem is formalized as relation extraction (RE), an important task in information extraction. RE aims to identify the semantic relationships between two entity mentions within the same sentences in text. Due to its important applications on many areas of natural language processing (e.g., question answering, knowledge base construction), RE has been actively studied in the last decade, featuring a variety of feature-based or kernel-based models for this problem (Zelenko et al., 2002; Zhou et al., 2005; Bunescu and Mooney, 2005; Sun et al., 2011; Chan and Roth, 2010; Nguyen et al., 2009). Recently, the introduction of deep learning has produced a new generation of models for RE with the state-of-the-art performance on many different benchmark datasets (Zeng et al., 2014; dos Santos et al., 2015; Xu et al., 2015; Liu et al., 2015; Zhou et al., 2016; Wang et al., 2016; Zhang et al., 2017, 2018b). The advantage of deep learning over the previous approaches for RE is the ability to automatically learn effective features for the sentences from data via various network architectures. The same trend has also been observed for RE in the biomedical domain where deep learning is gaining more and more attention from the research community (Mehryary et al., 2016; Björne and Salakoski, 2018; Nguyen and Verspoor, 2018; Verga et al., 2018).

The typical deep learning models for RE have involved Convolutional Neural Networks (CNN) (Zeng et al., 2014; Nguyen and Grishman, 2015b; Zeng et al., 2015; Lin et al., 2016; Zeng et al., 2017), Recurrent Neural Networks (RNN), (Miwa and Bansal, 2016; Zhang et al., 2017), Transformer (self-attention) networks (Verga et al., 2018), and Graph Convolutional Neural Networks (GCNN) (Zhang et al., 2018b). There are two major common components in such deep learning models for RE, i.e., the representation component and the pooling component. First, in the representation component, some deep learning architectures are employed to compute a sequence of vectors to represent an input sentence for RE for which each vector tends to capture the specific context information for a word in that sentence. Such word-specific representation sequence is then fed into the second pooling component (e.g., max pooling) that aggregates the representation vectors to obtain an overall vector to represent the whole input sentence for the classification problem in RE.

While there have been many work in the literature to compare different deep learning architectures for the representation component, the pos-

Proceedings of the 10th International Workshop on Health Text Mining and Information Analysis (LOUHI 2019), pages 18–27
Hong Kong, November 3, 2019. ©2019 Association for Computational Linguistics
https://doi.org/10.18653/v1/D19-62

sible methods for the pooling component of the deep learning models have not been systematically benchmarked for RE in general and for the biomedical domain in particular. Specifically, the prior work on relation extraction with deep learning has only assumed one form of pooling in the model without considering the possible alternatives for this component. In this work, we argue that the pooling mechanisms also have significant impact on the performance of the deep learning models for RE and it is important to understand how well different pooling methods perform in this case. Consequently, in this work, we conduct a comprehensive investigation on the effectiveness of different max pooling methods for the deep learning models of RE, focusing on the biomedical domain as the case study. Our goal is to determine the best pooling methods for the deep learning models in biomedical RE. We also want to emphasize the experiments where the pooling methods are compared in a compatible manner with the same representation components and resources for the biomedical RE models in this work. Such compatible comparison is unfortunately very rare in the current literature about deep learning for RE as new models are being intensively proposed, employing a diversity of options and resources (i.e., pre-trained word embeddings, optimizers, etc.). Therefore, this is actually the first work to compare different pooling methods for deep relation extraction on the same setting.

In the experiments, we find that syntactic information (i.e., dependency parsing) can be exploited to provide the best pooling strategies for biomedical RE. In fact, our experiments also suggest that it is more beneficial to apply the syntactic information in the pooling component of the deep learning models for biomedical RE than that in the representation component. This is different from most of the prior work on relation extraction that has only employed the syntactic information in the representation component of the deep learning models (Xu et al., 2016; Miwa and Bansal, 2016). Based on the syntax-based pooling mechanism, we achieve the state-of-the-art performance on two benchmark datasets for biomedical RE.

2 Model

Relation Extraction can be seen as a multi-class classification problem that takes a sentence and two entity mentions of interest in that sentence as the input. The goal is to predict the semantic relation between these two entity mentions according to some predefined set of relations. Formally, let $W = [w_1, w_2, \ldots, w_n]$ be the input sentence where n is the number of tokens and w_i is the i-th word/token in W. As entity mentions can span multiple consecutive words/tokens, let $[s_1, e_1]$ be the span of the first entity mention M_1 where s_1 and e_1 are the indexes for the first and last token of M_1 respectively. Similarly, we define $[s_2, e_2]$ as the span for the second entity mention M_2. For convenience, we assume that the entity mentions are not nested, i.e., $1 \leq s_1 \leq e_1 < s_2 \leq e_2 \leq n$.

2.1 Input Vector Representation

In order to encode the positions and the entity types of the two entity mentions in the input sentence, following (Zhang et al., 2018b), we first replace the tokens in the entity mentions M_1 and M_2 with the special tokens of format M_1-$Type_1$ and M_2-$Type_2$ respectively ($Type_1$ and $Type_2$ represent the entity types of M_1 and M_2 respectively). The purpose of this replacement is to help the models to abstract from the specific tokens/words of the entity mentions and only focus on their positions and entity types, the two most important pieces of information of the entity mentions for RE.

Given the enriched input sentence, the first step in the deep learning models for RE is to convert each word in the input sentence into a vector to facilitate the real-valued computation of the models. In this work, the vector v_i for w_i is obtained by concatenating the following two vectors:

1. The word embeddings of w_i: The embeddings for the special tokens are initialized randomly while the embeddings for the other words are retrieved from the pre-trained word embedding table provided by the *Word2Vec* toolkit with 300 dimensions (Mikolov et al., 2013).

2. The embeddings for the part-of-speech (POS) tag of w_i in W: We assign a POS tag for each word in the input sentence using the Stanford CoreNLP toolkit. The embedding for each POS tag is also randomly initialized in this case.

Note that both the word embeddings and the POS embeddings are updated during the training time of the models in this work. The word-to-vector conversion transforms the input sentence $W = [w_1, w_2, \ldots, w_n]$ into a sequence of vectors $V = [v_1, v_2, \ldots, v_n]$ (respectively) that would be used as the input for all the deep learning models considered in this work to ensure a compatible comparison. As mentioned in the introduction, the deep learning models for RE involves two major components, i.e., the representation component and the pooling component. We describe the options for such components in the following sections.

2.2 The Representation Component for RE

Given the input sequence of vectors $V = [v_1, v_2, \ldots, v_n]$, the next step in the deep learning models for RE is to transform this vector sequence into a more abstract vector sequence $A = [a_1, a_2, \ldots, a_n]$ so a_i would capture the underlying representation for the context information specific to the i-th word in the sentence. In this work, we examine the following typical architectures to obtain such an abstract sequence A for V:

1. *CNN* (Zeng et al., 2014; Nguyen and Grishman, 2015b; dos Santos et al., 2015): *CNN* is one of the early deep learning models for RE. It involves an 1D convolution layer over the input vector sequence V with multiple window sizes for the filters. *CNN* produces a sequence of vectors in which each vector capture some n-grams specific to a word in the sentence. This sequence of vectors is used as A for our purpose.

2. *BiLSTM* (Nguyen and Grishman, 2015a): In *BiLSTM*, two Long-short Term Memory Networks (LSTM) are run over the input vector sequence V in the forward and backward direction. The hidden vectors generated at the position i by the two networks are then concatenated to constitute the abstract vector a_i for this position. Due to the recurrent nature, a_i involves the context information over the whole input sentence W although a greater focus is put on the context of the current word.

3. *BiLSTM-CNN*: This models resembles the MASS model presented in (Le et al., 2018). It first applies a bidirectional LSTM layer over the input sequence V whose results are further processed by a Convolutional Neural Network (CNN) layer as in *CNN*. We also use the output of the CNN layer as the abstract vector sequence A for this model.

4. *BiLSTM-GCNN* (Zhang et al., 2018b): Similar to *BiLSTM-CNN*, *BiLSTM-GCNN* also first employs a bidirectional LSTM network to abstract the input vector sequence V. However, in the second step, different from *BiLSTM-CNN*, *BiLSTM-GCNN* introduces a Graph Convolutional Neural Network (GCNN) layer that consumes the LSTM hidden vectors and augments the representation for a word with the representation vectors of the surrounding words in the dependency trees. The output of the GCNN layer is also a sequence of vectors to represent the contexts for the words in the sentence and functions as the abstract sequence A in our case. *BiLSTM-GCNN* (Zhang et al., 2018b) is one of the current state-of-the-art models for RE in the literature.

Note that there are many other variants of such models for RE in the literature (Xu et al., 2016; Zhang et al., 2017; Verga et al., 2018). However, as our goal in this paper is to evaluate different pooling mechanisms for RE, we focus on these standard representation learning methods to avoid the confounding effect of the complicated models, thus better revealing the effectiveness of the pooling methods.

2.3 The Pooling Component for RE

The goal of the pooling component is to aggregate the representation vectors in the abstract sequence A to constitute an overall vector F to represent the whole input sentence W and the two entity mentions of interest (i.e., $F = \mathrm{aggregate}(A)$). The overall representation vector should be able to capture the most important features induced in A. The typical method to achieve such aggregation in the RE models is to apply the element-wise max-pooling operation over subsets of vectors in A whose results are combined to obtain the overall representation vector. While there are different methods to select the vector subsets for the max-pooling operation, the prior work for RE has only employed one particular selection method in their deep learning models (Nguyen and Grishman, 2015a; Zhang et al., 2018b; Le et al., 2018). This raises the question about the impact of the other subset selection methods for such prior RE models. Can these methods benefit from different pooling mechanisms? What are the best pooling methods for the deep learning models in RE? In order to answer these questions, besides the architectures for the representation component in the previous section, we investigate the following subset selection methods for the pooling component of the RE models in this work:

1. *ENT-ONLY*: In this pooling method, we use the subsets of the vectors corresponding to the words in the two entity mentions of interest in

A for the max-pooling operations (i.e., M_1 with the words in the range $[s_1, e_1]$ and M_2 with the words in the range $[s_2, e_2]$). This is motivated by the utmost importance of the two entity mentions of interest for RE and employed in some prior work (Nguyen and Grishman, 2015a; Zhang et al., 2018b):

$$F_{M_1} = \text{max-pool}(a_{s_1}, a_{s_1+1}, \ldots, a_{e_1})$$
$$F_{M_2} = \text{max-pool}(a_{s_2}, a_{s_2+1}, \ldots, a_{e_2})$$
$$F_{ENT-ONLY} = [F_{M_1}, F_{M_2}]$$

2. *ENT-SENT*: Besides the entity mentions, the other context words in the sentence might also involve important information for the relation prediction in RE. For instance, in the sentence "*Acetazolamide can elevate cyclosporine levels.*", the context word "*elevate*" is crucial to determine the semantic relations between the two entity mentions of interest "*Acetazolamide* and "*cyclosporine*". In order to capture such important contexts for pooling, the typical approach in the prior work for RE is to perform the max-pooling operation over the abstract vectors for every word in the sentence (i.e., the whole set A) (Zeng et al., 2014; dos Santos et al., 2015; Le et al., 2018). The rationale is to select the features of the abstract vectors in A with the highest values in each dimension to reveal the most important context for RE. The max-pooled vector over the whole set A is combined with the $F_{ENT-ONLY}$ vector in this method:

$$F_{SENT} = \text{max-pool}(a_1, a_2, \ldots, a_n)$$
$$F_{ENT-SENT} = [F_{ENT-ONLY}, F_{SENT}]$$

3. *ENT-DYM*: Similar to *ENT-SENT*, this method also seeks the important context information beyond the two entity mentions of interest. However, instead of taking the whole vector sequence A for the pooling, *ENT-DYM* divides A into three separate vector subsequences based on the start and end indexes of the first and second entity mentions (i.e., s_1 and e_2) respectively. The max-pooling operation is then applied over these three subsequences and the resulting vectors are combined to form an overall vector (i.e., dynamic pooling) (Zeng et al., 2015):

$$F_{LEFT} = \text{max-pool}(a_1, a_2, \ldots, a_{s_1-1})$$
$$F_{MIDDLE} = \text{max-pool}(a_{s_1}, a_{s_1+1}, \ldots, a_{e_2})$$
$$F_{RIGHT} = \text{max-pool}(a_{e_2+1}, a_{e_2+2}, \ldots, a_n)$$
$$F_{ENT-DYM} = [F_{LEFT}, F_{MIDDLE}, F_{RIGHT},$$
$$F_{ENT-ONLY}]$$

4. *ENT-DEP0*: The previous pooling methods have only relied on the sequential structures of the sentence where the chosen subsets of A for pooling always contain vectors for the consecutive words in the sentence. Unfortunately, such sequential pooling might introduce irrelevant words into the selected subsets of A, potentially causing noise in the pooling features and impeding the performance of the RE models. For instance, in the previous sentence example "*Acetazolamide can elevate cyclosporine levels.*", the *ENT-SENT* and *ENT-DYM* methods woulds also include the word "*levels*" in the pooling subsets that is not very important for the relation prediction in this case. Consequently, in *ENT-DEP0*, we explore the possibility to use the dependency parse tree of the input sentence W to filter out the irrelevant words for the pooling operation. In particular, instead of considering every word in the input sentence, *ENT-DEP0* only pools over the abstract vectors in A that correspond to the words along the shortest dependency path (SDP) between the two entity mentions M_1 and M_2 in the dependency tree for W (called $SDP0(M_1, M_2)$). Note that the shortest dependency paths have been shown to be able to select the important context words for RE in many previous work (Zhou et al., 2005; Chan and Roth, 2010; Xu et al., 2016). Similar to *ENT-SENT* and *ENT-DYM*, we also include $F_{ENT-ONLY}$ in this method:

$$F_{DEP0} = \text{max-pool}_{a_i \in SDP0(M_1, M_2)}(a_i)$$
$$F_{ENT-DEP0} = [F_{DEP0}, F_{ENT-ONLY}]$$

5. *ENT-DEP1*: This method is similar to *ENT-DEP0*. However, instead of directly pooling over the words in the shortest dependency path $SDP0(M_1, M_2)$, *ENT-DEP1* extends this path to also include every word that is connected to some word in $SDP0(M_1, M_2)$ via an edge in the dependency tree for W (i.e., one edge distance from $SDP0(M_1, M_2)$). We denote this extended word set by $SDP1(M_1, M_2)$ for which the corresponding abstract vectors in A would be chosen for

the max-pooling operation. The motivation for $SDP1(M_1, M_2)$ is that the representations of the words close to the shortest dependency path between M_1 and M_2 might also provide useful information to improve the performance for RE. In our experiments, we find that one edge is the optimal distance to enlarge the shortest dependency paths. Using larger distance for the pooling mechanism would hurt the performance of the deep learning models for RE:

$$F_{DEP1} = \text{max-pool}_{a_i \in SDP1(M_1, M_2)}(a_i)$$

$$F_{ENT-DEP1} = [F_{DEP1}, F_{ENT-ONLY}]$$

Once the overall representation vector F for the input sentence W and the two entity mentions of interest has been produced, we feed it into a feed-forward neural network with a softmax layer in the end to obtain the probability distribution $P(y|W, M_1, M_2) = \text{feed-forward}(F)$ over the possible relation types for our RE problem. This probability distribution would then be used for both making prediction (i.e., by taking the relation type with the highest probability) and training models (i.e., by optimizing the negative log-likelihood function).

3 Experiments

3.1 Datasets

In order to evaluate the performance of the models in this work, we employ the following biomedical datasets for RE in the experiments:

DDI-2013 (Herrero-Zazo et al., 2013): This dataset contains 730 documents from the Drugbank database, involving about 25,000 examples for the training and test sets (each example consists of a sentence and two entity mentions of interest for classification). There are 4 entity types (i.e., *drug*, *brand*, *group* and *brand_n*) and 5 relation types (i.e., *mechanism*, *advise*, *effect*, *int*, and *no_relation*) in this dataset. The *no_relation* is to indicate any example that does not belong to any relation types of interest. This dataset is severely imbalanced, containing 85% negative examples in the training dataset. In order to deal with such imbalanced data, we employ weighted sampling that equally distributes the selection probability for the positive and negative examples.

BB3 (Deléger et al., 2016). This dataset contains 95 documents; each of them involves a title and abstract from a document from the PubMed database. There are 800 examples in this dataset divided into two separate sets (i.e., the training set and the validation set). BB3 also include a test set; however, the relation types for the examples in this test set are not provided. In order to obtain the performance of the models on the test set, the performers need to submit their system outputs to an official API that would evaluate the output and return the model performance. We train the models in this work on the training data and employ the official API to obtain their test set performance to be reported in the experiments for this dataset.

Following the prior work on these datasets (Chowdhury and Lavelli, 2013; Lever and Jones, 2016; Zhou et al., 2018; Le et al., 2018), we use the micro-averaged F1 scores as the performance measure in the experiments to ensure a compatible comparison.

3.2 Parameters and Resources

As the DDI-2013 dataset does not involve a development set, we tune the parameters for the models in this work based on the validation data of the BB3 dataset and use the selected parameters for both datasets in the experiments. The best parameters from this tuning process include the learning rate of 0.5 and momentum of 0.8 for the stochastic gradient descent (SGD) optimizer with nesterov's momentum to optimize the models. In order to regularize the models, we apply dropout between layers with the drop rate for word embeddings set to 0.7 and other drop rates set to 0.5. We also employ the weight dropout *DropConnect* in (Wan et al., 2013) to regularize the hidden-to-hidden transition matrix within each bidirectional LSTM in the models (Merity et al., 2017). For all the models that involve bidirectional LSTMs (i.e., *BiLSTM*, *BiLSTM-CNN*, and *BiLSTM-GCNN*), two layers of bidirectional LSTMs are utilized with 300 hidden units for each LSTM network. For the models with CNN components (i.e., *CNN* and *BiLSTM-CNN*), we use one CNN layer with multiple window sizes of 2, 3, 4, and 5 for the filters (200 filters for each window size). For the *BiLSTM-GCN* model, two GCNN layers are employed with 300 hidden units in each layer. Finally, for the final feed-forward neural network to compute the probability distribution (i.e., feed-forward), we utilize two hidden layers for which 1000 hidden units are used for the first layer and the number of hidden units for the second layer is determined by the number of relation types in the datasets.

3.3 Evaluating the Pooling Methods for RE

This section evaluates the performance of different pooling methods when they are applied to the deep learning models for RE on the two datasets DDI-2013 and BB3. In particular, we integrate each of the pooling methods in Section 2.3 (i.e., *ENT-ONLY, ENT-SENT, ENT-DYM, END-DEP0*, and *END-DEP1*) into each of the deep learning models in Section 2.2 (i.e., *CNN, BiLSTM, BiLSTM-CNN*, and *BiLSTM-GCN*), resulting 20 different model combinations to be investigated in this section. For each model combination, we train five versions of the model with different random seeds for parameter initialization over the training datasets. The performance of such versions over the test sets is averaged to serve as the overall model performance on the corresponding dataset. Tables 1 and 2 report the performance of the models on the DDI-2013 dataset and BB3 dataset respectively.

Model	P	R	F1
CNN			
+ *ENT-ONLY*	52.7	43.1	47.4
+ *ENT-SENT*	75.8	60.7	67.3
+ *ENT-DYM*	66.5	70.6	68.5
+ *ENT-DEP0*	59.8	61.5	60.6
+ *ENT-DEP1*	67.6	65.1	66.3
BiLSTM			
+ *ENT-ONLY*	74.0	69.4	71.6
+ *ENT-SENT*	74.8	71.7	73.1
+ *ENT-DYM*	71.5	73.4	72.4
+ *ENT-DEP0*	72.8	69.4	71.1
+ *ENT-DEP1*	71.6	76.4	**73.9**
BiLSTM-CNN			
+ *ENT-ONLY*	69.6	72.3	70.9
+ *ENT-SENT*	69.4	74.9	72.0
+ *ENT-DYM*	71.0	69.7	71.8
+ *ENT-DEP0*	72.2	69.5	70.8
+ *ENT-DEP1*	71.0	74.3	72.6
BiLSTM-GCNN			
+ *ENT-ONLY*	69.3	71.4	70.4
+ *ENT-SENT*	72.2	71.9	72.0
+ *ENT-DYM*	69.7	73.9	71.7
+ *ENT-DEP0*	70.1	71.1	70.6
+ *ENT-DEP1*	72.7	72.9	72.8

Table 1: Results on DDI 2013

From the tables, we have the following observations about the effectiveness of the pooling methods for RE with deep learning:

1. Comparing *ENT-SENT, ENT-DYM* and *ENT-ONLY*, we see that the pooling methods over the whole sentence (i.e., *ENT-SENT* and *ENT-DYM*) are significantly better than *ENT-ONLY* that only focuses on the two entity mentions of interest in

Model	P	R	F1
CNN			
+ *ENT-ONLY*	54.2	65.7	59.1
+ *ENT-SENT*	55.0	62.5	59.1
+ *ENT-DYM*	54.6	53.3	53.5
+ *ENT-DEP0*	55.9	65.8	60.6
+ *ENT-DEP1*	55.7	67.7	61.1
BiLSTM			
+ *ENT-ONLY*	58.9	59.6	59.2
+ *ENT-SENT*	60.7	59.2	59.9
+ *ENT-DYM*	50.2	66.0	56.9
+ *ENT-DEP0*	51.6	78.0	61.9
+ *ENT-DEP1*	54.7	72.6	62.4
BiLSTM-CNN			
+ *ENT-ONLY*	56.4	66.2	60.8
+ *ENT-SENT*	53.6	69.2	60.5
+ *ENT-DYM*	47.1	78.0	58.7
+ *ENT-DEP0*	55.9	71.4	**62.5**
+ *ENT-DEP1*	54.1	74.7	62.4
BiLSTM-GCNN			
+ *ENT-ONLY*	62.7	56.1	58.9
+ *ENT-SENT*	58.4	58.7	58.5
+ *ENT-DYM*	56.8	58.4	56.6
+ *ENT-DEP0*	55.6	67.4	60.8
+ *ENT-DEP1*	54.4	71.1	61.5

Table 2: Results on BioNLP BB3

the DDI-2013 dataset. This is true across different deep learning models in this work. However, this comparison is reversed for the BB3 dataset where *ENT-ONLY* is in general better or comparable to *ENT-SENT* and *ENT-DYM* over different deep learning models. We attribute such phenomena to the fact that the BB3 dataset often contains many entity mentions and relations within a single sentence (i.e., overlapping contexts) while the sentences in DDI-2013 tend to involve only a single relation with few entity mentions. This make *ENT-SENT* and *ENT-DYM*) ineffective for BB3 as the pooling mechanisms over the whole sentence are likely to involve the contexts for the other entity mentions and relations in the sentences, causing the low quality of the resulting representations and the confusion of the model for the relation prediction. This problem is less severe in DDI-2013 as the context of the whole sentence (with a single relation) is more aligned with the important context for the relation prediction. We call the many entity mentions and relations in a single single sentence of BB3 as the multiple relation effect for convenient discussion in this paper.

2. Comparing *ENT-SENT* and *ENT-DYM*, their performance are comparable in DDI-2013 (except for *CNN* where *ENT-DYM* is better); however, in the BB3 dataset, *ENT-SENT* significantly outperforms *ENT-DYM* over all the models. This suggests the amplification of the multiple relation ef-

fect in BB3 due to *ENT-DYM* where the separation of the sentence context for pooling encourages the emergence of context information for multiple relations in the final representation vector and increases the confusion of the models.

3. Comparing the syntax-based pooling methods and the non-syntax pooling methods, the pooling based on dependency paths (i.e., *ENT-DEP0*) is worse than the non-syntax pooling methods (i.e., *ENT-SENT* and *ENT-DYM*) and perform comparably with *ENT-ONLY* in the DDI-2013 dataset over all the models (except for the *CNN* model where *ENT-ONLY* is much worse). These evidences suggest that the dependency paths themselves are not able to capture effective contexts for the pooling operation beyond the entity mentions for biomedical RE in DDI-2013. However, when we switch to the BB3 dataset, it turns out that *ENT-DEP0* is significantly better than all the non-syntax pooling methods (i.e., *ENT-ONLY*, *ENT-SENT* and *ENT-DYM*) for all the comparing models. This can be explained by the multiple relation effect in BB3 for which the dependency paths help to identify the most related context words for the two given entity mentions and filter out the confusing context words for the other relations in the sentences. The models would thus become less confused with different contexts for multiple relations as those in *ENT-SENT* and *ENT-DYM* for better performance in this case.

4. Finally, among all the pooling methods, we find that *ENT-DEP1* significantly outperforms the other pooling methods across different models and datasets (except the *CNN* model on DDI-2013 and *BiLSTM* on BB3). In particular, the performance improvement is substantial over the non-syntax pooling methods in BB3 where *ENT-DEP1* is up to 2% better than *ENT-SENT*, *ENT-DYM* and *ENT-ONLY* on the absolute F1 scores. This helps to demonstrate the benefits of *ENT-DEP1* for biomedical RE to both recognize the important context words for pooling in DDI-2013 and reduce the confusion effect of the multiple relations in single sentences for the models in BB3.

3.4 Comparing the Deep Learning Models for RE

Regarding the comparison among different deep learning models, the major observations from from Tables 1 and 2 include:

1. The performance of *CNN* is in general worse that the other models with the bidirectional LSTM components (i.e., *BiLSTM*, *BiLSTM-CNN* and *BiLSTM-GCN*) over different pooling methods and datasets. This illustrates the importance of bidirectional LSTMs to capture the effective feature representations for biomedical RE.

2. Comparing *BiLSTM* and *BiLSTM-CNN*, we find that *BiLSTM* is better in DDI-2013 while *BiLSTM-CNN* achieves better performance in BB3 (over different pooling methods). In other words, the CNN layer is only helpful for the *BiLSTM* model in the BB3 dataset. This can also be attributed to the multiple relation effect in BB3 where the CNN layer helps to further abstract the representations from *BiLSTM* to better reveal the underlying structures in such confusing and complicated contexts in the sentences of BB3 for RE.

3. Graph convolutions over the dependency trees are not effective for biomedical RE as incorporating it into the *BiLSTM* model hurts the performance significantly. In particular, *BiLSTM-GCNN* is significantly worse than *BiLSTM* no matter which pooling methods are applied and which datasets are used for evaluation.

4. Interestingly, comparing the *BiLSTM* model with the *ENT-DEP1* pooling method (i.e., *BiLSTM + ENT-DEP1*) and the *BiLSTM-GCN* model with the non-syntax pooling methods (i.e., *ENT-ONLY*, *ENT-SENT* and *ENT-DYM*), we see that *BiLSTM + ENT-DEP1* is significantly better with large performance gaps over both datasets DDI-2013 and BB3. For example, *BiLSTM + ENT-DEP1* is 1.9% better than *BiLSTM-GCNN + ENT-SENT* in the DDI-2013 dataset and 3.5% better than *BiLSTM-GCNN + ENT-ONLY* in BB3 with respect to the absolute F1 scores. In fact, *BiLSTM + ENT-DEP1* also achieves the best performance among the compared models in this section for both datasets. The major difference between *BiLSTM + ENT-DEP1* and *BiLSTM-GCN* with the non-syntax pooling methods lies at the specific component of the models where the syntactic information (i.e., the dependency trees) is applied. In *BiLSTM-GCN* with the non-syntax pooling methods, the syntactic information is employed in the representation learning component while in *BiLSTM + ENT-DEP*, the application of the syntactic information is postponed all the way to the pooling component. Our experiments thus demonstrate that it is more effective to utilize the syntactic information in the pooling component

than in the representation learning component of the deep learning models for biomedical RE. This is an interesting and unique observation given that the prior work for RE has only focused on using the syntactic information in the representation component and never explicitly investigated the effectiveness of the syntactic information for the pooling component of the deep learning models.

3.5 Comparing to the State-of-the-art Models

In order to further demonstrate the advantage of the syntactic information for the pooling component for biomedical RE, this section compares *BiLSTM + ENT-DEP1* (i.e., the best model with the *ENT-DEP1* pooling in this work) with the best reported models on the two datasets DDI-2013 and BB3. For a fair comparison between models, we select the previous single (non-ensemble) models for the comparison in this section. Tables 3 and 4 presents the model performance.

Models	P	R	F1
(Raihani and Laachfoubi, 2017)	73.6	70.1	71.8
(Zhang et al., 2018a)	74.1	71.8	72.9
(Zhou et al., 2018)	75.8	70.3	73.0
(Björne and Salakoski, 2018)	75.3	66.3	70.5
BiLSTM + ENT-DEP1	71.6	76.4	**73.9**

Table 3: Comparison with the state-of-the-art systems on the DDI-2013 test set

Models	P	R	F1
(Lever and Jones, 2016)	51.0	61.5	55.8
(Mehryary et al., 2016)	62.3	44.8	52.1
(Li et al., 2016)	56.3	58.0	57.1
(Le et al., 2018)	59.8	51.3	55.2
BiLSTM + ENT-DEP1	54.7	72.6	**62.4**

Table 4: Comparison with the state-of-the-art systems on the BB3 test set

The most important observation from the tables is that the *BiLSTM* model, once combined with the *ENT-DEP1* pooling method, significantly outperforms the previous models on DDI-2013 and BB3, establishing new state-of-the-art performance for these datasets. In particular, in the DDI-2013 dataset, *BiLSTM + ENT-DEP1* is 0.9% better than the current state-of-the-art model in (Zhou et al., 2018) while the performance improvement over the best reported model for BB3 in (Li et al., 2016) is 5.3% (over the absolute F1 scores). Such substantial improvement clearly demonstrates the ad-

vantages of the syntactic information and its delayed application in the pooling component of the deep learning models for biomedical RE.

4 Related Work

Traditional work on RE has mostly used feature engineering with syntactical information for statistical or kernel based classifiers (Zelenko et al., 2002; Zhou et al., 2005; Bunescu and Mooney, 2005; Sun et al., 2011; Chan and Roth, 2010). Recently, deep learning has been shown to advance many benchmark datasets for this RE problem due to its representation learning capacity. The typical architectures for such deep learning models involve CNN, LSTM, the attention mechanism and their variants (Zeng et al., 2014; dos Santos et al., 2015; Zhou et al., 2016; Wang et al., 2016; Nguyen and Grishman, 2015a; Miwa and Bansal, 2016; Zhang et al., 2017, 2018b). Deep learning has also been applied to biomedical RE in the last couple of years and started to demonstrate much potentials for this area (Mehryary et al., 2016; Björne and Salakoski, 2018; Nguyen and Verspoor, 2018; Verga et al., 2018).

Pooling is a common and crucial component in most of the deep learning models for RE. (Nguyen and Grishman, 2015b; dos Santos et al., 2015) apply the pooling operation over the whole sentence for RE while Zeng et al. (2015) proposes the dynamic pooling mechanism in the CNN models. However, none of these prior work systematically examines different pooling mechanisms for deep learning in RE as we do in this work.

5 Conclusion

We conduct a comprehensive study on the effectiveness of different pooling mechanisms for the deep learning models in biomedical relation extraction. Our experiments suggest that the pooling mechanisms have a significant impact on the performance of the deep learning models and a careful evaluation should be done to decide the appropriate pooling mechanism for the biomedical RE problem. From the experiments, we also find that syntactic information (i.e., dependency parsing) provides the best pooling methods for the models and biomedical RE datasets we investigate in this work (i.e., *ENT-DEP1*). We achieve the state-of-the-art performance for biomedical RE over the two datasets DDI-2013 and BB3 with such syntax-based pooling methods.

References

Jari Björne and Tapio Salakoski. 2018. Biomedical Event Extraction Using Convolutional Neural Networks and Dependency Parsing. In *Proceedings of the BioNLP 2018 Workshop*, pages 98–108.

Razvan Bunescu and Raymond Mooney. 2005. A Shortest Path Dependency Kernel for Relation Extraction. In *Proceedings of the EMNLP-HLT 2005*, pages 724–731.

Yee S. Chan and Dan Roth. 2010. Exploiting background knowledge for relation extraction. In *COLING*.

Md Faisal Mahbub Chowdhury and Alberto Lavelli. 2013. FBK-irst : A Multi-Phase Kernel Based Approach for Drug-Drug Interaction Detection and Classification that Exploits Linguistic Information. In *Proceedings of the Seventh International Workshop on Semantic Evaluation*, pages 351–355.

Louise Deléger, Robert Bossy, Estelle Chaix, Mouhamadou Ba, Arnaud Ferré, Philippe Bessières, and Claire Nédellec. 2016. Overview of the Bacteria Biotope Task at BioNLP Shared Task 2016. In *Proceedings of the BioNLP 2016 Workshop*, pages 12–22.

Cicero dos Santos, Bing Xiang, and Bowen Zhou. 2015. Classifying Relations by Ranking with Convolutional Neural Networks. In *Proceedings of the IJCNLP 2015*, pages 626–634.

María Herrero-Zazo, Isabel Segura-Bedmar, Paloma Martínez, and Thierry Declerck. 2013. The DDI corpus: An annotated corpus with pharmacological substances and drug–drug interactions. *Journal of Biomedical Informatics*, 46(5):914–920.

Hoang-Quynh Le, Duy-Cat Can, Sinh T. Vu, Thanh Hai Dang, Mohammad Taher Pilehvar, and Nigel Collier. 2018. Large-scale Exploration of Neural Relation Classification Architectures. In *Proceedings of the EMNLP 2018*, pages 2266–2277.

Jake Lever and Steven JM Jones. 2016. VERSE: Event and Relation Extraction in the BioNLP 2016 Shared Task. In *Proceedings of the BioNLP 2016 Workshop*, pages 42–49.

L. Li, and, and D. Huang and. 2016. Biomedical event extraction via Long Short Term Memory networks along dynamic extended tree. In *Proceedings of the IEEE-BIBM 2016*, pages 739–742.

Yankai Lin, Shiqi Shen, Zhiyuan Liu, Huanbo Luan, and Maosong Sun. 2016. Neural Relation Extraction with Selective Attention over Instances. In *Proceedings of the ACL 2016*, pages 2124–2133.

Yang Liu, Furu Wei, Sujian Li, Heng Ji, Ming Zhou, and Houfeng Wang. 2015. A dependency-based neural network for relation classification. *arXiv preprint arXiv:1507.04646*.

Farrokh Mehryary, Jari Björne, Sampo Pyysalo, Tapio Salakoski, and Filip Ginter. 2016. Deep Learning with Minimal Training Data: TurkuNLP Entry in the BioNLP Shared Task 2016. In *Proceedings of the BioNLP 2016 Workshop*, pages 73–81.

Stephen Merity, Nitish Shirish Keskar, and Richard Socher. 2017. Regularizing and Optimizing LSTM Language Models. In *Proceedings of the ICLR 2018*.

Tomas Mikolov, Kai Chen, Greg Corrado, and Jeffrey Dean. 2013. Efficient Estimation of Word Representations in Vector Space. *arXiv:1301.3781 [cs]*.

Makoto Miwa and Mohit Bansal. 2016. End-to-End Relation Extraction using LSTMs on Sequences and Tree Structures. In *Proceedings of the ACL 2016*, pages 1105–1116.

Dat Quoc Nguyen and Karin Verspoor. 2018. Convolutional neural networks for chemical-disease relation extraction are improved with character-based word embeddings. In *Proceedings of the BioNLP 2018 Workshop*, pages 129–136.

Thien Huu Nguyen and Ralph Grishman. 2015a. Combining neural networks and log-linear models to improve relation extraction. *arXiv preprint arXiv:1511.05926*.

Thien Huu Nguyen and Ralph Grishman. 2015b. Relation Extraction: Perspective from Convolutional Neural Networks. In *Proceedings of the 1st Workshop on Vector Space Modeling for Natural Language Processing*, pages 39–48.

Truc-Vien T. Nguyen, Alessandro Moschitti, and Giuseppe Riccardi. 2009. Convolution kernels on constituent, dependency and sequential structures for relation extraction. In *EMNLP*.

Anass Raihani and Nabil Laachfoubi. 2017. A Rich Feature-based Kernel Approach for Drug- Drug Interaction Extraction. In *In Proceedings of the IJACSA 2017*, volume 8.

Ang Sun, Ralph Grishman, and Satoshi Sekine. 2011. Semi-supervised relation extraction with large-scale word clustering. In *ACL*.

Patrick Verga, Emma Strubell, and Andrew McCallum. 2018. Simultaneously Self-Attending to All Mentions for Full-Abstract Biological Relation Extraction. In *Proceedings of the NAACL-HLT 2018*, pages 872–884.

Li Wan, Matthew Zeiler, Sixin Zhang, Yann Le Cun, and Rob Fergus. 2013. Regularization of Neural Networks using DropConnect. In *Proceedings of the ICML 2013*, pages 1058–1066.

Linlin Wang, Zhu Cao, Gerard de Melo, and Zhiyuan Liu. 2016. Relation Classification via Multi-Level Attention CNNs. In *Proceedings of the ACL 2016*, pages 1298–1307.

Yan Xu, Ran Jia, Lili Mou, Ge Li, Yunchuan Chen, Yangyang Lu, and Zhi Jin. 2016. Improved relation classification by deep recurrent neural networks with data augmentation. In *Proceedings of the COLING 2016*, pages 1461–1470.

Yan Xu, Lili Mou, Ge Li, Yunchuan Chen, Hao Peng, and Zhi Jin. 2015. Classifying Relations via Long Short Term Memory Networks along Shortest Dependency Paths. In *Proceedings of the EMNLP 2015*, pages 1785–1794.

Dmitry Zelenko, Chinatsu Aone, and Anthony Richardella. 2002. Kernel Methods for Relation Extraction. In *Proceedings of the EMNLP 2002*, pages 71–78.

Daojian Zeng, Kang Liu, Yubo Chen, and Jun Zhao. 2015. Distant Supervision for Relation Extraction via Piecewise Convolutional Neural Networks. In *Proceedings of the EMNLP 2015*, pages 1753–1762.

Daojian Zeng, Kang Liu, Siwei Lai, Guangyou Zhou, and Jun Zhao. 2014. Relation Classification via Convolutional Deep Neural Network. In *Proceedings of the COLING 2014*, pages 2335–2344.

Wenyuan Zeng, Yankai Lin, Zhiyuan Liu, and Maosong Sun. 2017. Incorporating Relation Paths in Neural Relation Extraction. In *Proceedings of the EMNLP 2017*, pages 1768–1777.

Yijia Zhang, Wei Zheng, Hongfei Lin, Jian Wang, Zhihao Yang, and Michel Dumontier. 2018a. Drug-Drug Interaction Extraction Via Hierarchical Rnns on Sequence and Shortest Dependency Paths. *Bioinformatics*, 34(5):828–835.

Yuhao Zhang, Peng Qi, and Christopher D. Manning. 2018b. Graph Convolution over Pruned Dependency Trees Improves Relation Extraction. In *Proceedings of the EMNLP 2018*, pages 2205–2215.

Yuhao Zhang, Victor Zhong, Danqi Chen, Gabor Angeli, and Christopher D. Manning. 2017. Position-aware Attention and Supervised Data Improve Slot Filling. In *Proceedings of the EMNLP 2017*, pages 35–45.

Deyu Zhou, Lei Miao, and Yulan He. 2018. Position-Aware Deep Multi-Task Learning for Drug–Drug Interaction Extraction. *Journal of Artificial Intelligence in Medicine*, 87:1–8.

Guodong Zhou, Jian Su, Jie Zhang, and Min Zhang. 2005. Exploring Various Knowledge in Relation Extraction. In *Proceedings of the ACL 2005*, pages 427–434.

Peng Zhou, Wei Shi, Jun Tian, Zhenyu Qi, Bingchen Li, Hongwei Hao, and Bo Xu. 2016. Attention-Based Bidirectional Long Short-Term Memory Networks for Relation Classification. In *Proceedings of the ACL 2016*, pages 207–212.

Syntax-aware Multi-task Graph Convolutional Networks
for Biomedical Relation Extraction

Diya Li[*], Heng Ji[†‡]

[*] Computer Science Department, Rensselaer Polytechnic Institute
lid18@rpi.edu
[†] Department of Computer Science [‡] Department of Electrical and Computer Engineering
University of Illinois at Urbana-Champaign
hengji@illinois.edu

Abstract

In this paper we tackle two unique challenges
in biomedical relation extraction. The first
challenge is that the contextual information
between two entity mentions often involves
sophisticated syntactic structures. We propose
a novel graph convolutional networks model
that incorporates dependency parsing and contextualized embedding to effectively capture
comprehensive contextual information. The
second challenge is that most of the benchmark data sets for this task are quite imbalanced because more than 80% mention pairs
are negative instances (i.e., no relations). We
propose a multi-task learning framework to
jointly model relation identification and classification tasks to propagate supervision signals from each other and apply a focal loss
to focus training on ambiguous mention pairs.
By applying these two strategies, experiments
show that our model achieves state-of-the-art
F-score on the 2013 drug-drug interaction extraction task.

1 Introduction

Recently relation extraction in biomedical literature has attracted increasing interests from medical language processing research community as
an important stage for downstream tasks such as
question answering (Hristovski et al., 2015) and
decision making (Agosti et al., 2019). Biomedical relation extraction aims to identify and classify relations between two entity mentions into
pre-defined types based on contexts. In this paper
we aim to extract drug-drug interactions (DDIs),
which occur when taking two or more drugs within
a certain period of time that alters the way one
or more drugs act in human body and may result
in unexpected side effects (Figure 1). Extracting
DDI provides important clues for research in drug
safety and human health care.

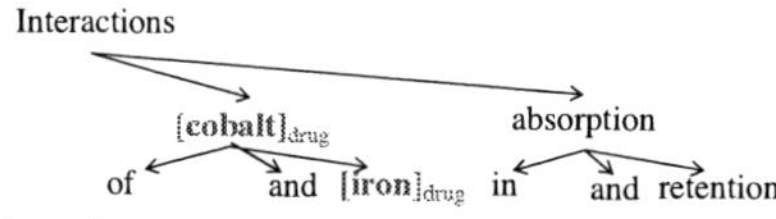

Figure 1: Example of drug-drug interaction on dependency tree.

Dependency parses are widely used in relation
extraction task due to the advantage of shortening the distance of words which are syntactically
related. As shown in Figure 1, the partial dependency path $\{iron \leftarrow cobalt \leftarrow interactions\}$
reveals that these two drugs are interactive, and
the path $\{interactions \rightarrow absorption \rightarrow retention\}$
further indicates the *mechanism* relation between
these two mentions. Therefore capturing the syntactic information involving the word *interaction*
on the dependency path $\{iron \leftarrow cobalt \leftarrow in-$
$teractions \rightarrow absorption \rightarrow retention\}$ can effectively help on the classification of the relation between these two mentions $\langle cobalt, iron \rangle$. In order to capture indicative information from wide
contexts, we adopt the graph convolutional networks (GCN) (Kipf and Welling, 2016; Marcheggiani and Titov, 2017) to obtain the syntactic information by encoding the dependency structure over
the input sentence with graph convolution operations. To compensate the loss of local context information in GCN, we incorporate the contextualized word representation pre-trained by the BERT
model (Devlin et al., 2019) in large-scale biomedical corpora containing over 200K abstracts from
PubMed and over 270K full texts from PMC (Lee
et al., 2019) .

Moreover, we notice that data imbalance is another major challenge in biomedical text as the distribution of relations among biomedical mentions

Proceedings of the 10th International Workshop on Health Text Mining and Information Analysis (LOUHI 2019), pages 28–33
Hong Kong, November 3, 2019. ©2019 Association for Computational Linguistics
https://doi.org/10.18653/v1/D19-62

are usually very sparse. Over 80% candidate mention pairs have no relation in DDI 2013 (Herrero-Zazo et al., 2013) training set. To tackle this problem, we propose a binary relation identification task as an auxiliary task to facilitate the main multi-classification task. For instance, the detection of drug interaction on dependency path {*iron* ← *cobalt* ← **interactions** → *absorption* → *retention*} will assist the prediction of the relation type *mechanism* by using the signals from binary classification as an inductive bias to avoid misclassifying it as no relation. We also exploit the focal loss (Lin et al., 2017) to potentially help the multi-class relation classification task by forcing the loss implicitly focus on ambiguous examples.

To recap, our contributions are twofold: First, we adopt the syntax-aware graph convolutional networks incorporating contextualized representation. Second, we further design an auxiliary task to solve the data imbalance problem, which achieves the state-of-the-art micro F-score on the DDIExtraction 2013 shared task.

2   Methods

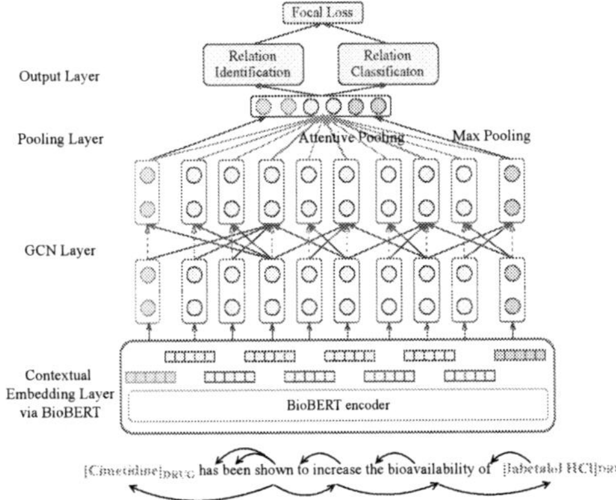

Figure 2: Framework of syntax-aware multi-task graph convolutional networks.

2.1   Contextual and Syntax-aware GCN

As a variant of the convolutional neural networks (LeCun et al., 1998), the graph convolutional networks (Kipf and Welling, 2016) is designed for graph data and it has been proven effective in modeling text data via syntactic dependency graphs (Marcheggiani and Titov, 2017).

We encode the tokens in a biomedical sentence of size n as $x = \{x_1, \ldots, x_n\}$, where x_i is a vector which concatenates the representation of the token i and the position embeddings corresponding to the relative positions from candidate mention pairs. We feed the token vectors into a L-layer GCN to obtain the hidden representations of each token which are directly influenced by its neighbors no more than L edges apart in the dependency tree. We apply the Stanford dependency parser (Chen and Manning, 2014) to generate the dependency structure:

$$h_i^{(l)} = \sigma\left(\sum_{j=1}^{n} \widetilde{A}_{ij} W^{(l)} h_j^{l-1} / d_i + b^{(l)}\right)$$

where $\widetilde{A} = A + I$ with A is the adjacent matrix of tokens in dependency tree, I is the identity matrix. $W^{(l)}$ is a linear transformation, $b^{(l)}$ is a bias term, and σ is a nonlinear function. Following Zhang et al. (2018), d_i is the degree of the token i in dependency tree with an additional self-loop.

We notice that some token representations are more informative by gathering information from syntactically related neighbors through GCN. For example, the representation of the token *interactions* from a 2-layer GCN operating on its two edges apart neighbors provides inductive information for predicting a *mechanism* relation. Thus, we adopt attentive pooling (Zhou et al., 2016) to achieve the optimal pooling:

$$\alpha = softmax(w^T \tanh(h))$$

$$h_{attentive} = h\alpha^T$$

where w is a trained parameter to assign weights based on the importance of each token representation.

We obtain the final representation by concatenating the sentence from attentive pooling and the mention representations from max pooling. We finally obtain the prediction of relation type by feeding the final representations into a fully connected neural network followed by a softmax operation.

Graph neural networks (Zhou et al., 2018b) can learn effective representations but suffer from the loss of local context information. We believe the local context information is also crucial for biomedical relation extraction. For example, in the following sentence *"The response to [Factrel]DRUG may be blunted by [phenothiazines]DRUG and [dopamine antagonists]DRUG "*, it's intuitive to tell *Factrel* and *phenothiazines* are interactive while *phenothiazines* and *dopamine antagonists* have no interaction according to the sentence order. However, GCNs treat the three drugs as interacting with each other as they are close in dependency structure with no order information.

BERT (Devlin et al., 2019) is a recently proposed model based on a multi-layer bidirectional Transformer (Vaswani et al., 2017). Using pre-trained BERT has been proven effective to create contextualized word embeddings for various NLP tasks (Han et al., 2019; Wang et al., 2019). The BioBERT (Lee et al., 2019) is a biomedical language representation model pre-trained on large-scale biomedical corpora. The output of each encoder layer of the input token can be used as a feature representation of that token. As shown in Figure 2, we encode the input tokens as contextualized embeddings by leveraging the last hidden layer of the corresponding token in BioBERT. As the BERT model uses WordPiece (Wu et al., 2016) to decompose infrequent words into frequent subwords for unsupervised tokenization of the input token, if the token has multiple BERT subword units, we use the first one. After getting the contextualized embedding of each token, we feed them into the GCN layer to make our model context-aware.

2.2 Auxiliary Task Learning with Focal Loss

In the DDIExtraction 2013 task, all possible interactions between drugs within one sentence are annotated, which means a single sentence with multiple drug mentions will lead to separate instances of candidate mention pairs (Herrero-Zazo et al., 2013). There are 21,012 mention pairs generated from 3,790 sentences in training set and over 80% have no relations. This data imbalance problem due to sparse relation distribution is a main reason for low recall in DDI task (Zhou et al., 2018a; Sun et al., 2019).

Here we address this relation type imbalance problem by adding an auxiliary task on top of the syntax-aware GCN model. To conduct the auxiliary task learning, we add a separate binary classifier for relation identification as shown in Figure 2. All classifiers share the same GCN representation and contextualized embeddings, and thus they can potentially help each other by propagating their supervision signals.

Additionally, instead of setting the objective function as the negative log-likelihood loss, here we optimize the parameters in training by minimizing a focal loss (Lin et al., 2017) which focuses on hard relation types. For instance, the *int* relation indicates drug interaction without providing any extra information (*e.g.*, *Some [anticonvulsants]DRUG may interact with [Mephenytoin]DRUG*). This relation type only accounts for 0.82% in training set and is often misclassified into other relation types. We denote t_i and p_i as the ground truth and the conditional probability value of the type i in relation types C, the focal loss can be defined as:

$$\mathcal{L} = -\sum_{i}^{C} (\alpha_i (1 - p_i)^\gamma t_i \log(p_i)) + \lambda ||\theta||^2$$

where α is a weighting factor to balance the importance of samples from various types, γ is the focusing parameter to reduce the influence of well-classified samples in the loss. λ is the *L2* regularization parameter and θ is the parameter set.

The auxiliary task along with the focal loss enhances our model's ability to handle imbalance data by leveraging the inductive signal from the easier identification task and meanwhile downweighting the influence of easy classified instances thus directing the model to focus on difficult relation types.

3 Experiments

3.1 Datasets and Task Settings

System	Prec	Rec	F1
CNN (Liu et al., 2016)	75.70	64.66	69.75
Multi Channel CNN (Quan et al., 2016)	75.99	65.25	70.21
GRU (Yi et al., 2017)	73.67	70.79	72.20
AB-LSTM (Sahu and Anand, 2018)	74.47	64.96	69.39
CNN-GCNN (Asada et al., 2018)	73.31	71.81	72.55
Position-aware LSTM (Zhou et al., 2018a)	75.80	70.38	72.99
RHCNN (Sun et al., 2019)	77.30	73.75	**75.48**
LSTM baseline	69.34	62.74	65.88
GCN baseline	71.96	67.14	69.47
–without attentive pooling	77.12	75.03	76.06
–without BioBERT	76.51	73.56	75.01
–without multi-task learning	76.01	71.92	73.91
Our Model	77.62	75.69	**76.64**

Table 1: Precision (Prec), recall (Rec) and micro F-score (F1) results on DDI 2013 corpus.

We evaluate our model on the DDIExtraction 2013 relation dataset (Herrero-Zazo et al., 2013). The corpus is annotated with drug mentions and their four types of interactions: *Mechanism* (pharmacokinetic mechanism of a DDI), *Effect* (effect of a DDI), *Advice* (a recommendation or advice regarding a DDI) and *Int* (a DDI simply occurs without extra information). We randomly choose 10% from the training dataset as the development set. Following previous work (Liu et al., 2016; Quan et al., 2016; Zhou et al., 2018a; Sun et al., 2019), we use a negative instance filtering strategy to filter out some negative drug pairs based on manually-formulated rules. Instances containing drug pair referring to the same thing and drug pair appearing in the same coordinate structure with more than two drugs (*e.g.*, drug1, drug2, and drug3) will be filtered. Entity mentions are masked with *DRUG* for better generalization and avoiding overfitting.

We train the model with GCN hidden state size of 200, the SGD optimizer with a learning rate of 0.001, a batch size of 30, and 50 epochs. Dropout is applied with a rate of 0.5 for regularization. The contextual embedding size from BioBERT is 768. The focusing parameter γ is set as 1. All hyperparameters are tuned on the development set.

3.2 Results and Analysis

The experiment results are reported from a 2-layer GCN which achieves the best performance and shown in Table 1. Our model significantly outperforms all previous methods at the significance level of 0.05. To analyze the contributions and effects of the various components in our model, we also perform ablation tests. The ablated GCN model outperforms the LSTM baseline by 3.6% F1 score, which demonstrates the effectiveness of GCN on modeling mention relations through dependency structure. The utilization of contextualized embedding from BioBERT which encodes the contextual information involving sequence order and word disambiguation implicitly helps the model to learn contextual relation patterns, therefore the performance is further improved. We obtain a significant F-score improvement (2.7%) by applying multi-task learning. As over 80% mention pairs are negative samples, the multi-task learning effectively solves the problem by jointly modeling relation identification and classification tasks and applying focal loss to focusing on ambiguous mention pairs, and thus we also gain 3.8% absolute score on recall. Specifically, the F1 score of *int* type is increased from 54.38% to 59.79%.

For the remaining errors, we notice that our model often fails to predict relations when the sentence are parsed poorly due to the complex content which suggests us to seek for more powerful parser tools. Besides, we also observe some errors occurring in extremely short sentences. For example, in the following sentence "*[Calcium]$_{DRUG}$ Supplements/[Antacids]$_{DRUG}$*", our model cannot capture informative representations as the mentions are masked with *DRUG* and the sentence is too concise to offer indicative evidence.

4   Related Work

Traditional feature/kernel-based models for biomedical relation extraction rely on engineered features which suffer from low portability and generalizability (Kim et al., 2015; Zheng et al., 2016; Raihani and Laachfoubi, 2017). To tackle this problem, recent studies apply Convolutional Neural Networks (CNNs) and Recurrent Neural Networks (RNNs) to automatically learn feature representations with input words encoded as pre-trained word embeddings (Zhao et al., 2016; Liu et al., 2016; Quan et al., 2016; Zhang et al., 2017; Zhou et al., 2018a; Sun et al., 2019). Learning representations of graphs are widely studied and several graph neural networks have been applied in the biomedical domain. Lim et al. (2018) proposed recursive neural network based model with a subtree containment feature. Asada et al. (2018) encoded drug pairs with CNNs and used external knowledge base to encode their molecular pairs with two graph neural networks. Here we directly apply syntax-aware GCNs on biomedical text to extract drug-drug interaction.

5   Conclusions and Future Work

We propose a syntax-aware multi-task learning model for biomedical relation extraction. Our model can effectively extract the drug-drug interactions by capturing the syntactic information through graph convolution operations and modeling context information via contextualized embeddings. An auxiliary task with focal loss is designed to mitigate the data imbalance by leveraging the inductive signal from binary classification and increasing the influence of decisive relation types. In the future, we plan to explore more informative parsers like the abstract meaning representation parser to create graph structure and consider leveraging external knowledge to further enhance the extraction quality.

Acknowledgments

This work was supported by the U.S. NSF No. 1741634 and Air Force No. FA8650-17-C-7715. The views and conclusions contained in this document are those of the authors and should not be interpreted as representing the official policies, either expressed or implied, of the U.S. Government. The U.S. Government is authorized to reproduce and distribute reprints for Government purposes notwithstanding any copyright notation here on.

References

Maristella Agosti, Giorgio Maria Di Nunzio, Stefano Marchesin, and Gianmaria Silvello. 2019. A relation extraction approach for clinical decision support. *arXiv preprint arXiv:1905.01257*.

Masaki Asada, Makoto Miwa, and Yutaka Sasaki. 2018. Enhancing drug-drug interaction extraction from texts by molecular structure information. In *Proc. ACL2018*.

Danqi Chen and Christopher Manning. 2014. A fast and accurate dependency parser using neural networks. In *Proc. EMNLP2014*.

Jacob Devlin, Ming-Wei Chang, Kenton Lee, and Kristina Toutanova. 2019. Bert: Pre-training of deep bidirectional transformers for language understanding. In *Proc. NAACL-HLT2019*.

Rujun Han, Mengyue Liang, Bashar Alhafni, and Nanyun Peng. 2019. Contextualized word embeddings enhanced event temporal relation extraction for story understanding. *arXiv preprint arXiv:1904.11942*.

María Herrero-Zazo, Isabel Segura-Bedmar, Paloma Martínez, and Thierry Declerck. 2013. The ddi corpus: An annotated corpus with pharmacological substances and drug–drug interactions. *Journal of biomedical informatics*, 46(5):914–920.

Dimitar Hristovski, Dejan Dinevski, Andrej Kastrin, and Thomas C Rindflesch. 2015. Biomedical question answering using semantic relations. *BMC bioinformatics*, 16(1):6.

Sun Kim, Haibin Liu, Lana Yeganova, and W John Wilbur. 2015. Extracting drug–drug interactions from literature using a rich feature-based linear kernel approach. *Journal of biomedical informatics*, 55:23–30.

Thomas N Kipf and Max Welling. 2016. Semi-supervised classification with graph convolutional networks. *arXiv preprint arXiv:1609.02907*.

Yann LeCun, Léon Bottou, Yoshua Bengio, Patrick Haffner, et al. 1998. Gradient-based learning applied to document recognition. *Proceedings of the IEEE*, 86(11):2278–2324.

Jinhyuk Lee, Wonjin Yoon, Sungdong Kim, Donghyeon Kim, Sunkyu Kim, Chan Ho So, and Jaewoo Kang. 2019. Biobert: pre-trained biomedical language representation model for biomedical text mining. *arXiv preprint arXiv:1901.08746*.

Sangrak Lim, Kyubum Lee, and Jaewoo Kang. 2018. Drug drug interaction extraction from the literature using a recursive neural network. *PloS one*, 13(1):e0190926.

Tsung-Yi Lin, Priya Goyal, Ross Girshick, Kaiming He, and Piotr Dollár. 2017. Focal loss for dense object detection. In *Proceedings of the IEEE international conference on computer vision*, pages 2980–2988.

Shengyu Liu, Buzhou Tang, Qingcai Chen, and Xiaolong Wang. 2016. Drug-drug interaction extraction via convolutional neural networks. *Computational and mathematical methods in medicine*, 2016.

Diego Marcheggiani and Ivan Titov. 2017. Encoding sentences with graph convolutional networks for semantic role labeling. *arXiv preprint arXiv:1703.04826*.

Chanqin Quan, Lei Hua, Xiao Sun, and Wenjun Bai. 2016. Multichannel convolutional neural network for biological relation extraction. *BioMed research international*, 2016.

Anass Raihani and Nabil Laachfoubi. 2017. A rich feature-based kernel approach for drug-drug interaction extraction. *International journal of advanced computer science and applications*, 8(4):324–3360.

Sunil Kumar Sahu and Ashish Anand. 2018. Drug-drug interaction extraction from biomedical texts using long short-term memory network. *Journal of biomedical informatics*, 86:15–24.

Xia Sun, Ke Dong, Long Ma, Richard Sutcliffe, Feijuan He, Sushing Chen, and Jun Feng. 2019. Drug-drug interaction extraction via recurrent hybrid convolutional neural networks with an improved focal loss. *Entropy*, 21(1):37.

Ashish Vaswani, Noam Shazeer, Niki Parmar, Jakob Uszkoreit, Llion Jones, Aidan N Gomez, Łukasz Kaiser, and Illia Polosukhin. 2017. Attention is all you need. In *Advances in neural information processing systems*, pages 5998–6008.

Haoyu Wang, Ming Tan, Mo Yu, Shiyu Chang, Dakuo Wang, Kun Xu, Xiaoxiao Guo, and Saloni Potdar. 2019. Extracting multiple-relations in one-pass with pre-trained transformers. *arXiv preprint arXiv:1902.01030*.

Yonghui Wu, Mike Schuster, Zhifeng Chen, Quoc V Le, Mohammad Norouzi, Wolfgang Macherey, Maxim Krikun, Yuan Cao, Qin Gao, Klaus Macherey, et al. 2016. Google's neural machine translation system: Bridging the gap between human and machine translation. *arXiv preprint arXiv:1609.08144*.

Zibo Yi, Shasha Li, Jie Yu, Yusong Tan, Qingbo Wu, Hong Yuan, and Ting Wang. 2017. Drug-drug interaction extraction via recurrent neural network with multiple attention layers. In *International Conference on Advanced Data Mining and Applications*, pages 554–566.

Yijia Zhang, Wei Zheng, Hongfei Lin, Jian Wang, Zhihao Yang, and Michel Dumontier. 2017. Drug–drug interaction extraction via hierarchical rnns on sequence and shortest dependency paths. *Bioinformatics*, 34(5):828–835.

Yuhao Zhang, Peng Qi, and Christopher D Manning. 2018. Graph convolution over pruned dependency trees improves relation extraction. *arXiv preprint arXiv:1809.10185*.

Zhehuan Zhao, Zhihao Yang, Ling Luo, Hongfei Lin, and Jian Wang. 2016. Drug drug interaction extraction from biomedical literature using syntax convolutional neural network. *Bioinformatics*, 32(22):3444–3453.

Wei Zheng, Hongfei Lin, Zhehuan Zhao, Bo Xu, Yijia Zhang, Zhihao Yang, and Jian Wang. 2016. A graph kernel based on context vectors for extracting drug–drug interactions. *Journal of biomedical informatics*, 61:34–43.

Deyu Zhou, Lei Miao, and Yulan He. 2018a. Position-aware deep multi-task learning for drug–drug interaction extraction. *Artificial intelligence in medicine*, 87:1–8.

Jie Zhou, Ganqu Cui, Zhengyan Zhang, Cheng Yang, Zhiyuan Liu, and Maosong Sun. 2018b. Graph neural networks: A review of methods and applications. *arXiv preprint arXiv:1812.08434*.

Peng Zhou, Wei Shi, Jun Tian, Zhenyu Qi, Bingchen Li, Hongwei Hao, and Bo Xu. 2016. Attention-based bidirectional long short-term memory networks for relation classification. In *Proc. ACL2016*.

BioReddit: Word Embeddings for User-Generated Biomedical NLP

Marco Basaldella and **Nigel Collier**
Department of Theoretical and Applied Linguistics
University of Cambridge
Cambridge, UK
{mb2313, nhc30}@domain

Abstract

Word embeddings, in their different shapes and evolutions, have changed the natural language processing research landscape in the last years. The biomedical text processing field is no stranger to this revolution; however, researchers in the field largely trained their embeddings on scientific documents, even when working on user-generated data. In this paper we show how training embeddings from a corpus collected from user-generated text from medical forums heavily influences the performance on downstream tasks, outperforming embeddings trained both on general purpose data or on scientific papers when applied to user-generated content.

1 Introduction

In the Natural Language Processing community, user-generated content, i.e. data from social media, user forums, review websites, and so on, has been the subject of many studies in the past years; the same holds for the biomedical domain, where there has been a great effort on the applications of NLP techniques for biomedical scientific publications, patient records, and so on. However, the intersection of the two fields is still in its infancy, even when dealing with relatively basic NLP tasks. For instance, in the field of user-generated biomedical natural language processing (hence UG-BioNLP), to the best of our knowledge there are no publicly available corpora for Named Entity Recognition (NER) akin in size and purpose e.g. to the CoNLL 2003 dataset. (Tjong Kim Sang and De Meulder, 2003), making it hard to compare systems effectively. Moreover, while there have been experiments on training word embeddings with biomedical data, we are not aware of any publicly available word embeddings trained on UG-BioNLP data.

For this reason, we decided to investigate the impact of using purpose-trained word embeddings in the Bio-UG field. In order to train such embeddings, we collected a dataset from Reddit, scraping posts from medical-themed subreddits, both on general health topics such as 'r/AskDocs', or on disease-specific subreddits, such as 'r/cancer', 'r/asthma', and so on. We then trained word embeddings on this corpus using different off-the-shelf techniques. Then, to evaluate the embeddings, we collected a second dataset of 4800 threads from the health forum *HealthUnlocked*, which was annotated for the NER task. Then, we analyzed the performance of the embeddings on the tasks of NER and of adverse effect mention detection. For NER, we used Conditional Random Fields as a baseline. We compared them against Bidirectional LSTM-CRFs (Lample et al., 2016), on which we analyzed the impact of using our custom-trained word embeddings against embeddings trained on general purpose data and scientific biomedical publications when evaluating on our purpose-built HealthUnlocked dataset and on the PsyTar and CADEC corpora. Finally, we evaluated the performance of a simple architecture for adverse reaction mention detection on the PsyTAR corpus. We conclude the paper explaining our intentions for future research, in other to obtain other results that confirm the preliminary findings we present in this work.

2 Related Work

The benefit of using in-domain embeddings for the biomedical domain has already been proven effective. For example, (Pakhomov et al., 2016) and (Wang et al., 2018) found that using clinical notes or biomedical articles for training word embeddings has generally a positive impact on down-

Proceedings of the 10th International Workshop on Health Text Mining and Information Analysis (LOUHI 2019), pages 34–38
Hong Kong, November 3, 2019. ©2019 Association for Computational Linguistics
https://doi.org/10.18653/v1/D19-62

stream NLP tasks. (Nikfarjam et al., 2015) trained embeddings on user-generated medical content and used them successfully on the pharmacovigilance task; however, they trained the embeddings an adverse reaction mining corpus, hence making them too task-specific to be considered useful on generic UG-BioNLP tasks.

3 Datasets

3.1 BioReddit

To train our embeddings on user-generated biomedical text, we choose to scrape data from the discussion website Reddit. The website is organized by forums, called *subreddits*, where the discussion is restricted to a topic, e.g. general news, computer science, and so on. There is a great number of health-themed subreddits, where users from all around the world discuss their health problems or ask for medical advice, which is ideal for training our embeddings.

We also evaluated the micro-blogging platform Twitter as a possible source for the embeddings, but we quickly discarded it due to its unstructured nature. On Twitter, in fact, information is not pre-aggregated by subject, and one has to search for the required posts by searching for keyword or *hashtag*. This, along with the restrictive limits imposed by Twitter APIs, makes it hard to find relevant content, so we decided to continue with Reddit instead.

We designed a scraping script that downloaded discussions from 68 health themed subreddits. We selected subreddits where users

- could ask for advice, e.g. `/r/AskDocs`, `/r/DiagnoseMe`, `r/AskaPharmacist`,
- discuss a specific illness, e.g. `r/cancer`, `r/migraine`, `r/insomnia`,
- can discuss on any health-related topic, e.g. `r/health`, `r/HealthIT`, `r/HealthInsurance`.

We collected all the posts from these subreddits from the beginning of 2015 to the end of 2018. After that, we cleaned the corpus for bot-generated content, e.g. bots automatically suggesting to seek professional medical advice. We obtained a corpus with 300 million tokens and a vocabulary size of 780,000 words. While the number of tokens is considerably lower than the size of other word embedding training datasets, which could be two orders of magnitude bigger, the vocabulary is

quite big; for example, GloVE (Pennington et al., 2014) was trained with a 1.2 million big vocabulary and 27 billion tokens when using Twitter, and on a 600,000 word vocabulary and 6 billion tokens when using Wikipedia.

3.2 HealthUnlocked

In order to evaluate our embeddings, as a first step, we decided to focus on the Named Entity Recognition task. We obtained 4800 forum threads from HealthUnlocked[1], a British social network for health where users can discuss their health with people with similar conditions and obtain advice from professionals.

We annotated the dataset by marking the entities belonging to seven categories, namely: *Phenotype, Disease, Anatomy, Molecule, Gene, Device*, and *Procedure*. We describe in detail the categories in Table 1.

Since the dataset is collected from patients' discussions, the language used is far from technical. For example[2],

- an user describes paresthesia of arm as *"a tickling sensation in my arms"*;
- another patient, to describe her swollen abdomen, writes that she *"looked six months pregnant"*,
- another user writes that *"her mood is low"*, to explain her depression.

All these phrases, while expressed in layman's language, describe very specific symptoms. For this reason, we developed a set of annotation guidelines where the annotators were asked to mark *any possible mention* of an entity belonging to the seven categories above, even if not expressed with technical language. After running a pilot annotation task on a small set of discussions, we fine tuned the annotation guidelines, and we asked PhD-qualified biomedical experts to annotate 4800 threads from the forums. After the annotation, the files were shuffled and split in train, test, and development set, obtaining 8750, 2526, and 1250 sentences respectively. The number of annotations per category and per set is described in Table 1.

3.3 PsyTAR

The PsyTAR dataset *"contains patients expression of effectiveness and adverse drug events as-*

---

[1] `https://healthunlocked.com/`

[2] Please note that we use feminine pronouns to preserve the privacy of the patients.

Category	Description	Train	Dev	Test
Anatomy	Any anatomical structure, organ, bodily fluids, tissues, etc.	1060	146	308
Device	Any medical device used in diagnosis, therapy or prevention.	276	26	82
Disease	Any disorder or abnormal condition.	1234	203	363
Gene	Any molecule carrying genetic information.	342	47	87
Molecule	Any chemical substance.	1791	240	544
Phenotype	Any abnormal morphology, physiology or behaviour.	2963	421	872
Procedure	Any medical procedure used in diagnosis, therapy or prevention.	1158	163	294

Table 1: Description and statistics of the HealthUnlocked dataset used for the experiments.

sociated with psychiatric medications." (Zolnoori et al., 2019). The dataset contains 6000 sentences annotated for mentions and spans (i.e. NER) of Adverse Drug Reactions, Withdrawal Symptoms, Drug Effectiveness, Drug Ineffectiveness, Sign/Symptoms/Illness, and Drug Indications. Each entity is grounded against the Unified Medical Language System (UMLS) and SNOMED Clinical Terms. The source of the corpus is the drug review website Ask a Patient[3]. The language used is very simple, without the use of specialist terms, and with no guarantee of grammatical/spelling correctness.

3.4 CADEC

The CADEC corpus (Karimi et al., 2015) is a corpus of consumer reviews for pharmacovigilance. It is sourced from Ask a Patient too and it is annotated for mentions of concepts such as drugs, adverse reactions, symptoms and diseases, which are linked against SNOMED and MedDRA.

4 Experiments

4.1 Embeddings

Using the dataset described in Section 3.1, we trained three word embedding models, namely GloVe (Pennington et al., 2014), ELMo (Peters et al., 2018), and Flair (Akbik et al., 2018). We choose these models due to their popularity, performance, and relative low resource requirements. In particular, GloVe requires just hours to be trained on a CPU, while ELMo and Flair obtained state-of-the-art results in the NER task at the time of their publication, and both models can be trained in relatively short time (~1 week) using 1 or 2 GPUs. As general purpose and PubMed embeddings, we use the ones provided or recommended by the respective architecture authors; unfortunately, we are not aware of any GloVe

---

[3] https://www.askapatient.com/

Algorithm	P	R	F
CRF	69.7	60.1	64.5
GloVe-Default	69.6	68.3	68.9
GloVe-BioReddit-50	68.7	65.7	67.2
GloVe-BioReddit-100	70.2	71.7	70.9
GloVe-BioReddit-200	72.1	70.3	71.2
ELMo-Default	72.3	72.8	72.5
ELMo-PubMed	73.7	73.7	73.7
ELMo-BioReddit	73.9	76.7	75.3
Flair-Default	75.0	75.8	75.4
Flair-PubMed	75.8	75.1	75.4
Flair-BioReddit	76.5	76.2	76.4

Table 2: Performance of different embeddings technique on NER, when trained and evaluated on the dataset described in Section 3.2.

PubMed pre-trainer embeddings available in the public domain. Using our BioReddit dataset, we trained all the embeddings with their default parameters, as described in their respective papers.

4.2 Named Entity Recognition

In order to evaluate our embeddings we use Conditional Random Fields and as a baseline, and then we evaluate our embeddings using a Bidirectional LSTM-CRF sequence tagging neural network (Lample et al., 2016). We refer the reader to the original paper for an explanation on how this architecture works, as the details are outside to the scope of the present paper.

We present our results in Table 2. As expected, all the neural architectures largely improve the results obtained by the CRF and, in line with the literature, Flair performs slightly better than ELMo, which in turn performs better than GloVe. Using our purpose-built embeddings, called *BioReddit* in the Table, we always obtain an improvement with respect to using embeddings trained on general-purpose data (*Default* in Table) or on PubMed, barring the smallest GloVe vectors.

Category	P	R	F
Anatomy	72.2	76.6	74.3
Device	67.2	50.0	57.3
Disease	76.8	80.2	78.4
Gene	80.4	85.0	82.7
Molecule	88.4	88.6	88.5
Phenotype	70.5	66.9	68.6
Procedure	76.6	80.2	78.4

Table 3: Performance on the NER task of the Flair-BioReddit on the HealthUnlocked dataset on the seven categories defined in Section 3.2.

Corpus	Task	Embedding	P	R	F
PsyTAR	NER	Default	65.3	59.7	62.4
		PubMed	65.0	55.3	59.8
		BioReddit	63.7	63.8	63.7
PsyTAR	ADR	Default	81.3	69.2	74.8
		PubMed	77.5	72.6	75.0
		BioReddit	79.5	73.7	76.5
CADEC	NER	Default	77.1	76.0	76.5
		PubMed	77.2	76.1	76.7
		Bioreddit	78.6	77.4	78.0

Table 4: Performance of the Flair embeddings on the NER and Adverse Reaction Mention Detection on the PsyTAR and CADEC corpora.

In Table 3 we provide a per-category breakdown of the best performing embeddings, i.e. Flair embeddings trained on our BioReddit corpus. It's interesting to note how the most difficult categories are *Device* and *Phenotype*. We explain this results by noting that the former is the least represented category in the corpus, while the latter was actually expected to be the hardest category. In fact, looking into the corpus, we found that users are relatively precise when talking about disease names, genes, molecules, and so on, while they don't necessarily describe their symptoms using "proper" medical language.

In Table 4 we see the results we obtain on the NER task on the PsyTAR and CADEC corpora while using Flair embeddings, where BioReddit embeddings always outperform general-purpose and PubMed trained ones. Interestingly, PubMed embeddings behave considerably worse than the others on the PsyTAR corpus, which seems to support the intuition that using a specialized scientific corpus is not always the guarantee of better performance.

4.3 Adverse Reaction Mention Detection

The task of Adverse Reaction Mention Detection (hence ADR) consists in detecting whether in a sentence a user mentions that he is experiencing/experienced an adverse reaction to a drug. For this task, we designed a simple neural architecture, where a bidirectional GRU (Cho et al., 2014) reads a sentence, and a softmax layer on its top performs the binary classification task of detecting wether the input sentence contains an ADR or not. When evaluating on the PsyTAR corpus we again obtain the best performance when using our BioReddit embeddings, followed by the PubMed trained ones and the default ones.

5 Conclusions

In this paper we showed how training ad-hoc embeddings for the task of user-generated biomedical text processing improves the results in the tasks of named entity recognition and adverse reaction mention detection. While preliminary, our results show a strong indication that embeddings trained on biomedical scientific literature only are not guaranteed to be effective when used on user-generated data, since people use "layman terms" which are seldom, if ever, used in scientific literature. As future work, we acknowledge the need to better investigate the results we present here. A good starting point would be to analyze other embedding techniques, in order to investigate if the performance improvement is due to embedding techniques themselves or to the datasets used. Moreover, we need to analyze the performance of our BioReddit embeddings on non-user generated content, as e.g. scientific abstracts, in order to investigate whether they are able to perform effectively on this domain too. Finally, we think that a manual investigation of the results of the downstream tasks is important, to investigate e.g. if the improvement in the ADR task is due to the embeddings helping to classify sentences with more colloquial language. Unfortunately, due to licensing and privacy issues, we are not allowed to release the HealthUnlocked corpus. However, we make available our BioReddit embeddings trained on GloVe, ELMo and Flair at `https://github.com/basaldella/bioreddit`. For the sake of reproducibility, we also we make available our PsyTAR preprocessed splits online at `https://github.com/basaldella/psytarpreprocessor`.

Acknowledgements

The authors would like to thank HealthUnlocked for providing the dataset used in this paper, Taher Pilhevar for his useful suggestions, and NVIDIA Corporation for the donation of the GPU cards used to train the models presented in this paper. The authors acknowledge support by the EPSRC grant EP/M005089/1.

References

Alan Akbik, Duncan Blythe, and Roland Vollgraf. 2018. Contextual string embeddings for sequence labeling. In *Proceedings of the 27th International Conference on Computational Linguistics*, pages 1638–1649, Santa Fe, New Mexico, USA. Association for Computational Linguistics.

Kyunghyun Cho, Bart van Merriënboer, Caglar Gulcehre, Dzmitry Bahdanau, Fethi Bougares, Holger Schwenk, and Yoshua Bengio. 2014. Learning phrase representations using RNN encoder–decoder for statistical machine translation. In *Proceedings of the 2014 Conference on Empirical Methods in Natural Language Processing (EMNLP)*, pages 1724–1734, Doha, Qatar. Association for Computational Linguistics.

Sarvnaz Karimi, Alejandro Metke-Jimenez, Madonna Kemp, and Chen Wang. 2015. CADEC: A corpus of adverse drug event annotations. *Journal of Biomedical Informatics*, 55:73 – 81.

Guillaume Lample, Miguel Ballesteros, Sandeep Subramanian, Kazuya Kawakami, and Chris Dyer. 2016. Neural architectures for named entity recognition. In *Proceedings of the 2016 Conference of the North American Chapter of the Association for Computational Linguistics: Human Language Technologies*, pages 260–270, San Diego, California. Association for Computational Linguistics.

A. Nikfarjam, A. Sarker, K. O'Connor, R. Ginn, and G. Gonzalez. 2015. Pharmacovigilance from social media: mining adverse drug reaction mentions using sequence labeling with word embedding cluster features. *J Am Med Inform Assoc*, 22(3):671–681.

Serguei VS Pakhomov, Greg Finley, Reed McEwan, Yan Wang, and Genevieve B Melton. 2016. Corpus domain effects on distributional semantic modeling of medical terms. *Bioinformatics*, 32(23):3635–3644.

Jeffrey Pennington, Richard Socher, and Christopher Manning. 2014. Glove: Global vectors for word representation. In *Proceedings of the 2014 Conference on Empirical Methods in Natural Language Processing (EMNLP)*, pages 1532–1543, Doha, Qatar. Association for Computational Linguistics.

Matthew Peters, Mark Neumann, Mohit Iyyer, Matt Gardner, Christopher Clark, Kenton Lee, and Luke Zettlemoyer. 2018. Deep contextualized word representations. In *Proceedings of the 2018 Conference of the North American Chapter of the Association for Computational Linguistics: Human Language Technologies, Volume 1 (Long Papers)*, pages 2227–2237, New Orleans, Louisiana. Association for Computational Linguistics.

Erik F. Tjong Kim Sang and Fien De Meulder. 2003. Introduction to the CoNLL-2003 shared task: Language-independent named entity recognition. In *Proceedings of the Seventh Conference on Natural Language Learning at HLT-NAACL 2003*, pages 142–147.

Yanshan Wang, Sijia Liu, Naveed Afzal, Majid Rastegar-Mojarad, Liwei Wang, Feichen Shen, Paul Kingsbury, and Hongfang Liu. 2018. A comparison of word embeddings for the biomedical natural language processing. *Journal of Biomedical Informatics*, 87:12 – 20.

Maryam Zolnoori, Kin Wah Fung, Timothy B. Patrick, Paul Fontelo, Hadi Kharrazi, Anthony Faiola, Nilay D. Shah, Yi Shuan Shirley Wu, Christina E. Eldredge, Jake Luo, Mike Conway, Jiaxi Zhu, Soo Kyung Park, Kelly Xu, and Hamideh Moayyed. 2019. The PsyTAR dataset: From patients generated narratives to a corpus of adverse drug events and effectiveness of psychiatric medications. *Data in Brief*, 24:103838.

Leveraging Hierarchical Category Knowledge for Data-Imbalanced Multi-Label Diagnostic Text Understanding

Shang-Chi Tsai   Ting-Yun Chang   Yun-Nung Chen

Department of Computer Science and Information Engineering

National Taiwan University, Taipei, Taiwan

{r06946004,r06922168}@ntu.edu.tw y.v.chen@ieee.org

Abstract

Clinical notes are essential medical documents to record each patient's symptoms. Each record is typically annotated with medical diagnostic codes, which means diagnosis and treatment. This paper focuses on predicting diagnostic codes given the descriptive present illness in electronic health records by leveraging domain knowledge. We investigate various losses in a convolutional model to utilize hierarchical category knowledge of diagnostic codes in order to allow the model to share semantics across different labels under the same category. The proposed model not only considers the external domain knowledge but also addresses the issue about data imbalance. The MIMIC3 benchmark experiments show that the proposed methods can effectively utilize category knowledge and provide informative cues to improve the performance in terms of the top-ranked diagnostic codes which is better than the prior state-of-the-art. The investigation and discussion express the potential of integrating the domain knowledge in the current machine learning based models and guiding future research directions.

1 Introduction

Electronic health records (EHR) usually contain clinical notes, which are free-form text generated by clinicians during patient encounters, and a set of metadata diagnosis codes from the International Classification of Diseases (ICD), which represent the diagnoses and procedures in a standard way. ICD codes have a variety of usage, ranging from billing to predictive modeling of the patient state (Choi et al., 2016). Automatic diagnosis prediction has been studied since 1998 (de Lima et al., 1998). Mullenbach et al. (2018) pointed out the main challenges of this task: 1) the large label space, with over 15,000 codes in the ICD-9 taxonomy, and over 140,000 codes in the newer ICD-

10 taxonomies (Organization et al., 2007), and 2) noisy text, including irrelevant information, misspellings and non-standard abbreviations, and a large medical vocabulary. Several recent work attempted at solving this task by neural models (Shi et al., 2017; Mullenbach et al., 2018).

However, most prior work considered the output labels independently, so that the codes with few samples are difficult to learn (Shi et al., 2017). Therefore, Mullenbach et al. (2018) proposed an attentional model to effectively utilize the textural forms of codes to facilitate learning. In addition to textual definitions of codes, the *category* domain knowledge may provide additional cues to allow the codes under same category to share parameters, so the codes with few samples can benefit from it. To effectively utilize the *category knowledge* from the ICD codes, this paper proposes several refined category losses and incorporate them into convolutional models and then evaluate the performance on both MIMIC-3 (Johnson et al., 2016) and our internal dataset. The experiments on MIMIC shows that the proposed knowledge integration model significantly improves the previous methods and achieves the state-of-the-art performance, and the improvement can also be observed in our internal dataset. The idea is similar to the prior work (Singh et al., 2018), which considered the keyword hierarchy for information extraction from medical documents, but our work focuses on leveraging domain knowledge for clinical code prediction. Our contributions are three-fold:

- This paper first leverages external domain knowledge for diagnostic text understanding.

- The paper investigates multiple ways for incorporating the domain knowledge in an end-to-end manner.

- The proposed mechanisms improve all prior

Proceedings of the 10th International Workshop on Health Text Mining and Information Analysis (LOUHI 2019), pages 39–43

Hong Kong, November 3, 2019. ©2019 Association for Computational Linguistics

https://doi.org/10.18653/v1/D19-62

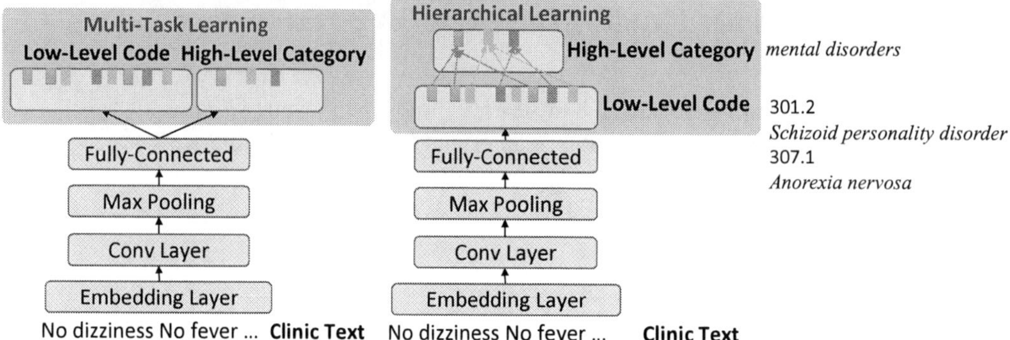

Figure 1: The architecture with the proposed category knowledge integration.

models and achieves the state-of-the-art performance on the benchmark MIMIC dataset.

2 Methodologies

Given each clinical record in EHR, the goal is to predict the corresponding diagnostic codes with the external hierarchical category information. This task is framed as a multi-label classification problem. The proposed mechanism is built on the top of various convolutional models to further combine with the category knowledge. Below we introduce the previously proposed convolutional models which are used for latter comparison in the experiment and detail the mechanism that leverages hierarchical knowledge.

2.1 Convolutional Models

There are various models for sequence-level classification, and this paper focuses on two types of convolutional models for investigation. The models are described as follows. Note that the proposed mechanism is flexible for diverse models.

TextCNN Let $x_i \in \mathbb{R}^k$ be the k-dimensional word embedding corresponding to the i-th word in the document, represented by the matrix $X = [x_1, x_2, ..., x_N]$, where N is the length of the document. TextCNN (Kim, 2014) applies both convolution and max-pooling operations in one dimension along the document length. For instance, a feature c_i is generated from a window of words $x_i, x_{i+1}, ..., x_{i+h}$, where h is the kernel size of the filters. The pooling operation is then applied over $c = [c_1, c_2, ..., c_{n-h+1}]$ to pick the maximum value $\hat{c} = \max(c)$ as the feature corresponding to this filter. We implement the model with kernel size = 3,4,5, considering different window sizes of words.

Convolutional Attention Model (CAML) Because the number of samples of each code is highly unbalanced, it is difficult to train each label with very few samples. To resolve this issue, the CAML model utilizes the descriptive definition of diagnosis codes, which additionally applies a per-label attention mechanism, where the additional benefit is that it selects the n-grams from the text that are most relevant to each predicted label (Mullenbach et al., 2018).

2.2 Knowledge Integration Mechanism

Considering the hierarchical property of ICD codes, we assume that using the higher level labels could learn more general concepts and thus improve the performance. For instance, the definitions of ICD-9 codes **301.2** and **307.1** are "*Schizoid personality disorder*" and "*Anorexia nervosa*" respectively. If we only use the labels given by the dataset, they are seen as two independent labels; however, in the ICD structure, both **301.2** and **307.1** belong to the same high-level category "*mental disorders*". The external knowledge shows that category knowledge provides additional cues to know code relatedness. Therefore, we propose four types of mechanisms that incorporate hierarchy category knowledge to improve the ICD prediction below.

Cluster Penalty Motivated by Nie et al. (2018), we compute two constraints to share the parameters of the ICD codes under the same categories. The between-cluster constraint, $\Omega_{between}$, indicates the total distance of parameters between mean of all ICD codes and the mean of each category.

$$\Omega_{between} = \sum_{k=1}^{K} \left\| \bar{\theta}_k - \bar{\theta} \right\|^2, \qquad (1)$$

where $\bar{\theta}$ is the mean vectors of all ICD codes, $\bar{\theta}_k$ is the mean vector of the k-th category. The within-cluster constraint, Ω_{within}, is the distance of parameters between the mean of each category and its low-level codes.

$$\Omega_{within} = \sum_{k=1}^{K} \sum_{i \in \mathcal{J}(k)} \left\| \theta_i - \bar{\theta}_k \right\|^2, \qquad (2)$$

where $\mathcal{J}(k)$ is a set of labels that belong to the k-th category. $\Omega_{between}$ and Ω_{within} are formulated as additional losses to enable the model to share parameters across codes with the same categories.

Multi-Task Learning Considering that the high-level category can be treated as another task, we apply a multi-task learning approach to leverage the external knowledge. This model focuses on predicting the low-level codes, y_{low}, as well as its high-level category, y_{high}, individually illustrated in Figure 1.

$$y_{high} = W_{high} \cdot h + b_{high} \qquad (3)$$

where $W_{high} \in \mathbb{R}^{N_{high} \times d}$, N_{high} means the number of high-level categories, and d is the dimension of hidden vectors derived from CNN.

Hierarchical Learning We build a dictionary for mapping our low-level labels to the corresponding high-level categories illustrated in Figure 1. To estimate the weights for high-level categories, y_{high}, two mechanisms are proposed:

- Average meta-label: The probability of the k-th high-level category can be approximated by the *averaged* weights for low-level codes that belong to the k-th category.

$$y_{high} = \frac{1}{k} \sum y_{low}^k \qquad (4)$$

- At-least-one meta-label: Motivated by Nie et al. (2018), meta labels are created by examining whether any disease label for the k-th category has been marked as tagged, where the high-level probability is derived from the low-level probability of disease labels.

$$y_{high} = 1 - \prod_k (1 - y_{low}^k) \qquad (5)$$

	MIMIC-3		Internal
	Full	**50**	**200**
# training documents	47,424	8,067	17,762
mean length of texts	1,485	1,530	50.35
vocabulary size	51,917	51,917	25,654
OOV rate	0.137	0.137	0.373
# labels	8,922	50	200
mean number of labels	15.9	5.7	1.7

Table 1: Dataset comparison and statistics. From the full set of the internal data (1495 labels) to 200, only 6.0% of data points are discarded.

2.3 Training

The knowledge integration mechanisms are built on top of the multi-label convolutional models, which treat each ICD label as a binary classification. The predicted values for high-level categories come from the proposed mechanisms. Considering that learning low-level labels directly is difficult due to the highly imbalanced label distribution, we add a loss term indicating the high-level category in order to learn the general concepts in addition to the low-level labels, and train the model in an end-to-end fashion. Note that the high-level loss is set as $loss_{high} = \Omega_{between} + \Omega_{within}$ for cluster penalty and the binary log loss for other methods.

$$loss = loss_{low} + \lambda \cdot loss_{high}, \qquad (6)$$

where λ is the parameter to control the influence of the knowledge category and we choose $\lambda = 0.1$.

3 Experiments

In order to measure the effectiveness of the proposed methods, the following experiments are conducted.

3.1 Setup

We evaluate our model on two datasets, one is the benchmark MIMIC-3 data and another is the dataset collected by National Taiwan University Hospital (NTUH). MIMIC-3 (Johnson et al., 2016) is a benchmark dataset, where the text and structured records from a hospital ICU. We use the same setting as the prior work (Mullenbach et al., 2018), where 47,724 discharge summaries is for training, with 1,632 summaries and 3,372 summaries for validation and testing, respectively. We also obtain a subdataset from original MIMIC3-Full, called MIMIC3-50, which has the top 50 high frequency labels. NTUH dataset is collected

MIMIC3-50	P@1	P@3	P@5	MAP	Macro-F	Micro-F	Macro-AUC	Micro-AUC
CNN (Shi et al., 2017)	82.8	71.2	61.4	72.4	57.9	63.0	88.2	91.2
+ Cluster Penalty	83.5†	71.9†	62.4†	73.1†	58.3†	63.7†	88.5†	91.3†
+ Multi-Task	83.5†	71.3†	61.9†	72.5†	57.6	62.8	88.1	91.1
+ Hierarchical *avg*	**84.5**†	**72.1**†	**62.4**†	**73.5**†	**58.6**†	**64.3**†	**88.9**†	**91.4**†
at-least-one	83.4†	72.1†	62.4†	73.4†	58.5†	63.8†	88.4†	91.3†
MIMIC3-Full	P@1	P@3	P@8	P@15	Macro-F	Micro-F	Macro-AUC	Micro-AUC
CNN (Shi et al., 2017)	80.5	73.6	59.6	45.4	3.8	42.9	81.8	97.1
+ Cluster Penalty	80.9†	74.0†	59.5	45.2	3.3	40.5	82.1†	97.0
+ Multi-Task	**82.8**†	**75.8**†	**61.5**†	**46.6**†	3.6	**43.9**†	**83.3**†	**97.3**†
+ Hierarchical *avg*	79.0	73.1	59.2	45.2	**4.3**†	42.7	83.0†	97.1
at-least-one	82.1†	74.3†	59.7†	44.9	2.6	42.0	80.3	96.7
CAML (Mullenbach et al., 2018)	89.6	83.4	69.5	54.6	6.1	51.7	88.4	98.4
+ Cluster Penalty	88.4	82.4	68.8	54.0	5.4	51.2	87.5	98.3
+ Multi-Task	**89.7**†	83.4	69.7†	54.8	6.9†	52.3†	88.8†	98.5†
+ Hierarchical *avg*	89.6	**83.5**†	**70.9**†	**56.1**†	**8.2**†	**53.9**†	**89.5**†	**98.6**†
at-least-one	89.4	83.3	69.5	54.8†	6.2†	51.7	88.3	98.4

Table 2: The results on MIMIC-3 data (%). † indicates the improvement over the baseline.

Data-200	Macro-F1	Micro-F1
CNN	7.6	39.8
+ Multi-Task	**11.7**†	41.6†
+ Hierarchical (avg)	9.2†	**44.1**†
CAML	6.2	42.6
+ Multi-Task	14.5†	44.7†
+ Hierarchical (avg)	**18.4**†	**45.7**†

Table 3: The results on NTUH data.

from an internal hospital, where each record includes narrative notes describing a patients stay and associated diagnostic ICD-9 codes. There are total 1,495 ICD-9 codes in the data, and the distribution is highly imbalanced. Our data is noisy due to typos and different writing styles, where the OOV rate is 0.373 based on the large vocabulary obtained from PubMed and PMC. As shown in Table 1, our data, Internal-200, is more challenging due to much shorter text inputs and higher OOV rate compared with the benchmark MIMIC-3 dataset. We split the whole set of 25,375 records from Internal-200 into 17,762 as training, 2,537 as validation, and 5,076 as testing.

All models use the same setting as the prior work (Kim, 2014; Mullenbach et al., 2018) and use skipgram word embeddings trained on PubMed[1] and PMC[2] (Mikolov et al., 2013). We evaluate the model performance using metrics for the multi-label classification task, including precision at K, mean average precision (MAP), and micro-averaged, macro-averaged F1 and AUC.

[1] https://www.ncbi.nlm.nih.gov/pubmed/
[2] https://www.ncbi.nlm.nih.gov/pmc/

3.2 Results

The baseline and the results of adding the proposed mechanisms are shown in Table 2. For MIMIC3-50, all proposed mechanisms achieve the improvement for almost all metrics, and the best one is from the hierarchical learning with average meta-label. The consistent improvement indicates that category knowledge provides informative cues for sharing parameters across low-level codes under the same categories. For MIMIC3-Full, our proposed mechanisms still outperform the baseline CNN model, and the best performance comes from the one with multi-task learning. The reason may be that multi-task learning has more flexible constraints compared with hierarchical learning, and it is more suitable for this more challenging scenario due to data imbalance. In addition, the proposed knowledge integration mechanisms using multi-task learning or hierarchical learning with average meta-label are able to improve the prior state-of-the-art model, CAML (Mullenbach et al., 2018), demonstrating the superior capability and the importance of domain knowledge.

To further investigate the model effectiveness, we perform the experiments on the NTUH dataset in Table 3. Due to shorter clinical notes and higher OOV rate, this dataset is more challenging and the results are lower than the ones in MIMIC-3. Nevertheless, the proposed methods still improve the performance by integrating category knowledge using multi-task learning or hierarchical learning with average meta-label. In sum, our proposed category knowledge integration mechanisms are capable of improving the text understanding performance by combining the domain knowledge with neural models and achieve the state-of-the-art results.

3.3 Qualitative Analysis

From our prediction results, we find that our proposed mechanisms tend to predict more labels than the baseline models for both CNN and CAML. Specifically, our methods can assist models to consider more categories from shared information in the hierarchy. The additional codes often contain the right answers and sometimes are in the correct categories but not exactly matched. Moreover, our mechanisms have the capability of correcting the wrong codes to the correct ones which are under the same category. The appendix provides some examples for reference.

4 Conclusion

This paper proposes multiple mechanisms using the refined losses to leverage hierarchical category knowledge and share semantics of the labels under the same category, so the model can better understand the clinical texts even if the training samples are limited. The experiments demonstrate the effectiveness of the proposed knowledge integration mechanisms given the achieved state-of-the-art performance and show the great generalization capability for multiple datasets. In the future, we plan to analyze the performance of each label, investigating which label can benefit more from the proposed approaches.

Acknowledgements

We would like to thank reviewers for their insightful comments on the paper. This work was financially supported from the Young Scholar Fellowship Program by Ministry of Science and Technology (MOST) in Taiwan, under Grant 108-2636-E002-003 and 108-2634-F-002-015.

References

Edward Choi, Mohammad Taha Bahadori, Andy Schuetz, Walter F Stewart, and Jimeng Sun. 2016. Doctor ai: Predicting clinical events via recurrent neural networks. In *Machine Learning for Healthcare Conference*, pages 301–318.

Alistair EW Johnson, Tom J Pollard, Lu Shen, H Lehman Li-wei, Mengling Feng, Mohammad Ghassemi, Benjamin Moody, Peter Szolovits, Leo Anthony Celi, and Roger G Mark. 2016. Mimic-iii, a freely accessible critical care database. *Scientific data*, 3:160035.

Yoon Kim. 2014. Convolutional neural networks for sentence classification. In *Proceedings of the 2014 Conference on Empirical Methods in Natural Language Processing (EMNLP)*, pages 1746–1751.

Luciano RS de Lima, Alberto HF Laender, and Berthier A Ribeiro-Neto. 1998. A hierarchical approach to the automatic categorization of medical documents. In *Proceedings of the seventh international conference on Information and knowledge management*, pages 132–139. ACM.

Tomas Mikolov, Ilya Sutskever, Kai Chen, Greg S Corrado, and Jeff Dean. 2013. Distributed representations of words and phrases and their compositionality. In *Advances in neural information processing systems*, pages 3111–3119.

James Mullenbach, Sarah Wiegreffe, Jon Duke, Jimeng Sun, and Jacob Eisenstein. 2018. Explainable prediction of medical codes from clinical text. In *Proceedings of the 2018 Conference of the North American Chapter of the Association for Computational Linguistics: Human Language Technologies*, volume 1, pages 1101–1111.

Allen Nie, Ashley Zehnder, Rodney L Page, Arturo L Pineda, Manuel A Rivas, Carlos D Bustamante, and James Zou. 2018. Deeptag: inferring all-cause diagnoses from clinical notes in under-resourced medical domain. *arXiv preprint arXiv:1806.10722*.

World Health Organization et al. 2007. International statistical classification of diseases and related health problems: tenth revision-version for 2007. *http://apps. who. int/classifications/apps/icd/icd10online/*.

Haoran Shi, Pengtao Xie, Zhiting Hu, Ming Zhang, and Eric P Xing. 2017. Towards automated icd coding using deep learning. *arXiv preprint arXiv:1711.04075*.

Gaurav Singh, James Thomas, Iain Marshall, John Shawe-Taylor, and Byron C. Wallace. 2018. Structured multi-label biomedical text tagging via attentive neural tree decoding. In *Proceedings of the 2018 Conference on Empirical Methods in Natural Language Processing*, pages 2837–2842. Association for Computational Linguistics.

Experiments with ad hoc ambiguous abbreviation expansion

Agnieszka Mykowiecka
ICS PAS
Jana Kazimierza 5
Warsaw, Poland
`agn@ipipan.waw.pl`

Małgorzata Marciniak
ICS PAS
Jana Kazimierza 5
Warsaw, Poland
`mm@ipipan.waw.pl`

Abstract

The paper addresses experiments to expand ad hoc ambiguous abbreviations in medical notes on the basis of morphologically annotated texts, without using additional domain resources. We work on Polish data but the described approaches can be used for other languages too. We test two methods to select candidates for word abbreviation expansions. The first one automatically selects all words in text which might be an expansion of an abbreviation according to the language rules. The second method uses clustering of abbreviation occurrences to select representative elements which are manually annotated to determine lists of potential expansions. We then train a classifier to assign expansions to abbreviations based on three training sets: automatically obtained, consisting of manual annotation, and concatenation of the two previous ones. The results obtained for the manually annotated training data significantly outperform automatically obtained training data. Adding the automatically obtained training data to the manually annotated data improves the results, in particular for less frequent abbreviations. In this context the proposed a priori data driven selection of possible extensions turned out to be crucial.

1 Introduction

Saving time and effort is a crucial reason for using abbreviations and acronyms in all types of texts. In informal texts like e-mails, communicator messages, and notes, it is very common to create ad hoc abbreviations, which are easy for the author (in the case of personal notes) or a reader to interpret in the context of a topic discussed by a group of people.

The time/effort saving principle is also valid for medical notes prepared by physicians during patient visits, hospital examinations, and for nursing notes. They are often written in a hurry, but have to be understandable for other people involved in the treatment of a patient. They cannot be completely hermetic, but usually, they are difficult to interpret both for patients and nonspecialists (Mowery et al., 2016b). Ad hoc abbreviations are also difficult for automatic data processing systems to handle. But proper understanding and normalization of all types of abbreviations is indispensable for the correct operation of information extraction systems (Pradhan et al., 2014), data classification (Névéol et al., 2016; Mowery et al., 2016a), question answering (Kakadiaris et al., 2018), and many other applications.

The interpretation of an abbreviation consists of two aspects: its recognition and expansion. The recognition of well established abbreviations and acronyms is usually done with the help of dictionaries. For the English medical domain, several dictionaries such as the resources of the U.S. National Library of Medicine are available. Ad hoc abbreviations are not present in dictionaries and they are mainly recognized as unknown words. Sometimes, they are ambiguous with correct full word forms listed in general language dictionaries. For example, *dept* might be an abbreviation of 'department' or 'deputy'; in Polish medical notes *temp* 'rate' is an abbreviation of *temperatura* 'temperature'. In informal texts, the period after abbreviations, required after some of them in Polish, English and many other languages, is often omitted, which makes the distinction between word and abbreviation more difficult. Ad hoc abbreviations are also ambiguous with standard language abbreviations, and their interpretation is different from those used in standard language: literature, papers or everyday use. For example, in many languages (e.g. English, German, Polish) the abbreviation *op* means the opus number in musical composition. In Polish medical notes, it can be *opa-*

Proceedings of the 10th International Workshop on Health Text Mining and Information Analysis (LOUHI 2019), pages 44–53
Hong Kong, November 3, 2019. ©2019 Association for Computational Linguistics
https://doi.org/10.18653/v1/D19-62

trunek 'dressing' especially in the context of 'gypsum dressing' of broken bones, *oko prawe* 'right eye' in the context of ophthalmic examinations or *opakowanie* 'package' of medications in recommendations. But it can have several other meanings e.g. *opuszek* 'fingertip' – unit of cervical dilation used by gynecologists. So, if we want to recognize ad hoc abbreviations in informal texts it is necessary not only to consider unknown strings but also short words and abbreviations recognized during morphologic analysis of text.

Our study focuses on expanding word abbreviations in medical notes in Polish. We are not systematically considering phrase abbreviations, usually called acronyms, as selecting candidates for their expansions requires different methods. In the paper we test two approaches for selecting candidates for word abbreviation expansions which are used for training a classifier to assign the appropriate expansion to an abbreviation, see (Pakhomov, 2002). In the first method, we check a hypothesis that full forms of ad hoc abbreviations are represented in texts of the same domain and type. So, for each candidate for abbreviation, we select all words which might be expansions for the abbreviation according to the language rules. We test if the occurrences of potential expansions in text can be sufficient for training the classifier. This method provides us with many suggestions which might never be used. To limit this number, we modified the method by selecting words with distributed representation close to the representation of the abbreviation. In the second method, we select candidates for abbreviation expansions based on annotation of selected elements of abbreviation occurrences clusters. Clustering is done by the Chinese whispers algorithm (Biemann, 2006) according to their contexts. For each cluster, we expand a manually randomly selected 2 to 6 elements of each cluster. This procedure gives us a short list of potential expansions.

2   Related Work

The problem of abbreviation recognition and expansion, has so far been addressed mainly for English data, e.g. (Nadeau and Turney, 2005) and (Moon et al., 2012) where supervised machine learning algorithms are used, and (Du et al., 2019) who describes a complex system for English data that recognizes many types of abbreviations. But, there are papers describing the problem for other languages too, e.g. Swedish (Kvist and Velupillai, 2014) – SCAN system based on lexicon, and German, e.g. (Kreuzthaler et al., 2016) where abbreviation identification linked to the disambiguation of period characters in clinical narratives is addressed.

Methods of dealing with acronyms are described among others in (Schwartz and Hearst, 2003) where the authors look for acronym definitions in data and identify them as a text in parentheses adjacent to the acronym/abbreviation; (Tengstrand et al., 2014) where experiments for Swedish are described; and (Spasic, 2018) where terminology extraction methods are applied.

Experiments in which similar to our data driven approach is tested, are described in (Oleynik et al., 2017). They used a method of abbreviation expansion based on N-grams and achieved an F1 score of 0.91, evaluated on a test set of 200 text excerpts.

3   Data

Medical reports which we used to carry out the experiments are an annomyzed sample of data collected by the company providing electronic health record services. The research is a part of a project, the purpose of which is, among other things, automatic preparation of statistics concerning data on diseases, symptoms and treatments. The statistics should be based on information extracted automatically from descriptions of patients visits. The data was collected in many clinics and concerned visits to doctors of various specialties. Identification information was removed from texts, and only descriptions of interrogation, examination and recommendations were processed.

As Polish is an inflectional language we preprocessed text to obtain base word forms and POS tags. Medical reports usually contain a limited dictionary but a lot of words are not present in general dictionaries, thus specialized medical taggers would be the most appropriate for performing this task. However, manually annotated data to train the medical tagger is not available for Polish, thus we had to processed texts with the general purpose morphological tagger. In this work we used Concraft2 – the new version of the Concraft (Waszczuk, 2012) tagger which cooperates with the general purpose morphological analyzer Morfeusz2 (Woliński, 2014) and also performs tokenization and sentence identification. Additionally, we ensured that line breaks were treated as sentence delimiters, as often a dot was not used at the end of a line, while the topic was changed. The quality of automatic tagging of medical texts in Polish is not high, see (Marciniak and Mykowiecka, 2011). Medical notes contain a large number of spelling errors, there are many acronyms/abbreviations[1] and specialized words not present in the general purpose morphological analyzer. Thus, we performed our experiments using both exact word forms and lemmas.

The entire dataset consists of about 10 million tokens and 15,000 different word forms. This number is larger than for English data of the same size as Polish is an inflectional language. It means that one word can have several forms, e.g.: *kropla* 'drop' is represented in our data as: *krople, kropli, kroplach, kroplami*. As we differentiate between capital and small letters we additionally have the following forms: *Krople, KROPLE*. The latter decision resulted from a desire to preserve information about acronyms for future work. For example, the form *PSA* is rather the acronym of an examination while the form *psa* might be interpreted as a 'dog' in the genitive (in medical texts it mainly occurs in the phrase *sierść psa* 'dog fur' in the context of allergens).

Around 7% of tokens are recognized as unknown words and this group of tokens consists of about 91,000 different elements: abbreviations; acronyms; proper names such as medications and illnesses containing proper names (e.g. Hashimoto's thyroiditis); and typos that occur in large numbers in medical notes. Some abbreviations are represented in Morfeusz2 but often their meaning is not appropriate for medical texts, e.g.: a string 'por' is recognized as an abbreviation of 'lieutenant' or 'compare' while in medical data it is 'clinic'.

Tokens which are not recognized by dictionaries, are natural candidates for being abbreviations. In many papers addressing the problem of abbreviation recognition, the authors limit themselves to considering such tokens, see (Park and Byrd, 2001), (Kreuzthaler et al., 2016). In our approach, when selecting potential abbreviations, we took into account all forms out of the dictionary, and short words (up to 5 letters) which were in the dictionary. As we wanted to use contexts in our experiment, we decided to consider forms which oc-

curred in the data more than 15 times. This limited the list of unknown tokens to 2808 and the list of word forms considered as potential abbreviations to 3152.

The data set was divided into ten parts, one was left for evaluation purposes and the remaining 9 were used as a training set and a source of information on the number and types of abbreviations used.

The test set consists of about 996.000 tokens and thousands of abbreviation occurrences. To make manual checking of the results feasible we decided to perform our experiment on a small subset of 15 abbreviations. This short list consists of abbreviations which seem to be ambiguous (a few likely interpretations) and are rather common – 3069 occurrences in test data which means 0.3% of tokens. Their proper recognition is, therefore, important for correct text interpretation. All occurrences of these 15 abbreviations in the test set were manually expanded by a person with experience in medical data processing. All difficult cases were consulted with a specialist. A fragment of the list together with exemplary variations is given in Table 1.

4 Language Models

On the basis of the entire data set, we trained four word2vec (Mikolov et al., 2013) versions of language models (the choice of specialized data seems to be straightforward, but was also supported by (Charbonnier and Wartena, 2018)). One pair of models was trained on the original (tokenized) texts – inflected forms of words. The second pair of models was trained on the lemmatized text (in Polish, nouns, verbs and adjectives have many different inflectional variants). In both pairs we calculated vectors of length 100; one model was trained on all tokens which occurred at least 5 times and the second one was trained on text in which all numbers were replaced by one symbol. In the final experiments form based models turned out to be the most efficient.

5 Baseline

We solve the problem of abbreviation expansion as the task of word sense disambiguation where a classifier is trained on all expansions represented in the data. As it is difficult to compare our approach to other work as the assumptions of the tasks related to abbreviation expansion were dif-

---

[1]Marciniak and Mykowiecka (2011) reported that around 6% of tokens in hospital records are acronyms and abbreviations.

Abr.	Variations	All	Possible Meanings in Test Data
fiz	FIZ, fiz, Fiz,	15	fizjologiczny, fizycznie, fizyczny, fizykalnie, fizykalny, fizykoterapia *Physiological, Physically, Physical, Physical, Physical Therapy*
cz	CZ, Cz	44	czerwień, czołowy, czynnik, czynność, częstość, część *redness, frontal, factor, activity, frequency, part*
gł	gł	22	głowa, główkowy, głównie, główny, głęboki *head, head(adj), mainly, main, deep*
op	OP Op	30	opak, opakowanie, opatrunek, opera, operacja, operacyjnie, operacyjny, oko prawy operowany, opieka, opis, opuszek, opór *awry, package, dressing, opera, operation, operationally, operational, right eye operated, care, description, pad, resistance*

Table 1: Four from the list of 15 abbreviations with variations, the number of all different longer words found in the training data.

	simulated train data			annotated	
abbr.	AL	SL	CL	train	test
cz	51022	46172	25642	96	137
fiz	10769	10684	9895	61	59
gł	15591	14460	9988	55	48
kr	37381	24349	20053	81	224
mies	9021	8949	6874	35	206
op	24677	21673	9285	410	1785
poj	4386	4035	3293	75	147
pow	22517	5037	17271	69	65
pr	88312	20386	57809	105	100
rodz	6459	6459	4903	26	52
śr	3894	2922	2316	61	65
wz	9942	6914	3345	42	31
zab	8670	8085	7755	69	90
zaw	3826	1296	2012	28	29
zał	1657	1544	717	18	31
total	298149	182965	181140	1231	3069

Table 2: Number of occurrences in train and test data. The three potential extensions lists for simulated training sets: AL – all words being potential expansions, SL – all the possible words in our distributional model whose similarity to a particular abbreviation was higher than 0.2 for a language model created on forms, CL – annotations of randomly selected cluster elements.

ferent, we suggest an artificial baseline, which consists of the most common interpretation of manually annotated abbreviations in the test set. Table 3 gives appropriate statistics. If we assign the most common interpretation of an abbreviation to all its occurrences we obtain the weighted precision equal to 0.568, the recall equal to 0.742 and the F1 measure equal to 0.64.

6 Methods for Determining Expansions

We checked two methods for determining potential ad hoc abbreviation expansions. The first one assumes that full versions of abbreviated forms are available somewhere in the data. So the problem can be seen as an attempt to determine which words from the text data can be abbreviated to a considered token and which of them correspond to an abbreviation in the given context. The second method uses clustering of abbreviation occurrences to select representative elements from each cluster to determine lists of potential expansions. This method allows a considered token that can abbreviate a phrase to be taken into account, while the first method is only oriented on word expansions.

6.1 All Words and Similar Word Methods

When we look for potential expansions of a selected token without any additional resources, we have to consider two cases. The first, is that we should leave the token unchanged as it could be a correct word or acronym. We do not address this problem. The second, is that we should select all words from the data that can be abbreviated to the considered token according to language rules. So, the list of potential expansions consists of all forms from the data which met the conditions of being an abbreviation in Polish. We analyse cases in which a token x might be an abbreviation of a word y if:

- the beginning of y is equal to x;

- the POS tag does not indicate an abbreviation or an unknown word (to avoid using incorrectly written words as potential extensions);

- the abbreviation does not cross Polish two-letter compounds ('rz', 'sz', 'cz', 'ch').

The first potential extensions list (AL) contains all words meeting the above conditions. It consists of 1345 elements. The AL list contains forms which are never shortened. Their usage should thus be different from that of the abbreviation itself. To eliminate such unlikely expansions and to limit the number of potential labels, we selected

abbr.	test anot.	expansions
cz	137	czerwień(1), czynnik(3), czynność(14), częstość(14) część(**102**)
		redness, factor, activity, frequency, part
fiz	59	fizjologiczny(2), fizycznie(1), fizyczny(5),fizykalnnie(**45**), fizykalny(6)
		physiological, physically, physical, physically, physical
gł	48	gładki(1), głowa(16), główkowy(4), głównie(**17**), główny(10)
		smooth, head, head, mainly, main
kr	224	krawędź(1), kreatynina(2), kropla(68), krople(4), kręgosłup(**149**)
		edge, creatinine, drop, drops, spine
mies	206	miesiąc(**187**), miesiączka(18), miesięczny(1)
		month, menstruation, monthly
op	1785	oko prawe (349), ostatni poród (1), opakowanie(**1384**), opatrunek(6), operacja(22), operacyjnie(3), operacyjny(10), operować(1), opieka(7), opuszek(2)
		right eye, last delivery, package, dressing, surgery, surgically, surgical, operate, care, fingertip
poj	147	pojawić(2), pojedynczy(**127**), pojemnik(18)
		appear, single, container
pow	65	powierzchnia(17), powiększony(6), powiększyć(8), powlekać(9), powyżej(**24**), powód(1)
		surface, enlarged, enlarge,coated, above, reason
pr	100	Pogotowie Ratunkowe(5), public relations(3), prawa ręka(1), PR(1), per rectum(5), prawidłowo(17), prawidłowy(**34**), prawy(20), preparat(2), prostata(4), przewód(1), przychodnia(1), próba(6)
		Emergency Service, public relations, right hand, PR(in ECG), per rectum, properly, normal, right, preparation, prostate, tract, clinic, test
rodz	52	rodzeństwo(8), rodzice(3), rodzina(4), rodzinne(1), rodzinnie(**20**), rodzinny(16)
		sibling, parents, family, family, family, family
wz	31	wziernik(**9**), wzrost(4)
		speculum, high
śr	65	średni(3), średnica(**47**), średnio(10), środa(1), środek(3), środkowy(1)
		medium, diameter, medium, Wednesday, middle, middle
zab	90	zabieg(14), zaburzenie(**76**)
		surgery, disorder
zaw	29	zawiesina(**23**), zawód(6)
		suspension, profession
zał	31	załamek(10), założyć(5), załączyć(**16**)
		crinkle, put on, attach

Table 3: Test set abbreviation expansions in numbers

from all the possible word forms in our distributional model, those whose similarity to a particular abbreviation was higher than 0.2 for a language model created on forms, see Section (4). These candidates form the second expansion list of 259 elements (SL). The numbers of occurrences of all expansions of these three lists in the training data are given in Table 2.

tation of the abbreviation, we established quite a high level of similarity between nodes in the initial graph. The similarity was counted as the cosine between vectors and we set it experimentally to 0.7 (it had no theoretical justification). Increasing the parameter of similarity we obtain more clusters and they represent higher granularity of abbreviation contexts.

6.2 Clustering and Manual Annotation

To check whether abbreviation usages form any differentiable clusters, we identified all their occurrences in the training data. For each such occurrence, we determined the context vector, which was equal to the average of vectors of surrounding tokens. In the experiment, we set the context as three tokens before and after each abbreviation. Then, we clustered occurrences of the abbreviation via the Chinese whispers algorithm (Biemann, 2006) which does not impose defining a priori a number of clusters. As we aimed to select examples of various interpretations of the same abbreviation and various usage of the same interpre-

For each cluster, we randomly selected from 2 to 6 elements (depending on the cluster size) and manually annotated them and the representative elements of the cluster pointed out by the algorithm, with proper expansions in the data. 85 elements used in this manual annotation constitutes the third list (CL) used in our experiments. In Table 2 the number of annotated examples in both train and test data are give. In test data a very high variance of abbreviations occurrences caused mainly by the big number of clusters obtained for the most frequent abbreviation (*op*) can be seen (from 29 to 1785).

6.3 Training Data

The core training set (SIM) is constructed via simulation by shortening word forms beginning with any of the abbreviation from the appropriate list processed in the exact experiment: AL, SL, and CL. The longest list AL contains 1345 potential expansions, SL limits the number of potential expansions to 259 elements while the manually created list CL has 85 elements. However, the SIM set may be biased as some of the words from these lists might be never shortened. What is more, in some typical places in which the chosen abbreviations occur, the full form may never or almost never be used. To check the real value of such simulated training set and, to test if a much smaller training set could be sufficient for this task, we also prepared manually annotated training set. It was built from all the manually annotated examples (a procedure described in 6.2). In Table 2, there is a comparison of the numbers of considered abbreviations in simulated (depending on the chosen expansion list) and manually annotated training and test data. As it turned out that nearly every abbreviation can also be an acronym, and one of them *oko prawy – op* occurs many times in our annotated data, to make comparisons more complete we also prepared a version of our training data (SIM-ac) in which two consecutive words recognized during manual annotation as a possible full form of an acronym, are abbreviated to the sequence of their first letters.

7 Neural Net Architecture

In the experiment, we used bidirectional LSTM nets as being most frequently judged as good for sequence processing. We formulated our task as a prediction task in which we predict a word on the basis of its context (and, optionally, on the basis of a representation of the abbreviation used instead of it). As clinical notes are short, concise and frequently change subject, we assign a label (which is a full word form) to a word on the basis of its left and right contexts of 3 or 5 words.

Input to a net consists of a subset of the following data (names given after features descriptions are used in Table 4 headings):

- word vectors form the models trained on the entire dataset,

- POS tags encoded as one-hot vector (pos) (31 most frequent categories),

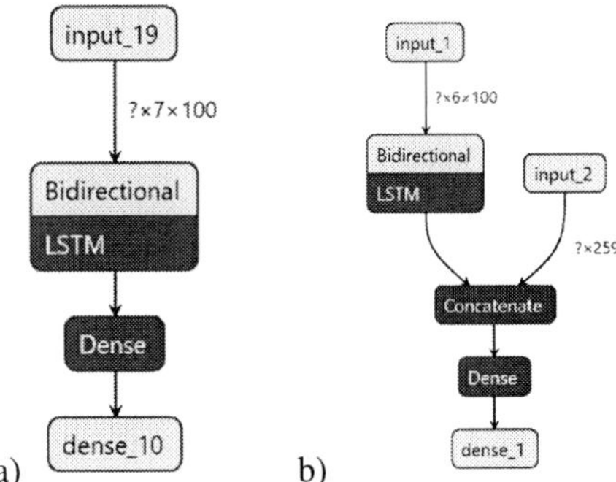

Figure 1: Two neural net architectures tested

- vector representing an abbreviation itself (padded with zeros if needed), (c),

- all possible longer versions of the particular abbreviations coded in a vector representing all possible forms of all the abbreviations taken into account. This vector was added as the additional information which was combined with the LSTM output layer (only one output value is considered), (cd), Figure 1b.

Two net architectures tested are given in Figure 1.[2] On the left side, a basic BiLSTM net with input consisting of seven word representation is shown (the central word is an abbreviation – being actually in the data or inserted in place of a full word form in the simulation variant). Representation consisted either of word embeddings only, or of embeddings concatenated with the POS one-hot encodings. We also tested variants in which only context words were used. The architecture given on the right takes only context words as input. The additional input vector represents all valid extensions of a given abbreviation (cd). In both cases, the last layer is a standard classification dense layer with a sigmoid activation function. Its size depends on the length of a particular extensions list. The implementation is done in Keras with Tensorflow backend. The Adam optimizer was used and the other settings are standard values used in Keras implementation.

8 Results

The net architecture for further experiments was chosen on the basis of the 10-fold cross validation results for one configuration, see Table 4.

---

[2]The picture was obtained using Netron `https://lutzroeder.github.io/netron/`

The number of epochs was established on the basis of validation on 1/10th of the training data while learning on the remaining training set to 2 for models which use big simulated data, and 5 for models which use only small annotated data. The batch size was equal to 32 and 1 respectively. Apart from the first model which does not take into account either the abbreviation representation (c) or the list of possible expansions (cd), other results do not vary much. We decided to choose the second best variant with the list of possible extensions added as an additional layer but with the three not five word context. As our annotated data is not big, we preferred a comparable model with fewer parameters.

The chosen set of features (second column in Table 4) was used for building models for all abbreviation lists and four variants of training sets (only annotated data – ANOT, only simulated – SIM (only word abbreviations) and SIM-ac (word abbreviations and acronyms), and the sum of both.

The results on cross validation (Table 5) are the best for simulated data, probably because of their size and repetitiveness. Only for the longest expansion list the results are the best on the smallest annotated set. The small number of examples could have just been memorized more easily. With this one exception it is also generally the rule that the bigger the training set the better, although adding annotated examples lowered the results slightly for SL list.

The results for the test set (Table 6) are better than those obtained for the cross validation on the training set in many cases. This is probably due to the small size of the test data set and many occurrences of the easy to resolve cases, for example the frequent occurrences of the 'op' abbreviation, which was correctly identified as 'opakowanie' *package*. However, models trained on the simulated set alone, performed significantly worse in terms of the recall (precision only deteriorated a little).

Using simulated data for training models has one important advantage – it saves time and effort. But there are also some disadvantages which have to be carefully analyzed. A few examples of miss-interpretation of 'pr' strings are given in Table 7. In our particular task, the possible problems can have different sources. First, some abbreviations are never (or almost never) expanded within the corpus. These are for example very common

acronyms (like *OP* – 'right eye') which are rarely written in the full form, or an abbreviation *meta* which is never used in its full form *metastaza* 'metastasis' in our corpus. We did not fully address the problem in this work and phrase extensions which were recognized during manual annotation were added manually to the expansion lists. The second problem is that some words are never abbreviated, but we automatically added them to our expansion lists making the problem harder to solve. However, the good results obtained for the AL list show that this situation was not very confusing for our models (which have access to annotated data). The third problem is the fact that the contexts in which the abbreviated form are used may differ from the contexts where the full form occurs. If in some contexts, only abbreviations are used and the full form never occurs, it is not possible to learn this pattern. For example, when prescribing the number of medicine packages, doctors always use *op* instead of *opakowanie* 'package', e.g. *Lantus (1 op. 30%)*. Our experiment confirmed that this is really the case. Results obtained by the models trained on simulated data only, although having very good cross validation results, have much worse recall on test data than models trained on annotated examples. However, adding annotated data to the simulated train set improved the results. For all but the AL list, the results obtained on the entire data even outperformed those obtained for the annotated data.

9 Conclusions

In the paper, we wanted to test if simulated abbreviations can be used to expand ambiguous ad hoc abbreviations in medical notes. Although simulation of the training data is a very useful practice, as manual data annotation is an expensive and time consuming process, our work shows that the obtained results are not always satisfactory. The F1 measure we obtained is below the artificially established baseline (the F1 measure equal to 0.64). Moreover, the experiments show that annotation of a small number of thoroughly selected examples of abbreviation occurrences gives satisfactory results for the task with the F1 measure equal to 0.92. It significantly outperforms the artificial baseline – the most common expansion e.g. the standard baseline for the word sense disambiguation task. However, the best results are obtained when the simulated data are combined with man-

	context=3, model based on forms					lemas	context=5
	pos-c-cd-	**pos-c-cd+**	pos-c+cd-	pos-cd+c+	pos+c-cd+	pos-c-cd+	pos-c-cd+
weighted precision	0.562	0.690	0.662	0.691	0.689	0.677	0.693
recall	0.652	0.770	0.757	0.764	0.764	0.764	0.778
F1	0.597	**0.721**	0.702	0.719	0.717	0.711	**0.726**
macro precision	0.367	0.469	0.458	0.471	0.472	0.475	0.484
recall	0.391	0.502	0.493	0.493	0.505	0.492	0.519
F1	0.368	0.476	0.465	0.472	0.478	0.473	0.491

Table 4: Results for 10-fold cross validation for different bidirectional LSTM settings for one training set (a subset of randomly selected cluster elements) and a chosen extension list. In all but the sixth case, word embeddings based on word forms were used. Additional information used in the models: pos – part of speech, c – vector representing an abbreviation itself, cd – vector codding possible extensions of the particular abbreviations (architecture from Figure 1b).

List Trainset	AL			SL			CL		
	P	R	F1	P	R	F1	P	R	F1
ANOT-rd	0.874	0.885	**0.875**	0.809	0.837	0.817	0.866	0.888	0.875
ANOT	0.854	0.869	0.853	0.855	0.855	0.855	0.884	0.901	0.891
SIM	0.864	0.872	0.866	0.896	0.901	**0.897**	0.968	0.969	**0.968**
ANOT+SIM	0.864	0.872	0.866	0.893	0.899	0.894	0.968	0.969	**0.968**

Table 5: Results for 10-fold cross validation of the selected net architecture for all extension lists and training set variants (notation explained in the text). The three potential extensions lists for simulated training sets: AL – all words being potential expansions, SL – all the possible words in our distributional model whose similarity to a particular abbreviation was higher than 0.2 for a language model created on forms, CL – annotations of randomly selected cluster elements. The best results for each expansion list are shown in bold.

List Model trained on	AL			SL			CL		
	P	R	F1	P	R	F1	P	R	F1
	weighted results								
ANOT	0.914	**0.926**	**0.917**	0.891	**0.906**	**0.893**	0.909	0.921	0.910
SIM	0.800	0.482	0.556	0.770	0.545	0.611	0.806	0.735	0.758
SIM-ac	0.907	0.386	0.441	0.888	0.472	0.537	0.930	0.748	0.777
ANOT+SIM	**0.947**	0.749	0.824	0.911	0.749	0.809	**0.944**	0.911	0.918
ANOT+SIM-ac	**0.947**	0.715	0.798	**0.915**	0.776	0.828	0.943	**0.928**	**0.930**
	macro results								
ANOT	0.516	0.514	0.500	0.492	0.513	0.479	0.495	0.540	0.489
SIM	0.308	0.314	0.286	0.317	0.359	0.304	0.460	0.539	0.469
SIM-ac	0.302	0.299	0.268	0.329	0.354	0.309	0.499	0.522	0.461
ANOT+SIM	0.357	0.363	0.343	0.372	0.409	0.370	0.571	0.616	0.570
ANOT+SIM-ac	0.384	0.383	0.369	0.366	0.406	0.366	0.546	0.588	0.546

Table 6: Results for the test set of the models trained on different datasets for all extension lists (notation explained in the text). The best results for each expansion list are shown in bold. The **artificial baseline results**, when we consider only those expansions which really occurred in the data and the most frequent expansion is taken as a solution, are (weighted) **P=0.568, R=0.742, F1=0.64**. Most of our results are well above this baseline and only models trained on simulated data gave lower results on two expansions lists.

ual annotation. Is it particularly important for less frequent expansions, as the increase of macro F1 is significantly greater than increase of the weighted one. This conclusion is somewhat in contradictions with a claim of (Oleynik et al., 2017) who suggested that the manual annotation is not necessary to obtain relatively high results. In this context, the suggested method of selecting extensions candidates turned out to be important – the results on the list of every possible word extension (the AL list) for the combined training set are much lower than for the SL and CL lists.

As the results obtained for the SL expansion list (a list of all words from the data whose dis-

Excerpt	Expansion	SIM	ANOT+SIM
Jama ustna, gardło: pr [line break] 'Mouth,throat: normal'	*prawidłowy* 'normal'	*prawy* 'right'	*prawidłowy* 'normal'
bez o. patologicznych, pr. Romberg [aprawidlowa] 'without pathological symptoms, Romberg's test [spelling error]'	*próba* 'test'	*prawidłowy* 'normal'	*prawidłowy* 'normal'
ogr. ruchomości kolana pr, przykurcz 'limitation of the right knee mobility, contracture'	*prawy* 'right'	*prawy* 'right'	*prawidłowy* 'normal'

Table 7: Examples of miss-interpretation of 'pr' for the CL list of potential expansions and for two training data: SIM and ANOT+SIM.

tributional similarity was higher than 0.2) and the ANOT+SIM training data are very good, it would be interesting to test how important the selection of annotated examples is and to test how many manually annotated data is necessary for obtaining satisfactory results. In the future work we want to test our method on a large set of abbreviations and include strings which are ambiguous between words and abbreviations.

Acknowledgments

This work was supported by the Polish National Science Centre project 2014/15/B/ST6/05186 and by EU structural funds as part of the Smart Growth Operational Programme POIR.01.01.01-00-0328/17

References

Chris Biemann. 2006. Chinese whispers: an efficient graph clustering algorithm and its application to natural language processing problems. In *Proceedings of the First Workshop on Graph Based Methods for Natural Language Processing*, TextGraphs-1, pages 73–80. Association for Computational Linguistics.

Jean Charbonnier and Christian Wartena. 2018. Using word embeddings for unsupervised acronym disambiguation. In *Proceedings of the 27th International Conference on Computational Linguistics, Santa Fe, New Mexico, USA*, pages 2610—-2619.

Xiaokun Du, Rongbo Zhu, Yanhong Li, and Ashiq Anjum. 2019. Language model-based automatic prefix abbreviation expansion method for biomedical big data analysis. *Future Generation Computer Systems*, (98):238–251.

Ioannis A. Kakadiaris, George Paliouras, and Anastasia Krithara. 2018. *Proceedings of the 6th BioASQ Workshop A challenge on large-scale biomedical semantic indexing and question answering*. Association for Computational Linguistics, Brussels, Belgium.

Markus Kreuzthaler, Michel Oleynik, Alexander Avian, and Stefan Schulz. 2016. Unsupervised abbreviation detection in clinical narratives. In *Proceedings of the Clinical Natural Language Processing Workshop (ClinicalNLP)*, pages 91–98. The COLING 2016 Organizing Committee.

Maria Kvist and Sumithra Velupillai. 2014. SCAN: A swedish clinical abbreviation normalizer - further development and adaptation to radiology. In *Information Access Evaluation. Multilinguality, Multimodality, and Interaction - 5th International Conference of the CLEF Initiative, CLEF 2014, Sheffield, UK, September 15-18, 2014. Proceedings*, volume 8685 of *Lecture Notes in Computer Science*, pages 62–73. Springer.

Małgorzata Marciniak and Agnieszka Mykowiecka. 2011. Towards Morphologically Annotated Corpus of Hospital Discharge Reports in Polish. In *Proceedings of BioNLP 2011*, pages 92–100.

Thomas Mikolov, Ilya Sutskever, Kai Chen, Greg Corrado, and Jeffrey Dean. 2013. Distributed representations of words and phrases and their compositionality. In *Advances in neural information processing systems*, pages 3111–3119.

Sungrim Moon, Serguei Pakhomov, and Genevieve B. Melton. 2012. Automated disambiguation of acronyms and abbreviations in clinical texts: window and training size considerations. In *AMIA Annu Symp Proc.*, pages 1310–9.

Danielle. Mowery, Brian Chapman, Mike Conway, Brett South, Erin Madden, Salomeh Keyhani, and Wendy Chapman. 2016a. Extracting a stroke phenotype risk factor from veteran health administration clinical reports: An information content analysis. *Journal of Biomedical Semantics*, 7(1).

Danielle L. Mowery, Brett R. South, Lee Christensen, Jianwei Leng, Laura-Maria Peltonen, Sanna Salanterä, Hanna Suominen, David Martinez, Sumithra Velupillai, Noémie Elhadad, Guergana Savova, Sameer Pradhan, and Wendy W. Chapman. 2016b. Normalizing acronyms and abbreviations to aid patient understanding of clinical texts: Share/clef ehealth challenge 2013, task 2. *Journal of Biomedical Semantics*, 7(1).

David Nadeau and Peter D. Turney. 2005. A supervised learning approach to acronym identification. In *Proceedings of the 18th Canadian Society Conference on Advances in Artificial Intelligence*, AI'05, pages 319–329, Berlin, Heidelberg. Springer-Verlag.

Aurélie Névéol, Cyril Grouin, Kevin B Cohen, Thierry Hamon, Thomas Lavergne, Liadh Kelly, Lorraine Goeuriot, Grégoire Rey, Aude Robert, Xavier Tannier, and Pierre Zweigenbaum. 2016. Clinical information extraction at the CLEF eHealth evaluation lab 2016. In *Proc of CLEF eHealth Evaluation lab*, pages 28–42, Evora, Portugal.

Michel Oleynik, Markus Kreuzthaler, and Stefan Schulz. 2017. Unsupervised abbreviation expansion in clinical narratives. In *MedInfo*, volume 245 of *Studies in Health Technology and Informatics*, pages 539–543. IOS Press.

Serguei Pakhomov. 2002. Semi-supervised maximum entropy based approach to acronym and abbreviation normalization in medical texts. In *Proceedings of 40th Annual Meeting of the Association for Computational Linguistics*, pages 160–167, Philadelphia, Pennsylvania, USA. Association for Computational Linguistics.

Youngja Park and Roy J. Byrd. 2001. Hybrid text mining for finding abbreviations and their definitions. In *Proceedings of the 2001 Conference on Empirical Methods in Natural Language Processing.*

Sameer Pradhan, Noémie Elhadad, Brett R South, David Martinez, Lee Christensen, Amy Vogel, Hanna Suominen, Wendy W Chapman, and Guergana Savova. 2014. Evaluating the state of the art in disorder recognition and normalization of the clinical narrative. *Journal of the American Medical Informatics Association*, 22(1):143–154.

Ariel S. Schwartz and Marti A. Hearst. 2003. A simple algorithm for identifying abbreviation definitions in biomedical text. In *Pacific Symposium on Biocomputing*, pages 451–462.

Irena Spasic. 2018. Acronyms as an integral part of multi-word term recognition - A token of appreciation. *IEEE Access*, 6:8351–8363.

Lisa Tengstrand, Beáta Megyesi, Aron Henriksson, Martin Duneld, and Maria Kvist. 2014. EACL - expansion of abbreviations in clinical text. In *PITR@EACL*, pages 94–103. Association for Computational Linguistics.

Jakub Waszczuk. 2012. Harnessing the CRF complexity with domain-specific constraints. the case of morphosyntactic tagging of a highly inflected language. In *In: Proceedings of COLING 2012, Mumbai, India.*

Marcin Woliński. 2014. Morfeusz reloaded. In *Proceedings of the Ninth International Conference on Language Resources and Evaluation, LREC 2014*, pages 1106–1111, Reykjavík, Iceland. ELRA.

Multi-Task, Multi-Channel, Multi-Input Learning for Mental Illness Detection using Social Media Text

Prasadith Buddhitha and **Diana Inkpen**
School of Electrical Engineering and Computer Science
University of Ottawa, Canada
`{pkiri056, diana.inkpen}@uottawa.ca`

Abstract

We investigate the impact of using emotional patterns identified by the clinical practitioners and computational linguists to enhance the prediction capabilities of a mental illness detection (in our case depression and post-traumatic stress disorder) model built using a deep neural network architecture. Over the years, deep learning methods have been successfully used in natural language processing tasks, including a few in the domain of mental illness and suicide ideation detection. We illustrate the effectiveness of using multi-task learning with a multi-channel convolutional neural network as the shared representation and use additional inputs identified by researchers as indicatives in detecting mental disorders to enhance the model predictability. Given the limited amount of unstructured data available for training, we managed to obtain a task-specific AUC higher than 0.90. In comparison to methods such as multi-class classification, we identified multi-task learning with multi-channel convolution neural network and multiple-inputs to be effective in detecting mental disorders.

1 Introduction

Social media platforms have revolutionized the way people interact as a society and have become an integral part of everyday life where many people have started sharing their day to day activities on these platforms. Such real-time data portraying one's daily life could reveal invaluable insights into one's cognition, emotion, and behavioral aspects. With its rapid growth among different demographics and being a source enriched with valuable information, social media can be a significant contributor to the process of mental disorder and suicide ideation detection.

In the domain of mental illness detection and especially when using social media text, lack of sufficiently-large annotated data and the inability to extract explanation on the derived outcome have restricted researchers to use traditional machine learning algorithms other than state-of-the-art methods such as deep neural networks. The proposed research explores the feasibility of applying the state-of-the-art processes in combination with features identified using manual feature engineering methods to enhance the prediction accuracy while maintaining low false negative and false positive rates. Also, this research looks into detecting multiple mental disorders at the same time by sharing lower level features among the different tasks. Intuitively, learning multiple mental disorders using a single neural network architecture in comparison to using a single network to identify only one mental illness is a logical approach considering the psychological and linguistic characteristics shared among the individuals susceptible of being diagnosed with different mental disorders.

Mental illness detection in social media using Natural Language Processing (NLP) methods is considered as a difficult task due to the complex nature of mental disorders. According to the American Psychiatric Association (2013), a mental disorder is a "syndrome characterized by a clinically significant disturbance in an individual's cognition, emotion regulation, or behavior that reflects a dysfunction in the psychological, biological, or developmental processes underlying mental functioning". Mental disorders have become and continue to be a global health problem. More than 300 million people from varied demographics suffer from depression (World Health Organization, 2018a), and have broader implications where 23% deaths in the world were caused due to mental and substance use disorders (World Health Organization, 2014). In Canada, one in every five Canadians has experienced some form of mental ill-

Proceedings of the 10th International Workshop on Health Text Mining and Information Analysis (LOUHI 2019), pages 54–64
Hong Kong, November 3, 2019. ©2019 Association for Computational Linguistics
https://doi.org/10.18653/v1/D19-62

ness (Canadian Mental Health Association, 2016). The adverse impact of mental disorders is prominent when looking into the number of Canadians who have committed suicide, where 90% of them were suffering from some form of a mental disorder (Mental Health Commission of Canada, 2016).

When considering the current treatment procedures for mental illnesses, the first step is to screen the individual for symptoms using questionnaires. With such an approach, the interviewee could be more vulnerable to memory bias and also could adapt to the guidelines prescribed by the assessor. Using such screening procedures might not expose the actual mental state of an individual, and hence, the prescribed treatments could be inadequate. Also, due to various socio-economic aspects, people with mental disorders have not been able to receive adequate treatments. The lack of sufficient treatments can be seen in countries with all types of income levels. For example, between 76% to 85% of the people from countries in low to middle-class income do not receive sufficient treatments for their mental illnesses, while 35% to 50% of the people from high-income countries do not receive adequate treatments (World Health Organization, 2018b). In addition to insufficient treatments, social stigma and discrimination have prohibited people from getting proper treatment and social support (World Health Organization, 2018b). Due to the constraints as mentioned above and social media becoming an integral part of everyday life for many individuals, researchers have identified the importance of using social media data as a source for ascertaining individuals susceptible of mental disorders (De Choudhury, 2013, 2014). Due to the rapid growth in social media users within different demographics (Statista, 2017), and the abundance of information that can be extracted about the users could bring invaluable insights that can be used to detect signs of mental illnesses and suicide ideation that can be challenging to obtain using structured questionnaires. Taking the factors mentioned above into consideration, we have proposed a solution that incorporates certain profound features manually engineered by researchers into a multi-task learning architecture, to enhance the model predictability to distinguish neurotypical users from users susceptible to mental disorders. We hope that our research will encourage other researchers to investigate further the possibilities of incorporating verified manually engineered features into architectures similar to the proposed one, to enhance the prediction accuracies in identifying users susceptible to mental disorders.

Key Contributions

- We demonstrate the applicability of a convolutional neural network with a multi-channel architecture on different classification tasks using unstructured and limited social media data.

- We illustrate the use of multi-task learning to predict users susceptible to depression and PTSD (Post Traumatic Stress Disorder).

- We built an emotion classifier to identify the emotion category (i.e., sad, anger, fear, joy) associated with the tweets posted by the users and used those categories as multiple inputs within the deep neural network architecture. We also explore the impact of using metadata (age and gender) as multiple inputs to enhance the model predictability.

2 Ethical Considerations

It is of greater importance to follow strict guidelines on ethical research conduct when the research data resembles vulnerable users who could be compromised. The researchers working with data that could be used to single out individuals must take adequate precautions to avoid further psychological distress. During our research, we have given thorough considerations to these ethical facets and have adopted strict guidelines to ensure the anonymity and privacy of the data. Similar to the guidelines proposed by Benton et al. (2017a), we have exercised strict hardware and software security measures. Our research does not involve any intervention and has focused mainly on the applicability of machine learning models in determining users susceptible to mental disorders.

3 Related work

As social media has become an integral part of ones' day-to-day-life, it will be insightful to identify to what extent an individual has disclosed her/his personal information and whether accurate and sufficient information is being published to determine whether or not a person has a mental disorder. Considering the Twitter platform, rather than just sharing depressed feelings, users are more likely to self-disclose to the extent where they reveal detailed information about their treatment history (Park et al., 2013). The same level of self-disclosure can be identified in the Reddit forums (Balani and De Choudhury, 2015) and specifically by users with anonymous accounts (Pavalanathan and De Choudhury, 2015). Also, it was identified that personality traits and meta-features such as age and gender could have a positive impact on the model performances when detecting users susceptible to PTSD and depression (Preot et al., 2015). Similarly, we have also identified that age and gender as multiple inputs have positively impacted model predictability when used with multi-task learning.

Text extracted from social media platforms such as Twitter, Facebook, Reddit, and other similar forums has been successfully used in various natural language processing (NLP) tasks to identify users with different mental disorders and suicide ideation. Social media text was used to classify users with insomnia and distress (Jamison-Powell et al., 2012; Lehrman et al., 2012), postpartum depression (De Choudhury et al., 2013a,b, 2014), depression (Resnik et al., 2015a, 2013, 2015b; Schwartz et al., 2014; Tsugawa et al., 2015), Post-Traumatic Stress Disorder (Coppersmith et al., 2014a,b), schizophrenia (Loveys et al., 2017) and many other mental illnesses such as Attention Deficit Hyperactivity Disorder (ADHD), Generalized Anxiety Disorder, Bipolar Disorder, Eating Disorders and obsessive-compulsive disorder (OCD) (Coppersmith et al., 2015a).

With the advancements in neural network-based algorithms, more research has been conducted successfully in detecting mental disorders, despite the limited amount of data. Kshirsagar et al. (2017) have used recurrent neural networks with attention to detect social media posts resembling crisis. Husseini Orabi et al. (2018) demonstrated that using convolution neural network-based architectures produces better results compared to recurrent neural network-based architectures when detecting users susceptible to depression. Even though our experiments are to categorize users into three classes (i.e., control, depression, PTSD), the proposed multi-channel architecture have produced comparable results to the ones presented by Husseini Orabi et al. (2018) using binary classification to distinguish users susceptible to depression.

4 Proposed solution

The proposed solution consists of two key components. The first identifies the type of emotion expressed by each user using the model trained on the WASSA 2017 shared task dataset. The identified emotion categories are used as multiple inputs within the multi-task learning environment. The second component is the model that predicts users susceptible to PTSD or depression. When structuring the two neural network models (i.e., for emotion classification and mental illness detection), a common base architecture is used. The base architecture is a multi-channel Convolutional Neural Network (CNN) with three different kernel sizes (i.e., 1, 2, and 3). Through experiments, we identified that the multi-channel CNN architecture manages to produce better validation accuracies compared to the accuracies produced by Recurrent Neural Network (RNN) based models, which are commonly used with sequence data.

4.1 Data

Emotion Classification We use the data from the 8th Workshop on Computational Approaches to Subjectivity, Sentiment & Social Media Analysis (WASSA-2017). The data was used in the shared task to identify emotion intensity (Mohammad and Bravo-Marquez, 2017). The tweets in the dataset were assigned with the labels: anger, fear, joy and sadness, and their associated intensities. Table 1 presents the detailed statistics of the dataset.

Emotion	Train	Test	Dev	Total
Anger	857	760	84	1701
Fear	1147	995	110	2252
Joy	823	714	79	1616
Sadness	786	673	74	1533
Total	3613	3142	347	7102

Table 1: The number of tweets under each emotion category

The dataset contains 194 tweets that belong to multiple emotion categories. For example, the tweet: "I feel like I am drowning. #depression #anxiety #falure #worthless" is associated with the labels 'fear' and 'sadness'. When training the model, we created a single training dataset by combining both the train and test data and tested the trained model on the development dataset. The main reason for using such an approach is to improve the neural network model training by providing as much data as possible so that model overfitting will be reduced while increasing the model generalization. During our training, we did not take into consideration the emotional intensity and expect to use it in our future research as an additional input.

Mental Illness Detection To detect whether a user is a neurotypical user or if the user is susceptible to having either PTSD or depression, we used the dataset from the Computational Linguistics and Clinical Psychology (CLPsych) 2015 shared task (Coppersmith et al., 2015b). Table 2 presents the detailed statistics of this dataset.

	Control	PTSD	Depressed
Number of users	572	246	327
Average age	24.4	27.9	21.7
Gender (female) distribution per class	74%	67%	80%

Table 2: CLPSych 2015 shared task dataset statistics

Preprocessing: All the URLs, @mentions, #hashtags, RTweets, emoticons, emojis, and numbers were removed. We removed a selected set of stopwords but kept first, second, and third person pronouns. It was discovered that users susceptible to mental disorders such as depression have frequently used the first-person singular pronouns compared to neurotypical users (Pennebaker et al., 2007). Also, the punctuation marks except for a selected few were removed. The full stop, comma, exclamation point, and the question mark were kept while removing all the other punctuation marks. The NLTK (Natural Language Toolkit) tweet tokenizer was used to tokenize the tweets. We selected 200,000 unique tokens to build the vocabulary, rather than choosing all the unique words, which could lead to sparse word vectors with high dimensionality.

Vocabulary Generation: To obtain an enriched dictionary containing the most relevant terms, we introduced a novel approach instead of the traditional approach used in many deep learning APIs (e.g., Keras deep learning high-level API[1]). Our approach takes into account the top 'K' terms based on their term frequency and inverse document frequency (TF-IDF) scores. To build the

dictionary, first, we calculated the TF-IDF values under each user (i.e., by considering all the train/validation tweets of a single user as one single document). Then we took the maximum score out from all the assigned TF-IDF scores given the word. The reason for taking the maximum is to extract the words identified as closely related to a given user. Based on the computed TF-IDF scores, we picked the top 'K' words (K=200,000) to construct the vocabulary. A dictionary created using the above approach allows the model to capture the underlying relationships between the critical words. In comparison to the word frequency-based approach, using the vocabulary based on the TF-IDF scores has produced relatively better results for the recorded matrices (refer Table 4). When analyzing the model's prediction accuracy and loss (i.e., on training and validation data) over five-fold cross-validation, we identified that the model trained using the TF-IDF based vocabulary has been more stable with less randomness compared to the model trained using the vocabulary based on word frequencies.

When choosing the maximum sequence length for the input data, it is essential to capture as much information as possible from each user, especially given consideration to the research domain of mental illness detection. Since we have concatenated all the individual tweets belonging to one user as a single string, a high variance in the sequence length was identified among users. On average, a single user has used around 15,800 tokens, where the maximum number of tokens used by an individual user is nearly 64,800. Rather than experimenting with different sequence lengths, we selected the maximum length for the sequence by adding three standard deviations to the average sequence length covering 99% of users with a sequence length of 46,200 tokens. The shorter sequences were padded with zeros (to the end of the sequence), and the longer sequences were truncated (from the end of the sequence).

Model Architecture The selected model architecture consists of three main components: multi-task learning, CNN with multi-channel, and multi-inputs. Multi-task learning is known to be successful when the data is noisy and limited so that when trying to learn one task, it could gain additional information from the other tasks to identify the most relevant features. Learning a shared representation so that individual tasks can benefit

---

[1] https://keras.io/preprocessing/text/

from one another (Caruana, 1997) can be considered as one of the most appropriate architectures when trying to detect multiple mental illnesses. Benton et al. (2017b) demonstrated the successful use of multi-task learning to recognize mental illnesses and suicide ideation. Different from their approach, we add multiple features discovered by researchers in the fields of computational linguistics and psychology, to enhance the model performances. We consider that it is important to identify the impact of manually engineered features on the model's predictability. We also recognized that using a CNN multi-channel architecture is best suited for tasks dealing with limited unstructured data compared to RNN architectures or multilayer perceptrons (MLP).

We used a multi-channel model as the base model in both emotion classification (i.e., to detect anger, sadness, joy, fear) and mental illness detection (i.e., to detect PTSD and depression). The multi-channel model uses three versions of a standard CNN architecture with different kernel sizes. We identified that using different kernel sizes (different n-grams sizes) with Global Maximum Pooling produces better results compared to a standard CNN architecture. The optimal validation accuracies for both emotion and mental illness detection models were derived using three channels with kernel sizes 1, 2, and 3. Increasing the kernel sizes or the number of channels reduced the validation accuracies.

For the emotion classification task, the CNN in each channel was tested with 64 filters, same padding and a stride of 1 (distance between successive sliding windows). We used Rectified Linear Unit (ReLU) as the activation function. To normalize the data and to reduce the impact of model overfitting, we used the batch normalization layer and used a dropout (Srivastava et al., 2014) as the regularization technique with a probability of 0.2. As the final layer in each channel, we used global maximum pooling to reduce the number of parameters needed to learn so that it could further reduce the impact of model overfitting. The outputs from each global maximum pooling layers (from each channel) were concatenated and fed into a fully connected layer with four hidden units that use sigmoid activation to generate the output. All the inputs were sent through trainable embedding layers (randomly initialized) with a dimension of 300 for the emotion classification task and 100 for

the mental illness detection task.

Throughout our research, we did not emphasize much on word embeddings as our primary objective was to identify the impact of merging features derived using deep learning methods with few of the notable features that were identified over the years by researchers on detecting mental illnesses. Even though our best results were obtained by instantiating the embedding layer weights with random numbers (refer Table 4), we conducted several preliminary experiments using word embeddings trained on the fastText (Joulin et al., 2017) algorithm. We decided to use fastText because given the unstructured nature of the twitter messages we could obtain a more meaningful representation by expressing a word as a vector constructed out of the sum of several vectors for different parts of the word (Bojanowski et al., 2017). One of the reasons for the low measurements could be due to the reason that we used fewer data to train our embeddings.

Mental illness detection When building the multi-task learning model to detect mental illnesses, the base model architecture (i.e., for the shared representation) has used a structure similar to the one used in emotion classification. The fundamental changes to the base model include: using 256 filters instead of 64 and using L1 kernel regularization in each convolution layer. We used the trained model on the emotion data to predict the emotion category of the individual Twitter messages in the CLPSych 2015 dataset. We grouped the predicted probabilities for each user under the different emotion categories by calculating the standard deviation. We used the predicted probabilities for each emotion category as multiple inputs when detecting neurotypical and depressed users, while age and gender were used as inputs when predicting users with PTSD. Before concatenating the multiple-inputs with the output from the multi-channel architecture, the multiple-inputs were transformed using a fully connected layer with 128 hidden units and ReLU activation. The output from the shared layers and the transformed multiple inputs were merged before being used as the input to the fully connected layer with three individual hidden units and sigmoid activation. Before applying multiple-inputs to the neural network architecture, all the relevant inputs were normalized using a minimum, maximum scaler initialized within the range 0 and 1. The neural network architecture used for the multi-task, multi-channel, multi-input model for mental illness detection is shown in Figure 1.

Model Training When training both the emotion and the mental illness detection models, we minimized the validation loss to learn the optimal neural network parameters. To train both the models, we used minibatch gradient descent with smaller batch sizes. Using a smaller batch size is known to stabilize model training while increasing the model generalizability. In many cases, the optimal results were obtained by using batch sizes smaller or equal to 32 (Masters and Luschi, 2018). In our experiments, we used batch sizes 32 and 8 respectively for training the emotion and mental illness detection models. Both models were trained for 15 epochs and used early stopping when the validation loss has stopped improving. The Adam optimizer (Kingma and Ba, 2014) was used when training both the models with the default learning rate of 0.001.

5 Results

5.1 Emotion classification

The emotion detection task was implemented as a multi-class, multi-label classification because the same Twitter message can belong to multiple classes. Since we would like to have independent probability values for each class rather than a probability distribution over the four classes, we used binary cross-entropy as the loss function. Having independent probability values is better-oriented towards the identification of independent emotion categories. Table 3 reports the emotion classification results obtained using multi-channel CNN (MCCNN), and in addition, results from several other experiments: CNN with max-pooling (CNNMax) and bidirectional Long Short Term Memory module (biLSTM) were reported for comparison.

	Acc (%)	F1(%)	P(%)	R(%)	P rank (%)
MCCNN	88.88	77.41	79.68	75.67	84.85
CNNMax	85.82	68.95	76.97	62.70	81.39
biLSTM	85.07	68.66	75.97	63.49	80.97

Table 3: Emotion multi-class, multi-label classification results

In Table 3, the recorded accuracy is based on the Keras API[2] accuracy calculation on the multi-class, multi-label models where it takes into account the individual label predictions rather than a complete match (i.e., if there is more than one label per instance). The F1-score, precision, and recall measures are calculated based on the 'macro' averaging on the exact match and hence the low percentages. We have also reported the label ranking average precision score, which averages over the individual ground truth label per instance. This metric is mainly used in tasks that involve multi-label ranking. From the reported results, it can be seen that the multi-channel CNN model has given the best results compared to the standard CNN model and the RNN based model. Based on the outcome, we have used the above trained multi-channel CNN model to make predictions on the CLPsych 2015 individual tweets.

5.2 Mental illness detection

For detecting mental illnesses, we used binary cross-entropy loss as the loss function and sigmoid activation as the final layer activation. The data was sampled using Stratified Shuffle Split to maintain class distribution between 80%/20% of training and validation data. Our models were evaluated only on the validation data because the CLPSych 2015 shared task test data labels were not made available. To ascertain the model reliability, we performed 5-fold cross-validation and discovered that having multi-inputs on a multi-task, multi-channel architecture does increase the model performances. The recorded model accuracies were averaged over five folds with a standard deviation of 0.01.

Table 4 demonstrates the model performances according to different combinations of the multi-task, multi-channel, and multi-input architectures. To demonstrate the effectiveness of the proposed approach, we conducted several experiments using variants of two deep learning architectures; Convolutional Neural Networks, and Recurrent Neural Networks and also to measure a baseline, we used the shallow learning approach; Support Vector Machine. The experiments: MtMcMi (Multi-task Multi-channel, Multi-input), MtMc (Multi-task, Multi-channel), MtMcMiFT(Multi-task, Multi-channel, Multi-input, using FastText word representations), MtMcMiFr (Multi-task, Multi-channel, Multi-input, using word Frequen-

---

[2]https://keras.io/metrics/

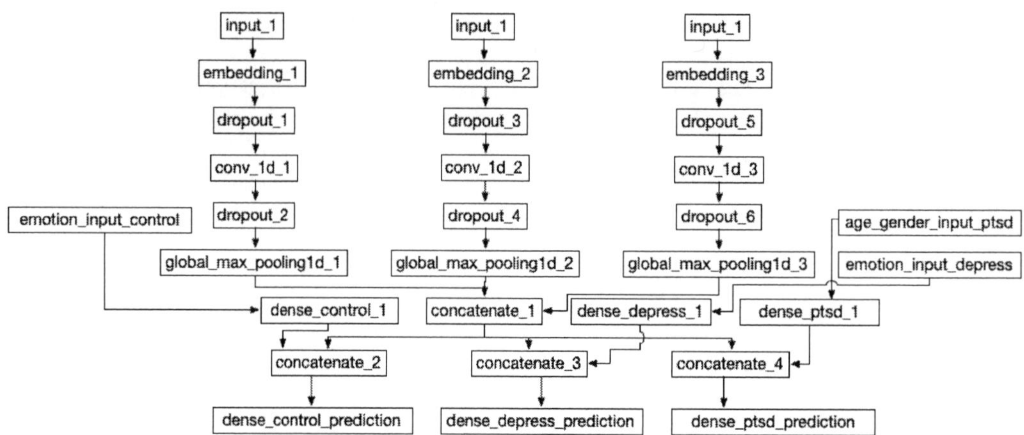

Figure 1: Multi-task, multi-channel, multi-input model for mental illness detection

	Accuracy(%)			Avg. F1(%)			Avg. Precision(%)			Avg. Recall(%)			Avg. AUC(%)		
	C	D	P	C	D	P	C	D	P	C	D	P	C	D	P
MtMcMi	89.08	**87.59**	91.35	**89.07**	**83.81**	86.06	**89.19**	**86.61**	89.81	89.07	**82.05**	83.50	**95.30**	**92.24**	**93.18**
MtMc	88.55	86.89	**91.96**	88.53	83.06	**87.02**	88.82	85.11	**90.54**	88.54	81.75	**84.64**	94.62	90.74	92.54
MtMcMiFT	85.50	86.28	91.26	85.45	82.73	85.90	85.91	84.06	89.70	85.49	81.97	83.30	93.88	91.01	91.91
MtMcMiFR	87.42	86.72	91.52	87.41	83.07	86.52	87.48	84.63	89.49	87.41	82.00	84.36	94.88	91.55	92.53
McMclass	**92.00**	75.69	76.73	89.05	76.83	81.22	86.34	78.25	86.93	**92.00**	75.69	76.73	92.44	84.55	86.32
biLSTMMtMi	51.35	71.61	78.60	41.27	41.73	44.00	41.92	35.80	39.30	51.52	50.00	50.00	56.87	59.89	60.28
biLSTMMt	52.48	72.31	78.60	47.48	46.40	44.00	47.84	59.81	39.30	52.57	52.06	50.00	56.32	59.96	54.89
svmMclass	81.73	52.91	42.85	75.82	57.01	46.87	70.76	62.11	51.97	81.73	52.91	42.85	81.18	79.12	77.70

Table 4: Mental illness detection using multi-task, multi-channel, multi-input architecture. The labels 'C', 'D' and 'P' denotes 'Control', 'Depressed' and 'PTSD'

cies), McMclass (Multi-channel, Multi-class), biLSTMMtMi (bidirectional Long Short Term Memory, Multi-task, Multi-input), biLSTMMt (bidirectional Long Short Term Memory, Multi-task), svmMclass (support vector machine, Multi-class) were conducted to identify a fitting approach to discover individuals who are susceptible of mental disorders.

The metrics used for the evaluation are accuracy, precision, recall, F1-score, and Area Under the Receiver Operating Characteristic curve (AUC). The AUC score is used to compare the model performances where the average AUC is calculated with standard deviation. The standard deviation is used as a mechanism to identify variance among model performances, which could bring insight into the reliability of the trained model. For each experiment, we recognized that the standard deviation is approximately around 0.01, which provides an empirical confirmation that the sampling using stratified shuffle splits provides an accurate representation of the complete dataset.

The "MtMcMi" architecture uses features based on emotion as multi-inputs on the control and depressed users while age and gender were used on the users with PTSD. In comparison to "MtMc" which is multi-task, multi-channel without multi-inputs, we could see that using multiple inputs have increased the average AUC score, as well as most of the other evaluation matrices (i.e., precision, recall, and F1-score). Even though the increase could not be considered as significant, the potential room for improvement is high if provided with more accurate emotion prediction and additional profound features identified by researchers. Concerning the emotion detection task, we could have obtained considerable improvements in prediction accuracies if provided with additional data when training the deep learning model.

When analyzing the result for "MtMcMiFr", which is using the Multi-task, Multi-channel, Multi-input architecture with the vocabulary created using the default word frequencies; we could identify that our proposed approach, which uses the vocabulary constructed using the weighted TF-IDF words has produced comparatively better results. Even though the gained improvements could not be considered as significant, the proposed method can be regarded as an effective approach when initiating the vocabulary, that provides a balance between the rare and frequently used words.

In the experiment "MtMcMiFT", where we used the fastText embeddings layer with the number of dimensions equal to 100 as an input to the Multi-channel convolutional neural network, it was observed that the results obtained using the randomly initialized embedding layer are higher than with the fastText pre-trained embeddings. This could be due to embeddings not being trained on a sufficiently large dataset (we trained them only on the CLPSych 2015 dataset). In future work, we will conduct further research to enhance the embedding layer word representation by using state-of-the-art language modeling approaches trained on larger datasets.

The effectiveness of using convolutional neural network models can be identified when evaluating the results obtained using Recurrent Neural Network (RNN) based architectures. The "biLSTMMtMi" method uses a bidirectional Long Short Term Memory ("biLSTM") model in a Multi-task, Multi-input design and comparatively has produced poor results for different combinations of hyperparameters. This could be due to several reasons such as the unstructured nature of the Twitter text as well as the non-existence of long-term dependencies. For example, our best results were obtained when using the kernel sizes (i.e., the number of consecutive tokens) one, two, and three and ones the kernel sizes are increased the overall model predictability decreases. When using a "biLSTM" model as the shared layer in multi-task learning without multiple inputs ("biLSTMMMt"), the results are somewhat better compared to when using multi-inputs.

To demonstrate the effectiveness of using multi-task learning to detect multiple mental disorders, we compared the proposed approach with multiclass classification to distinguish neurotypical users from users susceptible to having either PTSD or depression. First, we used a multi-channel convolutional neural network to predict the three classes (i.e., control, depress, and PTSD). In comparison to our proposed approach, we can identify that multiclass classification using CNN have produced slightly better results on two occasions, which is for average accuracy and recall under the control class. Through further analysis,

we see that average precision, F1-score, and AUC scores are higher for all three classes when using the proposed approach. Overall the multiclass classification task has produced low scores (especially for precision, F1-score, and AUC) when detecting users susceptible to depression and PTSD while the proposed approach has contributed significantly better results. The better results could be due to the reason that depression is commonly identified among individuals with PTSD, and the shared layer has managed to learn such common characteristics while the task-specific layers have learned the individual features unique to each disorder.

As a baseline, we used the linear SVM classifier with TF-IDF features (200,000 features) in a multiclass classification task (i.e., svmMclass). When sampling the data, five splits of 80% training and 20% testing were created using the Stratified Shuffle Split method to maintain class distribution. We also computed a majority class baseline, which classifies everything in the largest class (the control class in the training data). It achieved an accuracy of 50.21% on the test data. Overall, we can see that using limited unstructured data with an architecture based on CNN have produced better results compared to the solution based on RNN. Notably, the multi-task, multi-channel architecture with multiple-inputs has provided the best results and confirms that using multiple-inputs has a positive influence on the overall model performances. Also, the appropriateness of using multi-task learning instead of multiclass classification to detect multiple mental disorders is highlighted. Similar to the fact that certain mental disorders share specific common symptoms (American Psychiatric Association, 2013), multi-task learning has managed to learn such characteristics through a shared representation followed with task-specific layers to identify the unique attributes to differentiate multiple mental disorders.

6 Comparison to related work

Even though our work could not be directly compared with (Benton et al., 2017b), we can identify that our model has produced competitive results, especially when comparing the AUC score for detecting users with PTSD and depression. Our best model has scored an AUC score >0.90 in identifying all three individual classes (control, depression, PTSD) in comparison to an AUC score <0.80 for detecting PTSD and depression and an AUC score >0.90 for detecting the neurotypical users (Benton et al., 2017b). In the CLPSych 2015 shared task (Coppersmith et al., 2015b), Resnik et al. (2015a) have reported AUC scores of 0.86 (depression vs. control), 0.84 (depression vs. PTSD) and 0.89 (PTSD vs. control) and similarly Preotiuc-Pietro et al. (2015) have reported an average AUC score around 0.86 in differentiating neurotypical users from users susceptible to PTSD and depression. Even though we have produced better results using the validation dataset, we could not directly compare our results with the shared task participants as they have evaluated their models against the test dataset which was not made available to us. In our future work, we will conduct experiments using public forum post data extracted from platforms such as Reddit[3]. In comparison, the proposed approach can be tested with adequate adjustments to detect multiple mental disorders such as depression, anxiety, PTSD, and six others using the dataset introduced by Cohan et al. (2018). The authors have achieved an F1-score of 27.83% for multi-class classification and 53.56% and 57.60% respectively, when detecting depression and PTSD as binary classification tasks.

7 Conclusion

In this paper, we investigated the impact of merging features derived using deep neural network architectures with profound manually engineered features identified by researchers over the years using shallow learning to detect mental disorders using social media text. In particular, we have identified that by using a multi-channel convolutional neural network as a shared layer in a multi-task learning architecture with multiple-inputs (e.g., different emotion categories, age, and gender) have produced comparatively competitive results in detecting multiple mental disorders (in our case depression and PTSD). For future work, we will continue our research on suicide risk detection, and the temporal impact different mental disorders have on suicide ideation.

References

American Psychiatric Association. 2013. *Diagnostic and Statistical Manual of Mental Disorders*. Arlington.

Sairam Balani and Munmun De Choudhury. 2015. Detecting and Characterizing Mental Health Related Self-Disclosure in Social Media. In *Proceedings of the 33rd Annual ACM Conference Extended Abstracts on Human Factors in Computing Systems - CHI EA '15*, pages 1373–1378.

Adrian Benton, Glen Coppersmith, and Mark Dredze. 2017a. Ethical Research Protocols for Social Media Health Research. In *First Workshop on Ethics in Natural Language Processing,*, pages 94–102.

Adrian Benton, Margaret Mitchell, and Dirk Hovy. 2017b. Multi-Task Learning for Mental Health using Social Media Text. *CoRR*, abs/1712.0.

Piotr Bojanowski, Edouard Grave, Armand Joulin, and Tomas Mikolov. 2017. Enriching Word Vectors with Subword Information. *Transactions of the Association for Computational Linguistics*, 5:135—146.

Canadian Mental Health Association. 2016. Canadian Mental Health Association.

Rich Caruana. 1997. Multitask Learning. *Machine Learning*, pages 41–75.

Arman Cohan, Bart Desmet, Andrew Yates, Luca Soldaini, Sean MacAvaney, and Nazli Goharian. 2018. SMHD: A Large-Scale Resource for Exploring Online Language Usage for Multiple Mental Health Conditions. In *Proceedings of the 27th International Conference on Computational Linguistics*, pages 1485—1497, Santa Fe, New Mexico, USA. Association for Computational Linguistics.

Glen Coppersmith, Mark Dredze, and Craig Harman. 2014a. Measuring Post Traumatic Stress Disorder in Twitter. In *In Proceedings of the 7th International AAAI Conference on Weblogs and Social Media (ICWSM).*, volume 2, pages 23–45.

Glen Coppersmith, Mark Dredze, and Craig Harman. 2014b. Quantifying Mental Health Signals in Twitter. In *Proceedings of the Workshop on Computational Linguistics and Clinical Psychology: From Linguistic Signal to Clinical Reality*, pages 51–60.

Glen Coppersmith, Mark Dredze, Craig Harman, and Kristy Hollingshead. 2015a. From ADHD to SAD: Analyzing the Language of Mental Health on Twitter through Self-Reported Diagnoses. In *Computational Linguistics and Clinical Psychology*, pages 1–10.

Glen Coppersmith, Mark Dredze, Craig Harman, Hollingshead Kristy, and Margaret Mitchell. 2015b. CLPsych 2015 Shared Task: Depression and PTSD on Twitter. In *Proceedings of the 2nd Workshop on Computational Linguistics and Clinical Psychology: From Linguistic Signal to Clinical Reality*, pages 31–39.

[3]https://www.reddit.com/

Munmun De Choudhury. 2013. Role of Social Media in Tackling Challenges in Mental Health. In *Proceedings of the 2nd International Workshop on Socially-Aware Multimedia (SAM'13)*, pages 49–52.

Munmun De Choudhury. 2014. Can social media help us reason about mental health? In *23rd International Conference on World Wide Web*, Cdc, pages 1243–1244.

Munmun De Choudhury, Scott Counts, and Eric Horvitz. 2013a. Major Life Changes and Behavioral Markers in Social Media : Case of Childbirth. In *Computer Supported Cooperative Work (CSCW)*, pages 1431–1442.

Munmun De Choudhury, Scott Counts, and Eric Horvitz. 2013b. Predicting postpartum changes in emotion and behavior via social media. In *Proceedings of the SIGCHI Conference on Human Factors in Computing Systems - CHI '13*, page 3267.

Munmun De Choudhury, Scott Counts, Eric J. Horvitz, and Aaron Hoff. 2014. Characterizing and predicting postpartum depression from shared facebook data. *Cscw*, pages 626–638.

Ahmed Husseini Orabi, Prasadith Buddhitha, Mahmoud Husseini Orabi, and Diana Inkpen. 2018. Deep Learning for Depression Detection of Twitter Users. In *Fifth Workshop on Computational Linguistics and Clinical Psychology*, pages 88–97.

Susan Jamison-Powell, Conor Linehan, Laura Daley, Andrew Garbett, and Shaun Lawson. 2012. "I can't get no sleep": discussing #insomnia on Twitter. In *Conference on Human Factors in Computing Systems - Proceedings*, pages 1501–1510.

Armand Joulin, Edouard Grave, Piotr Bojanowski, and Tomas Mikolov. 2017. Bag of Tricks for Efficient Text Classification. In *15th Conference of the European Chapter of the Association for Computational Linguistics*, pages 427—-431.

Diederik P. Kingma and Jimmy Lei Ba. 2014. Adam: A Method for Stochastic Optimization. In *Proceedings of the 3rd International Conference on Learning Representations (ICLR)*, pages 1–15.

Rohan Kshirsagar, Robert Morris, and Samuel Bowman. 2017. Detecting and Explaining Crisis. In *Proceedings of the Fourth Workshop on Computational Linguistics and Clinical Psychology — From Linguistic Signal to Clinical Reality*, pages 66–73, Vancouver. Association for Computational Linguistics.

Michael Thaul Lehrman, Cecilia Ovesdotter Alm, and Rubén A. Proaño. 2012. Detecting Distressed and Non-distressed Affect States in Short Forum Texts. In *Second Workshop on Language in Social Media*, Lsm, pages 9–18, Montreal.

Kate Loveys, Patrick Crutchley, Emily Wyatt, and Glen Coppersmith. 2017. Small but Mighty: Affective Micropatterns for Quantifying Mental Health from Social Media Language. *Proceedings of the Fourth Workshop on Computational Linguistics and Clinical Psychology*, pages 85–95.

Dominic Masters and Carlo Luschi. 2018. Revisiting Small Batch Training for Deep Neural Networks. *CoRR*, pages 1–18.

Mental Health Commission of Canada. 2016. Mental Health Commission of Canada.

Saif M. Mohammad and Felipe Bravo-Marquez. 2017. WASSA-2017 Shared Task on Emotion Intensity. In *8th Workshop on Computational Approaches to Subjectivity, Sentiment and Social Media Analysis*.

Minsu Park, David W Mcdonald, and Meeyoung Cha. 2013. Perception Differences between the Depressed and Non-depressed Users in Twitter. *Proceedings of the 7th International AAAI Conference on Weblogs and Social Media (ICWSM)*, pages 476–485.

Umashanthi Pavalanathan and Munmun De Choudhury. 2015. Identity Management and Mental Health Discourse in Social Media Identity in Online Communities. In *WWW'15 Companion: 24th International World Wide Web Conference*, pages 18–22.

James W Pennebaker, Cindy K Chung, Molly Ireland, Amy Gonzales, and Roger J Booth. 2007. The Development and Psychometric Properties of LIWC2007 The University of Texas at Austin. Technical Report 2, The University of Texas at Austin, Austin, Texas.

Daniel Preot, Johannes Eichstaedt, Gregory Park, Maarten Sap, Laura Smith, Victoria Tobolsky, H Andrew Schwartz, and Lyle Ungar. 2015. The Role of Personality , Age and Gender in Tweeting about Mental Illnesses. In *Proceedings of the Workshop on Computational Linguistics and Clinical Psychology: From Linguistic Signal to Clinical Reality*, pages 21–30.

Daniel Preotiuc-Pietro, Maarten Sap, H. Andrew Schwartz, and Lyle Ungar. 2015. Mental Illness Detection at the World Well-Being Project for the CLPsych 2015 Shared Task. In *Proceedings of the 2nd Workshop on Computational Linguistics and Clinical Psychology: From Linguistic Signal to Clinical Reality*, pages 40–45.

Philip Resnik, William Armstrong, Leonardo Claudino, and Thang Nguyen. 2015a. The University of Maryland CLPsych 2015 Shared Task System. In *CLPsych 2015 Shared Task System*, c, pages 54–60.

Philip Resnik, William Armstrong, Leonardo Claudino, Thang Nguyen, Viet-an Nguyen, and Jordan Boyd-graber. 2015b. Beyond LDA : Exploring Supervised Topic Modeling for Depression-Related

Language in Twitter. In *Proceedings of the 2nd Workshop on Computational Linguistics and Clinical Psychology: From Linguistic Signal to Clinical Reality*, volume 1, pages 99–107.

Philip Resnik, Anderson Garron, and Rebecca Resnik. 2013. Using Topic Modeling to Improve Prediction of Neuroticism and Depression in College Students. In *Proceedings of the 2013 Conference on Empirical Methods in Natural Language Processing (EMNLP)*, October, pages 1348–1353.

H Andrew Schwartz, Johannes Eichstaedt, Margaret L Kern, Gregory Park, Maarten Sap, David Stillwell, Michal Kosinski, and Lyle Ungar. 2014. Towards Assessing Changes in Degree of Depression through Facebook. In *Proceedings of the Workshop on Computational Linguistics and Clinical Psychology: From Linguistic Signal to Clinical Reality*, pages 118–125.

Nitish Srivastava, Geoffrey Hinton, Alex Krizhevsky, Ilya Sutskever, and Ruslan Salakhutdinov. 2014. Dropout: A Simple Way to Prevent Neural Networks from Overfitting. *Journal of Machine Learning Research*, 15:1929–1958.

Statista. 2017. Number of social media users worldwide from 2010 to 2021.

Sho Tsugawa, Yusuke Kikuchi, Fumio Kishino, Kosuke Nakajima, Yuichi Itoh, and Hiroyuki Ohsaki. 2015. Recognizing Depression from Twitter Activity. In *Proceedings of the 33rd Annual ACM Conference on Human Factors in Computing Systems - CHI '15*, pages 3187–3196.

World Health Organization. 2014. WHO — Mental health: a state of well-being.

World Health Organization. 2018a. Depression.

World Health Organization. 2018b. Mental disorders.

Extracting relevant information from physician-patient dialogues for automated clinical note taking

Serena Jeblee[1,2], Faiza Khan Khattak[1,2,3], Noah Crampton[3],
Muhammad Mamdani[3], Frank Rudzicz[1,2,3,4]

[1]Department of Computer Science, University of Toronto, Toronto, Ontario, Canada
[2]Vector Institute for Artificial Intelligence, Toronto, Ontario, Canada
[3]Li Ka Shing Knowledge Institute, St Michael's Hospital, Toronto, Ontario, Canada
[4]Surgical Safety Technologies, Toronto, Ontario, Canada
sjeblee@cs.toronto.edu, faizakk@cs.toronto.edu,
cramptonn@unityhealth.to, muhammadm@unityhealth.to, frank@spoclab.com

Abstract

We present a system for automatically extracting pertinent medical information from dialogues between clinicians and patients. The system parses each dialogue and extracts entities such as medications and symptoms, using context to predict which entities are relevant. We also classify the primary diagnosis for each conversation. In addition, we extract topic information and identify relevant utterances. This serves as a baseline for a system that extracts information from dialogues and automatically generates a patient note, which can be reviewed and edited by the clinician.

1 Introduction

In recent years, electronic medical record (EMR) data have become central to clinical care. However, entering data into EMRs is currently slow and error-prone, and clinicians can spend up to 50% of their time on data entry (Sinsky et al., 2016). In addition, this results in inconsistent and widely variable clinical documentation, which present challenges to machine learning models.

Most existing work on information extraction from clinical conversations does not differentiate between entities that are relevant to the patient (such as experienced symptoms and current medications), and entities that are not relevant (such as medications that the patient says were taken by someone else).

In this work, we extract clinically relevant information from the transcript of a conversation between a physician and patient (and sometimes a caregiver), and use that information to automatically generate a clinical note, which can then be edited by the physician. This automated note-taking will save clinicians valuable time and allow them to focus on interacting with their patients

rather than the EMR interface. We focus on linguistic context and time information to determine which parts of the conversation are medically relevant, in order to increase the accuracy of the generated patient note. In addition, the automatically generated notes can provide cleaner and more consistent data for downstream machine learning applications, such as automated coding and clinical decision support.

Figure 1 shows a synthetic example of the kind of medical conversation where context and time information are important.

> **DR:** Are you currently taking [Adderall]$_{Med.}$?
> **PT:** No, but I took it [a few years ago]$_{TIMEX3}$.
> **DR:** And when was that?
> **PT:** Um, around [2015 to 2016]$_{TIMEX3}$.
> **DR:** And did you ever take [Ritalin]$_{Med.}$?
> **PT:** I dont think so.
> **Typical output:** *Adderall, Ritalin.*
> **Expected output:**
> *Medications: Adderall (2015-2016), no Ritalin*

Figure 1: Synthetic conversation example.

2 Related Work

Previous studies have shown that current EMR data are difficult to use in automated systems because of variable data quality (Weiskopf and Weng, 2013; Thiru et al., 2003; Roth et al., 2003). Weiskopf and Weng (2013) showed that EMR data is frequently incomplete, and is often not evaluated for quality. In addition, the variance in documentation style, abbreviations, acryonyms, etc. make it difficult for algorithms to interpret the text.

Some recent work on machine learning methods for EMR data includes predicting mortality

Proceedings of the 10th International Workshop on Health Text Mining and Information Analysis (LOUHI 2019), pages 65–74
Hong Kong, November 3, 2019. ©2019 Association for Computational Linguistics
https://doi.org/10.18653/v1/D19-62

and discharge diagnoses (Rajkomar et al., 2018), predicting unplanned hospital readmissions for 5k patients by encoding EMR data with a convolutional neural network (Nguyen et al., 2018), and predicting diagnosis codes along with text explanations (Mullenbach et al., 2018).

Although there is some existing work on generating text from structured data (Dou et al., 2018; Lebret et al., 2016), very little work has been done in the clinical domain. Liu (2018) generated patient note templates with a language model, which was able to approximate the organization of the note, but no new information from the patient encounter was used.

Du et al. (2019) introduced a system for extracting symptoms and their status (experienced or not) from clinical conversations using a multi-task learning model trained on 3,000 annotated conversations. However, their model was trained on a limited set of 186 symptoms and did not address other medically relevant entities.

A latent Dirichlet allocation (LDA) model (Blei et al., 2003) is a topic modeling technique and has been applied to clinical text to extract underlying useful information. For example, Bhattacharya et al. (2017) applied LDA on structured EMR data such as age, gender, and lab results, showing that the relevance of topics obtained for each medical diagnosis aligns with the co-occurring conditions. Chan et al. (2013) applied topic modeling on EMR data including clinical notes and provided an empirical analysis of data for correlating disease topics with genetic mutations.

3 Dataset

Primary diagnosis	Dyads
ADHD	100
Depression	100
COPD	101
Influenza	100
Osteoporosis	87
Type II diabetes	86
Other	226

Table 1: Data distribution (ADHD: Attention Deficit Hyperactivity Disorder; COPD: Chronic Obstructive Pulmonary Disorder)

For training and testing our models, we use a dataset of 800 patient-clinician dialogues (dyads) purchased from Verilogue Inc.[1], which includes demographic information about the patient as well as the primary diagnosis. The data consist of audio files and human-generated transcripts with speaker labels. Table 1 shows the distribution of diagnoses in the dataset.

Since these data are proprietary, we also use a few transcripts of staged clinical interviews from YouTube as examples. [2]

4 Annotation

First, the conversation transcripts are automatically annotated for time phrases using Heidel-Time, a freely available rule-based time phrase tagger (Strötgen and Gertz, 2010), as well as a limited set of common medical terms.

Two physicians then conduct manual annotation by correcting the automatic annotations and making any necessary additions, using a custom-developed annotation interface. The following types of entities are annotated: anatomical locations, diagnoses, symptoms, medications, reasons for visit, referrals, investigations/therapies, and time phrases. A total of 476 conversations are annotated by a unique physician, and inter-annotator agreement is calculated using DKPro Statistics[3] on 30 conversations which were annotated by both physicians. The agreement across all entity types is 0.53 Krippendorffs alpha (Krippendorff, 2004) and 0.80 F_1 (partial match).

We developed a custom annotation interface for labeling entities and their attributes in the transcripts, shown in Figure 3. The software includes the ability to add new annotation types and attributes, edit and delete previous annotations, and view the entire conversation for context.

---

[1] `http://www.verilogue.com`

[2] YouTube videos of simulated patient encounters were sourced by searching for the following terms: "medical history", "patient interview", and "clinical assessment". Our clinician team member watched potential videos in the search list and selected only the ones that met the following criteria: 1) clinician asking a patient questions in simulated clinical scenarios; 2) a subjective perception of adequate fidelity to real clinical encounters. The audio for these dialogues were transcribed by a professional transcriptionist. Examples used in this paper:
1: `https://www.youtube.com/watch?v=O2qYU8n4VsA`, 2: `https://www.youtube.com/watch?v=CUSxC-XHT2A`, 3: `https://www.youtube.com/watch?v=5_jIcAk1XeA`

[3] `https://dkpro.github.io/dkpro-statistics`

DR: *It's a shame how good the Blue Jays were a couple of seasons ago compared to now.* **PT:** *Yeah, I'm still not sure we should have got rid of Alex Anthopoulos.* **DR:** *Yeah, that was the turning point, eh? Anyways, you're here to review your [diabetes]*Diagnosis *right?* **PT:** *That's right.* **DR:** *How's the [numbness in your toes]*Sign/Symptom*/[toes]*Anatomical Location *?* **PT:** *The same. I'm used to it by now, and I'm grateful it's not getting worse.* **DR:** *Okay, that's good. Let's keep you on the [same dose of Metformin]*Medication *[for now]*TIMEX3 *then we'll check your [a1c]*Investigation/Therapy *again [in three months]*TIMEX3 *, and then I'll [see you back here after that]*Disposition plan*.* **PT:** *That makes sense to me.*	**DR:** *It's a shame how good the [Blue]*Medication *Jays were a couple of seasons ago compared to [now]*TIMEX3*.* **PT:** *Yeah, I'm still not sure we should have got rid of Alex Anthopoulos.* **DR:** *Yeah, that was the turning point, eh? Anyways, you're here to review your [diabetes]*Diagnosis *right?* **PT:** *That's right.* **DR:** *How's the numbness in your [toes]*Anatomical Location *?* **PT:** *The same. I'm used to it by [now]*TIMEX3*,and I'm grateful it's not getting worse.* **DR:** *Okay, that's good. Let's keep you on the same [dose]*Medication *of [Metformin]*Medication *for [now]*TIMEX3 *then we'll check your a1c again in [three months]*TIMEX3 *, and then I'll see you back here after that.* **PT:** *That makes sense to me.*

Figure 2: Example dialogue: (Left) Human annotation, (Right) automatic annotation. In both tables, highlight indicates the annotated entities; darker highlights indicate overlap between human and automatic annotations. Subscripts indicate the entity type.

5 Methods and experiments

The automated pipeline currently includes the following components: preprocessing, utterance type classification (questions, answers, statements, etc.), entity extraction (medications, symptoms, diagnoses, etc.), attribute classification (modality and pertinence), primary diagnosis classification, SOAP classification, and note generation. In this section we discuss each component in detail, including methods and results. See Figure 4 for a diagram of the system components.

5.1 Preprocessing and data splitting

Before passing the data to our models, the text of the transcripts is lowercased, and punctuation is separated from words using WordPunctTokenizer from NLTK (Steven Bird and Loper, 2009). For the utterance type and attribute classification tasks, each word in an utterance is represented as a word embedding. In this work, we use publicly available ELMo embeddings (Peters et al., 2018) trained on PubMed abstracts, as well as word2vec embeddings trained on PubMed[4].

Of the 476 annotated conversations, we randomly select 50 to use as a test set for entity extraction and attribute classification.

5.2 Utterance type classification

In order to understand the conversational context, it may be useful to know whether an utterance is a question or answer. To this end, we classify each utterance as one of the following types: question, statement, positive answer, negative answer, backchannel (such as 'uh-huh' or 'yeah') or excluded (incomplete or vague utterance).

The utterance type classification model is a two-layer bidirectional gated recurrent unit (GRU) neural network (Cho et al., 2014), implemented in PyTorch, with the architecture shown in Figure 5. We augment the training data with two external, publicly available datasets: the Switchboard corpus (Calhoun et al., 2010), and the AMI corpus[5]. We map the utterance labels from the AMI and Switchboard corpora to our set of labels, and add these data to our training set.

We evaluate the utterance type classifier on a set of 20 conversations, annotated independently by 2 annotators with inter-annotator agreement of 0.77 (Cohen's kappa).

Table 2 shows the classification results by utterance type. As the most frequent type, statements are the easiest for the model to identify. The low performance of infrequent classes indicates that we could potentially improve performance by using an oversampling or regularization method.

5.3 Entity extraction

5.3.1 Time phrase extraction

In order to determine clinical relevance, it is important to know the time and duration of events in the patient history. We use HeidelTime to identify

---

[4]http://evexdb.org/pmresources/
vec-space-models/

[5]http://groups.inf.ed.ac.uk/ami/
corpus/

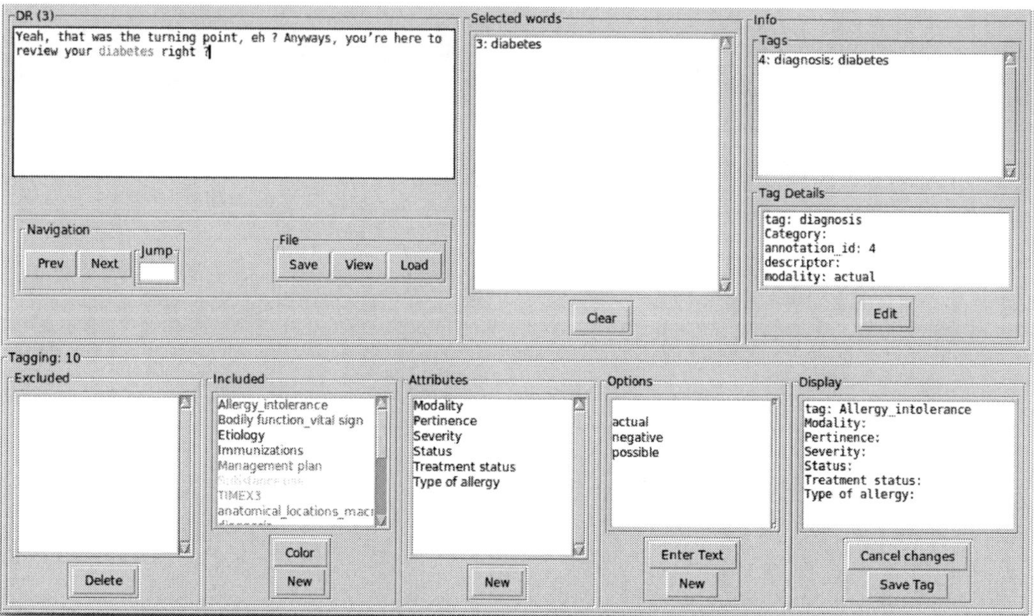

Figure 3: Annotation interface

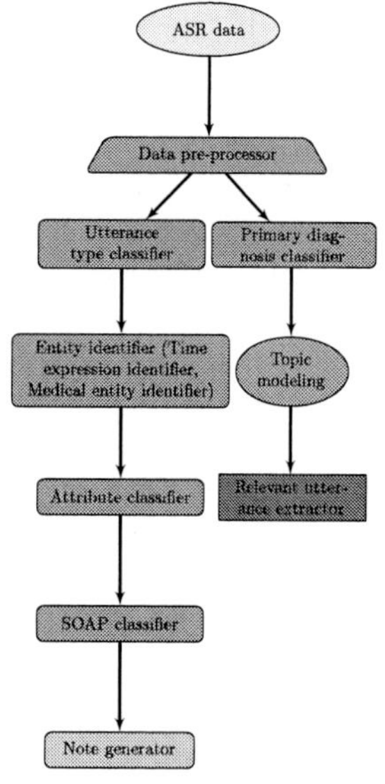

Figure 4: System components and data flow

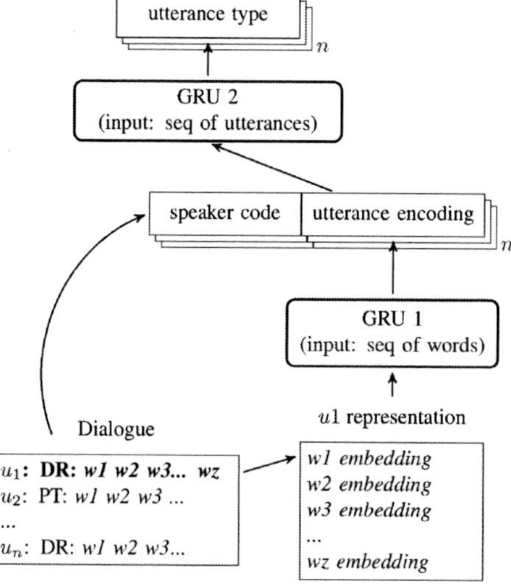

Figure 5: Utterance type classification model

time phrases in the transcripts, including times, dates, durations, frequencies, and quantities.

5.3.2 Clinical entity extraction

In addition to time phrases, we identify the following clinical concept types: anatomical locations, signs and symptoms, diagnoses, medications, referrals, investigations and therapies, and reasons for visit. To identify these entities, we search the transcript text for entities from a variety of medical lexicons, including the BioPortal Symptom lexicon [6], SNOMED-CT [7], the Consumer Health Vocabulary (CHV) [8], and RxNorm (a database of

[6] https://bioportal.bioontology.org/ontologies

[7] http://www.snomed.org/

[8] https://www.nlm.nih.gov/research/umls/sourcereleasedocs/current/CHV/

Type	Instances	P	R	F_1
Question	539	0.72	0.49	0.59
Statement	2,347	0.82	0.83	0.82
AnswerPositive	195	0.36	0.41	0.38
AnswerNegative	82	0.74	0.34	0.47
Backchannel	494	0.56	0.76	0.64
Excluded	131	0.20	0.16	0.18
Average	3,788	0.72	0.72	0.71

Table 2: Utterance type classification results

normalized medication names) [9].

Entity identification is currently limited to the terms present in our reference lists, which are large but cannot cover all possible expressions of relevant entities. There may be many valid variations of these entities that we hope to be able to identify in the future, potentially using a more sophisticated tagging method such as named entity recognition (NER).

Type	Instances	P	R	F_1
Anatomical locations	328	0.79	0.45	0.57
Diagnosis	346	0.88	0.62	0.72
Investigation or therapy	239	0.42	0.24	0.31
Medication	579	0.55	0.79	0.65
Referral	61	0.11	0.11	0.11
Sign/symptom	650	0.82	0.38	0.52
Time expression	1286	0.98	0.64	0.77
Average	3489	0.80	0.56	0.64

Table 3: Entity extraction results

5.3.3 Attribute classification

After extracting relevant entities, we classify them according to several attributes, including modality (i.e., whether the events were actually experienced or not) and pertinence (i.e., to which disease the entities are relevant, if any). For example, a patient might mention a medication that they have not actually taken, so we would not want to record that medication as part of the patient's history. In these

[9]https://www.nlm.nih.gov/research/umls/rxnorm/

Type	Instances	P	R	F_1
Actual	504	0.87	0.80	0.83
Negative	144	0.63	0.64	0.64
Possible	5	0.09	0.40	0.14
None	91	0.59	0.71	0.65
Average	744	0.78	0.76	0.77

Table 4: Modality classification results

Type	Instances	P	R	F_1
ADHD	126	0.54	0.41	0.28
COPD	22	0.20	0.45	0.28
Depression	32	0.27	0.81	0.41
Influenza	246	0.72	0.83	0.77
Other	312	0.79	0.51	0.62
None	6	0.32	1.00	0.48
Average	744	0.68	0.61	0.62

Table 5: Pertinence classification results

cases, the context of the conversation, as well as time information, is crucial to recording the patient's information accurately.

The attribute classifier is a support vector machine (SVM) trained with stochastic gradient descent using scikit-learn (Pedregosa et al., 2011). Each entity is represented as the average word embedding, concatenated with the word embeddings for the previous and next 5 words. We also include the speaker code of the utterance in which the entity appears. We train the model on 252 conversations and test on 50 for which we have human-assigned modality and pertinence labels.

We classify entities into the following modality categories: actual, negative, possible, or none. Table 4 shows the results of modality classification on the test set of 50 conversations. Since the majority of entities have a modality of 'actual', the model performs the best on this class. Entities are also classified as pertinent to one of the disease categories, or none. Table 5 shows the results of pertinence classification. Again we see that the classifier performs the best on the classes with more examples.

Modality classification performs fairly well with a context window of 5, likely because the relevant information can be found nearby in the text. However, pertinence classification is not as accurate, perhaps because it requires more global information about what conditions the patient has. In some cases, pertinence may be purely determined by a clinicians medical knowledge, not the information present in the text.

In the future we hope to have more annotated data on which to train, which should improve the overall performance, especially for the smaller classes.

5.4 Clinical note generation

In the note generation phase, we convert the structured data from the previous steps (i.e., entities and their attributes) into a free text clinical note that resembles what a physician would have written. This involves organizing the entities according to a structured note organization and, finally, generating the text of the note.

5.4.1 SOAP entity classification

After extracting clinical entities, we classify them according to the traditional four sections of a clinical note: subjective (S), objective (O), assessment (A), plan (P) (Bickley and Szilagyi, 2013). We also add a 'none' category, which means that the given entity should not be included in the note.

The SOAP classifier is a neural network trained on each word of the entity, the previous and next five words, the speaker code of the corresponding utterance, and the type of entity. The text and context are represented as word embeddings using the PubMed word2vec model. Since the note generation requires special annotations, we currently only have 58 conversations for training, and 20 for test.

Table 6 shows the results of SOAP classification. The model is the most accurate at determining which information to exclude from the note.

Type	Instances	P	R	F_1
S	299	0.52	0.56	0.54
O	51	0.44	0.43	0.44
A	55	0.35	0.16	0.22
P	66	0.22	0.15	0.18
None	708	0.69	0.72	0.70
Average	1189	0.59	0.61	0.60

Table 6: SOAP classification results

Class	P	R	F_1
ADHD	0.84	0.84	0.83± 0.05
Depression	0.80	0.64	0.71 ± 0.08
Osteoporosis	0.81	0.78	0.78 ± 0.04
Influenza	0.91	0.95	0.93 ± 0.04
COPD	0.75	0.65	0.68 ± 0.14
Type II Diabetes	0.81	0.75	0.76± 0.05
Other	0.71	0.82	0.76± 0.05
Average	0.79	0.78	0.78± 0.04

Table 7: Primary diagnosis classification results. 800 dyads using 5-fold cross-validation (Train: 80%, Test: 20%). F_1 score is the mean ± variance.

5.4.2 SOAP note generation

Our current note generation step organizes the entities into the SOAP sections, and lists them along with their attributes. Actually generating full sentences that more closely resemble a physician-generated note is the next step for our future work.

5.5 Primary diagnosis classification

We classify the primary diagnosis for each conversation. The purpose of this task is to automatically identify the main diagnosis for billing codes. We train and test the models on a 5-fold cross validation of the 800 dyads. We apply *tf-idf* on the cleaned text of each dyad and then use logistic regression, SVMs with various parameter settings, and random forest models for classification. The F_1 score is calculated based on the human-assigned labels available in the transcription.

The primary diagnosis classifier performs reasonably well even without labeled entity features. The results for influenza achieve almost 0.90 F_1 score, while the results for COPD and depression obtain an F_1 score of approximately 0.70. By inspecting the conversations, we find that visits with a primary diagnosis of depression mostly consist of general discussions related to daily routine, family life, and mood changes, which often result in misclassification probably because no medical terms are mentioned. By contrast, in patient visits where the primary diagnosis is influenza, the discussion is more focused on the disease.

The top words used by the classifier were *H1N1, ache, temperature, sore, sick, symptom, swine, body,* and *strep*, which possibly makes it easier to classify. On the other hand, COPD is misclassified mostly as the category 'other', which includes diseases such as asthma, CHF (Congestive heart failure), hypercholesterolemia, atopic dermatitis, HIV/AIDS, prenatal visit, hypercholesterolemia. That is, the COPD dyads may be misclassified because of the presence of other respiratory diseases in the 'other' category. We plan to extend the diagnosis classifier to multi-label classification.

5.6 Topic modeling

Topic modeling is an unsupervised machine learning technique used to form k topics (i.e., clusters of words) occurring together, where k is usually chosen empirically. We perform topic modeling with LDA using the open-source gensim package (Řehůřek and Sojka, 2010) with varying numbers of topics $k =$(5, 10, 12, 15, 20, 25, 30, and 40).

Due to their colloquial nature, patient-clinician conversations contain many informal words and non-medical terms. We remove common words, including stop words from NLTK (Steven Bird and Loper, 2009), backchannel words (like '*uh-huh*'), and words with frequencies above 0.05% of the total number words in all the documents (to reduce the influence of more generic words).

The topic modeling results are shown in Table 8; we choose k=12 topics because they provided the best topic distribution and coherence score. The words in each topic are reported in decreasing order of importance.

A manual analysis shows that topic 0 captures words related to ADHD and depression, while topic 1 is related to asthma and flu, and topic 3 is related to women's health, and so on. This qualitative evaluation of topics shows that topic modeling can be helpful in extracting important information and identifying the dominant topic of a conversation. In our future work, we also plan to do a quantitative evaluation of topic modeling results using state-of-the-art methods such as the methodology proposed by Wallach et al. (2009).

We see the potential use of topic modeling to keep track of the focus of each visit, the distribution of word usage, categorization, and to group patients together using similarity measures. We also use it for relevant text extraction in the next section.

5.7 Relevant utterance extraction

Identifying the key parts of the doctor-patient conversation can be helpful in finding the relevant information. In the previous section, we observe that topic modeling can be helpful in identifying the

Topic#	Topic words
0	focus, sleeping, depressed, asleep, attention, mind, cymbalta, appetite, psychiatrist, energy
1	ache, h1n1, treat, asthma, temperature, diarrhea, anybody, mucinex, chill, allergic
2	period, knee, birth, heavy, ultrasound, iron, metoprolol, pregnancy, pregnant, history,
3	meal, diabetic, lose, unit, mail, deal, crazy, card, swelling, pound
4	cymbalta, lantus, cool, cancer, crazy, allergy, sister, attack, nurse, wow
5	referral, trazodone, asked, shingle, woman, medicare, med, friend, clinic, form
6	breo, cream, puff, rash, smoking, albuterol, skin, allergy, proair, allergic
7	fosamax, allergy, tramadol, covered, plan, calcium, bladder, kept, alcohol, ache
8	metformin, x-ray, nerve, knee, lasix, bottle, lantus, hurting, referral, switch
9	lantus, looked, injection, botox, changed, flare, happening, cream, salt, sweating
10	generic, triumeq, cost, farxiga, physical, therapy, gosh, fracture, increase, invokana
11	unit, list, appreciate, therapy, difference, counter, report, lasix, lantus, endocrinologist

Table 8: Topic Modeling: Top 10 words for 12 topics.

underlying topics of the dyads. We also use topic modeling to extract the utterances relevant to the primary disease diagnosis.

We apply the following steps adapted from a publicly available text summarization method[10]:

1. Fit the LDA model to all dyads.
2. Pass the dyads for each class to the LDA model to determine the class-wise topic distribution.
3. Select the dominant topics for each class using the topic weight matrix.
4. For each dyad within this subset:

---

[10]`https://github.com/g-deoliveira/TextSummarization`

D: What brings you in today?
P: I've been having trouble with a cough and my ribs are really hurting, and I'm concerned about it.
D: And when did that start?
P: This has been about three weeks.
D: And is there anything that makes the coughing worse?
P: It's not the coughing so much, it's the pain that goes with the cough. So, I guess the coughing has been kind of consistent, but my ribs hurt more and more. So, it's with the cough and when I take a deep breath, that's when I get the pain.
D: I see, okay. So, your ribs hurt more when you're coughing and when you take a deep breath.
P: Correct.
D: Is there anything that makes the pain better?
P: A little bit of Ibuprofen will mellow it a little bit, but like I said, I'm concerned. It's three weeks in and I figured I could do something about this.
D: Yeah, it's good you came in.
P: Thanks.
D: And can you describe the quality of the pain? Is it sharp, stabbing?
P: It's sharp, like you said, yeah, it's getting to … in the ribs area, yeah, it's a sharp pain.
D: If you had to point with one finger to where it hurts the most?
P: I guess, it's hard to just with one because it kind of is in this area.

D: Hello, Louise. Hi, do you want to sit down? What can I do for you today?
P: Well, I was wondering about going on the pill. I'm after some contraception, and that sounded like the best one for me from what I've found out.
D: And why is it that you're needing contraception? Are you using something currently?
P: No, I'm not in a relationship at the moment, but I'm going away to uni, I just want to make sure that I've got a plan.
D: Oh, right, okay. So, I understand. That's quite sensible, really. And what's made you think the pill would be a good idea?
P: Well, I don't know much about it, but what I've heard from friends and things around me is that it helps periods a little bit, and I wondered if that might be right for me.
D: Yeah, it can help your periods. Have you got a bit of trouble with your periods?
P: Well, they're a bit heavy, and they can be quite painful.
D: And is that throughout the period?
P: It's usually at the beginning. Just the first few days, really.
D: And does that trouble you quite a lot?
P: It does a bit. I think it'd be an added bonus if it wasn't there.
D: And are your periods regular?
P: Yeah, yeah.

Figure 6: (Left) Presenting problem: *Cough and rib pain.* (Right) Presenting problem: *Women's health and contraception.* Extracted utterances are highlighted.

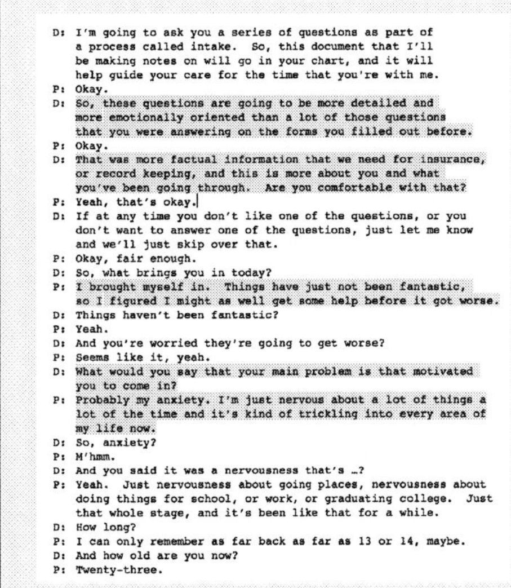

Figure 7: Presenting problem: *Anxiety.* Extracted utterances are highlighted.

(a) Split the conversation into sentences, using the NLTK (Steven Bird and Loper, 2009) sentence tokenizer.

(b) Determine the topic distribution of each sentence using LDA.

(c) Filter out the sentences whose dominant topic is not equal to the dominant topic of that dyad. What is left is a subset of sentences that reflect the given topic.

We conduct experiments on all 800 dyads and the 11 dyads from YouTube. Topic modeling is performed exactly as described in the previous section, with 12 topics. The results are shown in Table 6 and 7. The three dyads shown are from open-source YouTube data focusing on (a) cough and rib pain, (b) women's health and contraception, and (c) anxiety, respectively.

The results indicate a reasonable quality of relevant text extraction despite the limited amount of data. We can see that many of the utterances discussing the presenting problem are extracted. Since we do not have labels for the true relevance of the sentences to the disease, we are unable to provide any quantitative metrics, which is the subject of future work.

6 Conclusion & future work

The cumulative output of these models constitutes the initial automated system. Although for these experiments we used manual transcriptions, in practice the input would be from automatic speech recognition (ASR). Future research will include using ASR to record transcripts in real time, as well as expanding the types of entities we extract, identifying quantity, quality, and severity.

Diagnosis classification currently handles 6 classes only, and does not account for conditions other than the primary diagnosis that may be discussed in the conversation. We will also expand diagnosis classification to handle more classes, and to predict multiple diagnoses.

We have presented a system for extracting clinically relevant entities from physician-patient dialogues using linguistic context. The results show that clinical note-taking can be at least partially automated, saving clinicians valuable time. This system can result in a streamlined data entry process and a cleaner EMR note that can be used for analytics and automated decision making.

References

Moumita Bhattacharya, Claudine Jurkovitz, and Hagit Shatkay. 2017. Identifying patterns of co-occurring medical conditions through topic models of electronic health records. In *AMIA, iHealth 2017 Clinical Informatics Conference*.

Lynn S. Bickley and Peter G. Szilagyi. 2013. *Bates' pocket guide to physical examination and history taking 7th ed.* Wolters Kluwer Health/Lippincott Williams Wilkins.

David M Blei, Andrew Y Ng, and Michael I Jordan. 2003. Latent Dirichlet allocation. *Journal of machine Learning research*, 3(Jan):993–1022.

Sasha Calhoun, Jean Carletta, Jason M. Brenier, Neil Mayo, Dan Jurafsky, Mark Steedman, and David Beaver. 2010. The NXT-format Switchboard corpus: A rich resource for investigating the syntax, semantics, pragmatics and prosody of dialogue. *Language Resources and Evaluation 2010*, 44:387–419.

Katherine Redfield Chan, Xinghua Lou, Theofanis Karaletsos, Christopher Crosbie, Stuart Gardos, David Artz, and Gunnar Rätsch. 2013. An empirical analysis of topic modeling for mining cancer clinical notes. In *2013 IEEE 13th International Conference on Data Mining Workshops*, pages 56–63. IEEE.

Kyunghyun Cho, Bart van Merriënboer, Caglar Gulcehre, Dzmitry Bahdanau, Fethi Bougares, Holger Schwenk, and Yoshua Bengio. 2014. Learning phrase representations using RNN encoder–decoder for statistical machine translation. In *Proceedings of the 2014 Conference on Empirical Methods in Natural Language Processing (EMNLP)*, pages 1724–1734, Doha, Qatar. Association for Computational Linguistics.

Longxu Dou, Guanghui Qin, Jinpeng Wang, Jin-Ge Yao, and Chin-Yew Lin. 2018. Data2Text studio: Automated text generation from structured data. In *Proceedings of the 2018 Conference on Empirical Methods in Natural Language Processing: System Demonstrations*, pages 13–18, Brussels, Belgium. Association for Computational Linguistics.

Nan Du, Kai Chen, Anjuli Kannan, Linh Tran, Yuhui Chen, and Izhak Shafran. 2019. Extracting symptoms and their status from clinical conversations. In *Proceedings of the 57th Annual Meeting of the Association for Computational Linguistics*, pages 915–925, Florence, Italy. Association for Computational Linguistics.

Klaus Krippendorff. 2004. *Content Analysis: An Introduction to Its Methodology, Chapter 11*. Sage, Beverly Hills, CA, USA.

Rémi Lebret, David Grangier, and Michael Auli. 2016. Neural text generation from structured data with application to the biography domain. In *Proceedings of the 2016 Conference on Empirical Methods in Natural Language Processing*, pages 1203–1213, Austin, Texas. Association for Computational Linguistics.

Peter J Liu. 2018. Learning to write notes in electronic health records. *ArXiv eprint 1808.02622v1*.

James Mullenbach, Sarah Wiegreffe, Jon Duke, Jimeng Sun, and Jacob Eisenstein. 2018. Explainable prediction of medical codes from clinical text. In *Proceedings of NAACL-HLT 2018*, pages 1101–1111.

Phuoc Nguyen, Truyen Tran, Nilmini Wickramasinghe, and Svetha Venkatesh. 2018. Deepr: A convolutional net for medical records. *IEEE Journal of Biomedical and Health Informatics*, 21:22–30.

F. Pedregosa, G. Varoquaux, A. Gramfort, V. Michel, B. Thirion, O. Grisel, M. Blondel, P. Prettenhofer, R. Weiss, V. Dubourg, J. Vanderplas, A. Passos, D. Cournapeau, M. Brucher, M. Perrot, and E. Duchesnay. 2011. Scikit-learn: Machine learning in Python. *Journal of Machine Learning Research*, 12:2825–2830.

Matthew E. Peters, Mark Neumann, Mohit Iyyer, Matt Gardner, Christopher Clark, Kenton Lee, and Luke Zettlemoyer. 2018. Deep contextualized word representations. In *Proceedings of NAACL-HLT 2018*, pages 2227–2237.

Alvin Rajkomar, Eyal Oren, Kai Chen, Andrew M. Dai, Nissan Hajaj, Michaela Hardt, Peter J. Liu, Xiaobing Liu, Jake Marcus, Mimi Sun, Patrik Sundberg, Hector Yee, Kun Zhang, Yi Zhang, Gerardo Flores, Gavin E. Duggan, Jamie Irvine, Quoc Le, Kurt Litsch, Alexander Mossin, Justin Tansuwan, De Wang, James Wexler, Jimbo Wilson, Dana Ludwig, Samuel L. Volchenboum, Katherine Chou, Michael Pearson, Srinivasan Madabushi, Nigam H. Shah, Atul J. Butte, Michael D. Howell, Claire Cui, Greg S. Corrado, and Jeffrey Dean. 2018. Scalable and accurate deep learning for electronic health records. *NPJ Digital Medicine*, 2018:1–10.

Radim Řehůřek and Petr Sojka. 2010. Software Framework for Topic Modelling with Large Corpora. In *Proceedings of the LREC 2010 Workshop on New Challenges for NLP Frameworks*, pages 45–50, Valletta, Malta. ELRA.

Carol P Roth, Yee-Wei Lim, Joshua M. Pevnick, Steven M. Asch, and Elizabeth A. McGlynn. 2003. The challenge of measuring quality of care from the electronic health record. *American Journal of Medical Quality*, 24:385–394.

Christine Sinsky, Lacey Colligan, Ling Li, Mirela Prgomet, Sam Reynolds, Lindsey Goeders, Johanna Westbrook, Michael Tutty, and George Blike. 2016. Allocation of physician time in ambulatory practice: A time and motion study in 4 specialties. *Annals of Internal Medicine*, 165:753–760.

Ewan Klein Steven Bird and Edward Loper. 2009. *Natural Language Processing with Python*. OReilly Media.

Jannik Strötgen and Michael Gertz. 2010. Multilingual and cross-domain temporal tagging. *Language Resources and Evaluation 2013*, 47:269–298.

Krish Thiru, Alan Hassey, and Frank Sullivan. 2003. Systematic review of scope and quality of electronic patient record data in primary care. *BMJ*, 326:1070–1072.

Hanna M Wallach, Iain Murray, Ruslan Salakhutdinov, and David Mimno. 2009. Evaluation methods for topic models. In *Proceedings of the 26th annual international conference on machine learning*, pages 1105–1112. ACM.

Nicole Gray Weiskopf and Chunhua Weng. 2013. Methods and dimensions of electronic health record data quality assessment: Enabling reuse for clinical research. *Journal of the American Medical Informatics Association*, 20:144–151.

Biomedical Relation Classification by Single and Multiple source Domain Adaptation

Sinchani Chakraborty [1] **Sudeshna Sarkar** [2] **Pawan Goyal** [2] **Mahanandeeshwar Gattu** [3]

[1,2]Department of Computer Science and Engineering, IIT Kharagpur
[3]Excelra Knowledge Solutions Pvt Ltd, Hyderabad, India
[1] sinchanichakraborty@gmail.com
[2]{sudeshna, pawang}@cse.iitkgp.ac.in

Abstract

Relation classification is crucial for inferring semantic relatedness between entities in a piece of text. These systems can be trained given labelled data. However, relation classification is very domain-specific and it takes a lot of effort to label data for a new domain. In this paper, we explore domain adaptation techniques for this task. While past works have focused on single source domain adaptation for bio-medical relation classification, we classify relations in an unlabeled target domain by transferring useful knowledge from one or more related source domains. Our experiments with the model have shown to improve state-of-the-art F1 score on 3 benchmark biomedical corpora for single domain and on 2 out of 3 for multi-domain scenarios. When used with contextualized embeddings, there is further boost in performance outperforming neural-network based domain adaptation baselines for both the cases.

1 Introduction

In the biomedical domain, a relation can exist between various entity types like protein-protein, drug-drug, chemical-protein etc. Detecting relationships is a fundamental sub-task for automatic Information Extraction to overcome efforts of manual inspection, especially for growing biomedical articles. However, existing supervised systems are highly data-driven. This poses a challenge since manual labelling is a costly and time-consuming process and there is a dearth of labelled data in the biomedical domain covering all tasks and for new datasets. A system trained on a specific dataset[1] may perform poorly on another, for the same task (Mou et al., 2016), due to dataset variance which can arise owing to sample selection bias (Rios et al., 2018).

---

[1]Note: We use the terms dataset and domain interchangeably.

Domain Adaptation aims at adapting a model trained on a source domain to another target domain that may differ in their underlying data distributions. Past work on domain adaptation for bio-medical relation classification has focused on single-source adaptation (Rios et al., 2018). However, multiple sources from related domains can prove to be beneficial for classification in a low-resource scenario.

In this paper, we perform domain adaptation for biomedical binary relation classification at the sentence-level. For single-source single target (SSST) we transfer between different datasets of protein-protein interaction, along with drug-drug interaction. We also explore multi-source single target (MSST) adaptation to incorporate more richness in the knowledge transferred by using additional smaller corpora for protein-protein relation and multiple labels for chemical-protein relation respectively. Given an unlabeled target domain, we transfer common useful features from related labelled source domains using adversarial training (Goodfellow et al., 2014). It helps to overcome the sampling bias and learn common indistinguishable features, promoting generalization, using min-max optimization. We adopt the Multinomial Adversarial Network integrated with the Shared-Private model (Chen and Cardie, 2018) which was originally proposed for the task of Multi-Domain Text Classification. It can handle multiple source domains at a time which is in contrast to traditional binomial adversarial networks. The Shared-Private model (Bousmalis et al., 2016) consists of a split representation where the private space learns specific features related to a particular domain while a shared space learns features common to all the domains. Such representation promotes non-contamination of the two spaces preserving their uniqueness. The contributions of our approach are as follows:

Proceedings of the 10th International Workshop on Health Text Mining and Information Analysis (LOUHI 2019), pages 75–80
Hong Kong, November 3, 2019. ©2019 Association for Computational Linguistics
https://doi.org/10.18653/v1/D19-62

1) We show that using a shared-private model along with adversarial training improves SSST adaptation compared to neural network baselines. When multiple source corpora from similar domains are used it leads to further performance enhancement. Moreover, use of contextualized sentential embeddings leads to better performance than exisitng baseline methods for both MSST and SSST.

2) We explore the generalizability of our framework using two prominent neural architectures: CNN (Nguyen and Grishman, 2015) and Bi-LSTM (Kavuluru et al., 2017), where we find the former to be more robust across our experiments.

2 Methodology

For every labeled sources and a single unlabeled target we have set of NER tagged sentences, each of which is represented as: X = $\{e_1, e_2, w_1...w_n\}$ where e_1 and e_2 are two tagged entities and w_j is the jth word in the sentence . A labelled source instance is accompanied by the relation label (True or False). In this section we discuss the input representation followed by model description.

2.1 Input Representation

We form word and position embeddings for every word in an NER tagged sentence. We use the PubMed-and-PMC-w2v[2] to generate word embeddings. The size being $(|V| \cdot d_w)$, where d_w is the word embedding dimension which is 200 and $|V|$ is the vocabulary size. The position embedding vector for j^{th} word in a sentence relative to two tagged entities e_1 and e_2 is represented as a tuple: $(p_{e1(j)}, p_{e2(j)})$ where, $p_{e1(j)}$ and $p_{e2(j)} \in \mathbb{R}^e$.

2.2 Model

Fig 1 shows the adaptation of MAN framework whose various components are discussed below.

Shared & Domain feature extractor (F_s, F_{d_i}) The input representation is fed to both F_{d_i} and F_s for labeled source domains whereas for unlabeled target instances it is fed only to F_s. For SSST the model is trained on a single labeled source domain and tested on a unlabeled target domain. For MSST we do not combine the sources as a single corpus since that leads to a number of false negatives. We make two different assumptions to consider multiple sources: 1) Following Nguyen et al., (2014) we consider multiple labels

[2]http://evexdb.org/pmresources/vec-space-models/

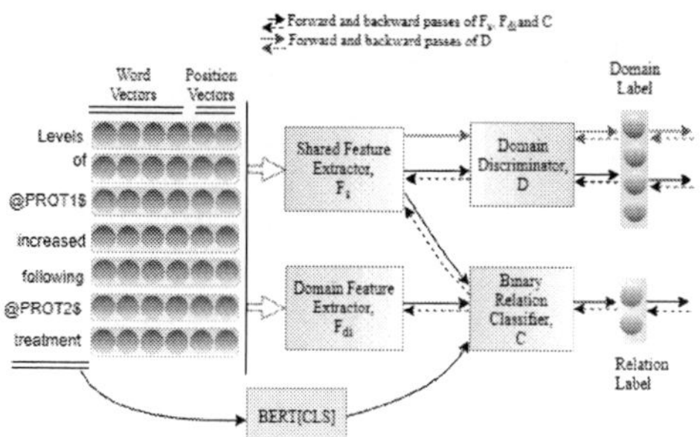

Figure 1: MAN for Domain Adaptation of Binary Relation Classification. The figure shows the training flow given a sentence from a labeled source domain. D is trained separately than rest of the network

from single corpus as different sources, 2) We use additional smaller corpora from a similar domain as multi-source. The *Shared Feature space* (F_s) learns domain agnostic representations and *Private Feature space* (F_{d_i}) learns domain specific features for every i^{th} labeled domain. We apply two different architectures for both F_s and F_{d_i} to analyze the changes in performance of the approach for the task : Convolutional neural network (Nguyen and Grishman, 2015) (MAN CNN) and Bi-LSTM (Kavuluru et al., 2017) (MAN Bi-LSTM). We have performed detailed experiments on each of these in Section 5.

Domain discriminator, D is a fully-connected layer with softmax that predicts multiple domain probabilities using *Multinomial Adversarial Network*. The output from F_s is fed to D which is adversarially trained separately from the entire network using L2 loss described as follows:

$$L_D(\hat{d}, d) = \sum_{i=1}^{N} (\hat{d}_i - 1_{\{d=i\}})^2$$

where, d is the index assigned for a domain and $\hat{d}$ is the prediction. It is generalized as $\sum_{i=1}^{N} \hat{d}_i = 1$ and $\forall_i : \hat{d}_i \geq 0$. F_s tries to fool D so that it can not correctly guess the domain from where a sample instance is coming from. Thus F_s learns indistinguishable features in the process.

Relation Classifier C is a fully-connected layer with a softmax, used to predict the class probabilities. We use Bio-BERT (Lee et al., 2019) embeddings for every sentence as features (Geeticka Chauhan, 2019) BERT[CLS] that have shown to improve performance in many downstream tasks. This is concatenated with the fixed size sentence representation from F_s and F_{d_i}, together

Datasets	Entity Pair	# of Sent	# of Positive	# of Negative
AiMed		1995	1000	4834
BioInfer		1100	2534	7132
LLL	PPI	77	164	166
HPRD50		145	163	270
IEPA		486	355	482

Table 1: Protein Protein Interaction Dataset statistics.

Datasets	Entity Pair	# of Train	# of Valid	# of Test
DDI	Drug-Drug	27779	-	5713
CPR: 3		768	550	665
CPR: 4		2254	1094	1661
CPR: 5	Chem-Prot	173	116	195
CPR: 6		235	199	293
CPR: 9		727	457	644

Table 2: Drug-Drug Interaction and Chemical-Protein Dataset statistics.

they serve as input to C. For unlabeled target, during test no domain specific features are generated from F_{d_i} and that part is set to zero vector. For binary classification we adopt Negative Log Likelihood Loss for C described below:

$$L_c(\hat{y}, y) = -log P(\hat{y} = y)$$

where, y is the true relation label and $\hat{y}$ is the softmax label. The objective of F_{d_i} is same as that of C and it relies only on labeled data. On the other hand the objective of the Shared Feature Extractor F_s is represented as follows:

$$Loss\,of\,F_s = Classifier\,loss + \lambda\,Domain\,loss$$

It consists of two loss components: improve performance of C and enhance learning of invariant features across all domains. A hyper parameter λ is used to balance both of them.

3 Datasets

The dataset statistics is summarized in Table 1 and Table 2. A 10-fold cross validation was performed for the Protein-Protein Interaction dataset. For given set of entities E in a sentence, it is split into $\binom{E}{2}$ instances. All positive instances of datasets with more than two relation types are merged and assigned True labels while negative instances are assigned False labels. Unlabeled data is formed by removing labels from development and test datasets.

4 Experiments

Pre-processing: We anonymize the named entities in the sentence by replacing them with predefined tags like @PROT1\$, @DRUG\$ (Bhasuran and Natarajan, 2018).

4.1 Single source single target (SSST)

A thorough experiment is conducted using all possible combinations of the three benchmark data-sets AiMed (Bunescu et al., 2005), BioInfer (Pyysalo et al., 2006), DDI (Herrero-Zazo et al., 2013) whose results are discussed in Table 3

4.2 Multi-source single target (MSST)

The experiments with two different assumptions to consider multiple sources are as follows:

Multiple smaller corpora from similar domain: For Protein Protein Interaction there are three smaller standard corpora in literature, namely, LLL (Nedellec, 2005), IEPA (Ding et al., 2001), HPRD50 (Fundel et al., 2007). All three were considered as additional sources to transfer knowledge. AiMed (AM) and BioInfer (BI) were alternately selected as the unlabeled target in 2 different experiments while the remaining 4 denoted as 4P are considered as source corpus.

Multiple labels from single corpora: For ChemProt corpora we consider various labels as different sources following Nguyen et al., (2014) The five positive labels of ChemProt are: CPR: 3, CPR: 4, CPR: 5, CPR: 6, CPR: 9 which stand for upregulator, downregulator, agonist, antagonist and substrate, respectively. We predict the classification performance for unlabeled targets CPR:6 and CPR:9 taking multi-source labeled input denoted as 3C from three sources- CPR: 3, CPR: 4, CPR: 5 as positive instances and remaining as negative.

4.3 Baselines

We compare our approach with different baselines which are mentioned as follows:

- **BioBERT (Rios et al., 2018):** For SSST we train it on one dataset and test on another. For MSST we combine the multiple sources as a single source and test on labeled target.

- **CNN+DANN (Lisheng Fu, 2017) :** A variant of adversarial training which is gradient reversal (RevGrad) is used with CNN (Nguyen and Grishman, 2015).

Method	BioInfer $\rightarrow$ AiMed	AiMed $\rightarrow$ BioInfer	BioInfer $\rightarrow$ DDI	DDI $\rightarrow$ BioInfer	AiMed $\rightarrow$ DDI	DDI $\rightarrow$ AiMed
CNN	45.22	36.72	39.75	22.13	15.83	27.93
Bi-LSTM	46.88	29.59	40.87	17.21	18.58	25.80
BioBERT*	76.48	69.23	67.89	57.84	51.22	54.83
CNN + DANN*	45.98	42.01	41.58	34.37	28.66	28.90
Bi-LSTM + RevGrad	46.41	40.11	39.41	37.20	27.72	35.29
Adv-CNN	48.79	54.13	44.19	48.53	45.96	44.71
Adv - Bi-LSTM	48.51	56.54	44.47	44.90	46.21	43.44
MAN CNN **	50.23	55.04	47.63	49.51	46.97	42.36
MAN Bi-LSTM **	49.19	58.69	46.77	46.28	47.84	41.53
MAN CNN + BERT[CLS] **	**53.08**	57.89	49.33	50.79	47.01	46.38
MAN Bi-LSTM + BERT[CLS] **	52.74	**61.01**	48.03	45.12	**50.19**	44.01

Table 3: F1 scores for SSST experiment on test set of target (RHS of $\rightarrow$) . **: Our model. *: Our implementation. Bold text: Best domain adaptation model for a dataset.

- **Adv Bi-LSTM + Adv CNN (Rios et al., 2018):** Conducts two-step training: pre-training with source followed by adversarial training with target. For MSST experiment we compare our method with Adv CNN and Adv Bi-LSTM by combining multiple sources.

5   Results and Discussions

In Table 3 we see that BioInfer generalizes well to AiMed and DDI corpora using vanilla LSTM or CNN architecture. However, with MAN and contextual embeddings, we do not see prominent gains as much as the other datasets. This can be due to the class imbalance in data (positive to negative instance ratio 1:5.9) (Hsu et al., 2015; Rios et al., 2018). For AiMed and BioInfer, we find that the knowledge transfer among themselves gives the best performance thus strengthening the fact that datasets from the same domain can contribute to performance enhancement justifying the performance gains in MSST experiments. Our model outperforms other baselines just with the use of adversarial training which might be attributed to joint learning better representation from shared and private feature extractors. The use of contextual BERT[CLS] tokens leads to increase in performance scores since they encode important relations between words in a sentence (Vig, 2019; Hewitt and Manning, 2019).

In Table 4, BioBERT is seen to perform well for ChemProt. We hypothesize that this may be due to the same underlying dataset being used during train and test. Though we use different labels as multi-source, that may not contribute to generating enough variance in sources since they

Method	3C $\rightarrow$ CPR:9	3C $\rightarrow$ CPR:6	4P $\rightarrow$ AM	4P $\rightarrow$ BI
BioBERT*	**69.27**	**73.50**	43.01	52.98
Adv-CNN*	58.23	56.69	45.30	51.79
Adv-BiLSTM*	56.30	57.13	42.01	52.67
MAN CNN**	59.69	58.30	52.33	57.21
MAN Bi-LSTM**	57.01	59.71	53.64	59.37
MAN CNN + BERT-[CLS]**	64.23	65.41	56.75	**64.83**
MAN Bi-LSTM + BERT-[CLS]**	62.07	64.09	**57.09**	63.92

Table 4: F1 scores for MSST experiment on test set of target (RHS of $\rightarrow$). **: Our model. *: Our implementation trained with unified labeled multi-source. Bold text: Best model for a dataset..

were from the same dataset. For AiMed and BioInfer, however, three different smaller corpora were used, where the proposed method outperforms BioBERT. When compared across all the six SSST experiments, the Bi-LSTM based model lacks in performance may be due to absence of any attention mechanism which would have helped in selecting more relevant context (Chen and Cardie, 2018). We observe that adversarial training along with contextualized BERT sentence embeddings leads to performance gains across all datasets.

6 Conclusions

Our proposed model significantly outperformed the existing neural network based domain adaptation baselines for SSST. Among the two MSST experiments, we showed that the system gains when multiple source corpora are used. We also experiment with two architectures out of which CNN is seen to perform marginally better compared to Bi-LSTM. Our analysis on Section 5 further explains the effect of sources, adversarial training and use of contextualized BERT sentential embeddings.

Acknowledgments

This work has been supported by the project Effective Drug Repurposing through literature and patent mining, data integration and development of systems pharmacology platform sponsored by MHRD, India and Excelra Knowledge Solutions, Hyderabad. Besides, the authors would like to thank the anonymous reviewers for their valuable comments and feedback.

References

Balu Bhasuran and Jeyakumar Natarajan. 2018. Automatic extraction of gene-disease associations from literature using joint ensemble learning. In *PloS one*.

Konstantinos Bousmalis, George Trigeorgis, Nathan Silberman, Dilip Krishnan, and Dumitru Erhan. 2016. Domain separation networks. In *NIPS*.

Razvan C. Bunescu, Ruifang Ge, Rohit J. Kate, Edward M. Marcotte, Raymond J. Mooney, Arun K. Ramani, and Yuk Wah Wong. 2005. Comparative experiments on learning information extractors for proteins and their interactions. *Artificial intelligence in medicine*, 33 2:139–55.

Xilun Chen and Claire Cardie. 2018. Multinomial adversarial networks for multi-domain text classification. In *Proceedings of NAACL-HLT 2018*, page 12261240.

Jing Ding, Daniel Berleant, Dan Nettleton, and Eve Syrkin Wurtele. 2001. Mining medline: Abstracts, sentences, or phrases? *Pacific Symposium on Biocomputing. Pacific Symposium on Biocomputing*, pages 326–37.

Katrin Fundel, Robert Küffner, and Ralf Zimmer. 2007. Relex - relation extraction using dependency parse trees. *Bioinformatics*, 23 3:365–71.

Peter Szolovits Geeticka Chauhan, Matthew B. A. McDermott. 2019. Reflex: Flexible framework for relation extraction in multiple domains. In *Proceedings of the BioNLP 2019 workshop*, page 3047.

Ian J. Goodfellow, Jean Pouget-Abadie, Mehdi Mirza, Bing Xu, David Warde-Farley, Sherjil Ozair, Aaron C. Courville, and Yoshua Bengio. 2014. Generative adversarial nets. In *NIPS*.

María Herrero-Zazo, Isabel Segura-Bedmar, Paloma Martínez, and Thierry Declerck. 2013. The ddi corpus: An annotated corpus with pharmacological substances and drug-drug interactions. *Journal of biomedical informatics*, 46 5:914–20.

John Hewitt and Christopher D. Manning. 2019. A structural probe for finding syntax in word representations. In *NAACL-HLT*.

Tzu-Ming Harry Hsu, Wei-Yu Chen, Cheng-An Hou, Yao-Hung Tsai, Yi-Ren Yeh, and Yu-Chiang Frank Wang. 2015. Unsupervised domain adaptation with imbalanced cross-domain data. *2015 IEEE International Conference on Computer Vision (ICCV)*, pages 4121–4129.

Ramakanth Kavuluru, Anthony Rios, and Tung Tran. 2017. Extracting drug-drug interactions with word and character-level recurrent neural networks. *2017 IEEE International Conference on Healthcare Informatics (ICHI)*, pages 5–12.

Jinhyuk Lee, Wonjin Yoon, Sungdong Kim, Donghyeon Kim, Sunkyu Kim, Chan Ho So, and Jaewoo Kang. 2019. Biobert: a pre-trained biomedical language representation model for biomedical text mining. *ArXiv*, abs/1901.08746.

Bonan Min Ralph Grishman Lisheng Fu, Thien Huu Nguyen. 2017. Domain adaptation for relation extraction with domain adversarial neural network. In *Proceedings of the The 8th International Joint Conference on Natural Language Processing*, page 425429.

Lili Mou, Zhao Meng, Rui Yan, Ge Li, Yuning Xu, Lu Zhang, and Zhi Jin. 2016. How transferable are neural networks in nlp applications? *ArXiv*, abs/1603.06111.

Claire Nedellec. 2005. Learning language in logic - genic interaction extraction challenge.

Minh Luan Nguyen, Ivor Wai-Hung Tsang, Kian Ming Adam Chai, and Hai Leong Chieu. 2014. Robust domain adaptation for relation extraction via clustering consistency. In *ACL*.

Thien Huu Nguyen and Ralph Grishman. 2015. Relation extraction: Perspective from convolutional neural networks. In *VS@HLT-NAACL*.

Sampo Pyysalo, Filip Ginter, Juho Heimonen, Jari Björne, Jorma Boberg, Jouni Järvinen, and Tapio Salakoski. 2006. Bioinfer: a corpus for information extraction in the biomedical domain. *BMC Bioinformatics*, 8:50 – 50.

Anthony Rios, Ramakanth Kavuluru, and Zhiyong Lu.
2018. Generalizing biomedical relation classification with neural adversarial domain adaptation. volume 34, pages 2973–2981.

Jesse Vig. 2019. Visualizing attention in transformer-based language representation models. *ArXiv*, abs/1904.02679.

Assessing the Efficacy of Clinical Sentiment Analysis and Topic Extraction in Psychiatric Readmission Risk Prediction

Elena Álvarez-Mellado[1,2], Eben Holderness[1,2], Nicholas Miller[1,2], Fyonn Dhang[1],
Philip Cawkwell[1], Kirsten Bolton[1], James Pustejovsky[2], and **Mei-Hua Hall[1]**
[1]Psychosis Neurobiology Laboratory, McLean Hospital, Harvard Medical School
[2]Department of Computer Science, Brandeis University
{ealvarezmellado, eholderness, mhall}@mclean.harvard.edu
nicholas.anthony.miller@gmail.com
{pcawkwell, fdhang, kbolton}@partners.org
{jamesp}@cs.brandeis.edu

Abstract

Predicting which patients are more likely to be readmitted to a hospital within 30 days after discharge is a valuable piece of information in clinical decision-making. Building a successful readmission risk classifier based on the content of Electronic Health Records (EHRs) has proved, however, to be a challenging task. Previously explored features include mainly structured information, such as sociodemographic data, comorbidity codes and physiological variables. In this paper we assess incorporating additional clinically interpretable NLP-based features such as topic extraction and clinical sentiment analysis to predict early readmission risk in psychiatry patients.

1 Introduction and Related Work

Psychotic disorders affect approximately 2.5-4% of the population (Perälä et al., 2007) (Bogren et al., 2009). They are one of the leading causes of disability worldwide (Vos et al., 2015) and are a frequent cause of inpatient readmission after discharge (Wiersma et al., 1998). Readmissions are disruptive for patients and families, and are a key driver of rising healthcare costs (Mangalore and Knapp, 2007) (Wu et al., 2005). Assessing readmission risk is therefore critically needed, as it can help inform the selection of treatment interventions and implement preventive measures.

Predicting hospital readmission risk is, however, a complex endeavour across all medical fields. Prior work in readmission risk prediction has used structured data (such as medical comorbidity, prior hospitalizations, sociodemographic factors, functional status, physiological variables, etc) extracted from patients' charts (Kansagara et al., 2011). NLP-based prediction models that extract unstructured data from EHR have also been developed with some success in other medical fields (Murff et al., 2011). In Psychiatry, due to the unique characteristics of medical record content (highly varied and context-sensitive vocabulary, abundance of multiword expressions, etc), NLP-based approaches have seldom been applied (Vigod et al., 2015; Tulloch et al., 2016; Greenwald et al., 2017) and strategies to study readmission risk factors primarily rely on clinical observation and manual review (Olfson et al., 1999) (Lorine et al., 2015), which is effort-intensive, and does not scale well.

In this paper we aim to assess the suitability of using NLP-based features like clinical sentiment analysis and topic extraction to predict 30-day readmission risk in psychiatry patients. We begin by describing the EHR corpus that was created using in-house data to train and evaluate our models. We then present the NLP pipeline for feature extraction that was used to parse the EHRs in our corpus. Finally, we compare the performances of our model when using only structured clinical variables and when incorporating features derived from free-text narratives.

2 Data

The corpus consists of a collection of 2,346 clinical notes (admission notes, progress notes, and discharge summaries), which amounts to 2,372,323 tokens in total (an average of 1,011 tokens per note). All the notes were written in English and extracted from the EHRs of 183 psychosis patients from McLean Psychiatric Hospital in Belmont, MA, all of whom had in their history at least one instance of 30-day readmission.

The age of the patients ranged from 20 to 67 (mean = 26.65, standard deviation = 8.73). 51% of the patients were male. The number of admis-

Proceedings of the 10th International Workshop on Health Text Mining and Information Analysis (LOUHI 2019), pages 81–86
Hong Kong, November 3, 2019. ©2019 Association for Computational Linguistics
https://doi.org/10.18653/v1/D19-62

sions per patient ranged from 2 to 21 (mean = 4, standard deviation = 2.85). Each admission contained on average 4.25 notes and 4,298 tokens. In total, the corpus contains 552 admissions, and 280 of those (50%) resulted in early readmissions.

3 Feature Extraction

The readmission risk prediction task was performed at the admission level. An admission consists of a collection of all the clinical notes for a given patient written by medical personnel between inpatient admission and discharge. Every admission was labeled as either 'readmitted' (i.e. the patient was readmitted within the next 30 days of discharge) or 'not readmitted'. Therefore, the classification task consists of creating a single feature representation of all the clinical notes belonging to one admission, plus the past medical history and demographic information of the patient, and establishing whether that admission will be followed by a 30-day readmission or not.

45 clinically interpretable features per admission were extracted as inputs to the readmission risk classifier. These features can be grouped into three categories (See Table 1 for complete list of features):

- Sociodemographics: gender, age, marital status, etc.

- Past medical history: number of previous admissions, history of suicidality, average length of stay (up until that admission), etc.

- Information from the current admission: length of stay (LOS), suicidal risk, number and length of notes, time of discharge, evaluation scores, etc.

The Current Admission feature group has the most number of features, with 29 features included in this group alone. These features can be further stratified into two groups: 'structured' clinical features and 'unstructured' clinical features.

3.1 Structured Features

Structure features are features that were identified on the EHR using regular expression matching and include rating scores that have been reported in the psychiatric literature as correlated with increased readmission risk, such as *Global Assessment of Functioning*, *Insight* and *Compliance*:

Sociodemographics
Age
Gender
Race
Marital status
Veteran
Past medical history
History of Suicidality
Number of past admissions
Average length of stay (previous)
Average # days between admissions
Previous 30-day readmission (Y/N)
Number of past readmissions
Readmission ratio
Average GAF at admission
Average GAF at discharge
Mode of past insight values
Mode of past medication compliance
Current admission
Structured features
Number of notes
Number of tokens
Number of tokens in discharge summary
Average note length
GAF at admission
GAF at discharge
GAF admission/discharge difference
Mean GAF (all notes for visit)
Insight (good, fair, poor)
Medication Compliance
Estimated length of stay
Actual length of stay
Difference b/w Estimated & Actual LOS
Is first admission (Y/N)
Unstructured features
Number of sentences (Appearance)
Number of sentences (Mood)
Number of sentences (Thought Content)
Number of sentences (Thought Process)
Number of sentences (Substance Use)
Number of sentences (Interpersonal)
Number of sentences (Occupation)
Clinical sentiment (Appearance)
Clinical sentiment (Mood)
Clinical sentiment (Thought Content)
Clinical sentiment (Thought Process)
Clinical sentiment (Substance Use)
Clinical sentiment (Interpersonal)
Clinical sentiment (Occupation)

Table 1: Extracted features by category.

Global Assessment of Functioning (GAF): The psychosocial functioning of the patient ranging from 100 (extremely high functioning) to 1 (severely impaired) (AAS, 2011).

Insight: The degree to which the patient recognizes and accepts his/her illness (either *Good*, *Fair* or *Poor*).

Compliance: The ability of the patient to comply with medication and to follow medical advice (either *Yes*, *Partial*, or *None*).

These features are widely-used in clinical practice and evaluate the general state and prognosis of

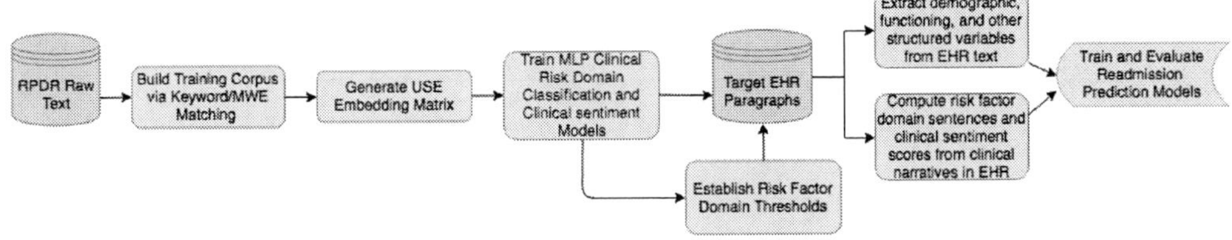

Figure 1: NLP pipeline for feature extraction.

the patient during the patient's evaluation.

3.2 Unstructured Features

Unstructured features aim to capture the state of the patient in relation to seven risk factor domains (Appearance, Thought Process, Thought Content, Interpersonal, Substance Use, Occupation, and Mood) from the free-text narratives on the EHR. These seven domains have been identified as associated with readmission risk in prior work (Holderness et al., 2018).

These unstructured features include: 1) the relative number of sentences in the admission notes that involve each risk factor domain (out of total number of sentences within the admission) and 2) clinical sentiment scores for each of these risk factor domains, i.e. sentiment scores that evaluate the patients psychosocial functioning level (positive, negative, or neutral) with respect to each of these risk factor domain.

These sentiment scores were automatically obtained through the topic extraction and sentiment analysis pipeline introduced in our prior work (Holderness et al., 2019) and pretrained on in-house psychiatric EHR text. In our paper we also showed that this automatic pipeline achieves reasonably strong F-scores, with an overall performance of 0.828 F1 for the topic extraction component and 0.5 F1 on the clinical sentiment component.

The clinical sentiment scores are computed for every note in the admission. Figure 1 details the data analysis pipeline that is employed for the feature extraction.

First, a multilayer perceptron (MLP) classifier is trained on EHR sentences (8,000,000 sentences consisting of 340,000,000 tokens) that are extracted from the Research Patient Data Registry (RPDR), a centralized regional data repository of clinical data from all institutions in the Partners HealthCare network. These sentences are automatically identified and labeled for their respec-

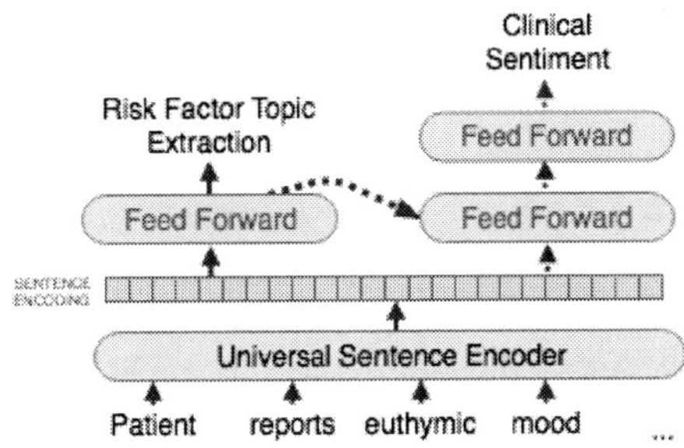

Figure 2: Model architecture for USE embedding generation and unstructured feature extraction. Dotted arrows indicate operations that are performed only on sentences marked for 1+ risk factor domain(s). USE top-layer weights are fine-tuned during training.

tive risk factor domain(s) by using a lexicon of clinician identified domain-related keywords and multiword expressions, and thus require no manual annotation. The sentences are vectorized using the Universal Sentence Encoder (USE), a transformer attention network pretrained on a large volume of general-domain web data and optimized for greater-than-word length sequences.

Sentences that are marked for one or more of the seven risk factor domains are then passed to a suite of seven clinical sentiment MLP classifiers (one for each risk factor domain) that are trained on a corpus of 3,500 EHR sentences (63,127 tokens) labeled by a team of three clinicians involved in this project. To prevent overfitting to this small amount of training data, the models are designed to be more generalizable through the use of two hidden layers and a dropout rate (Srivastava et al., 2014) of 0.75.

The outputs of each clinical sentiment model are then averaged across notes to create a single value for each risk factor domain that corresponds to the patient's level of functioning on a -1 to 1 scale (see Figure 2).

4 Experiments and Results

We tested six different classification models: Stochastic Gradient Descent, Logistic Regression, C-Support Vector, Decision Tree, Random Forest, and MLP. All of them were implemented and fine-tuned using the scikit-learn machine learning toolkit (Pedregosa et al., 2011). Because an accurate readmission risk prediction model is designed to be used to inform treatment decisions, it is important in adopting a model architecture that is clinically interpretable and allows for an analysis of the specific contribution of each feature in the input. As such, we include a Random Forest classifier, which we also found to have the best performance out of the six models.

To systematically evaluate the importance of the clinical sentiment values extracted from the free text in EHRs, we first build a baseline model using the structured features, which are similar to prior studies on readmission risk prediction (Kansagara et al., 2011). We then compare two models incorporating the unstructured features. In the "Baseline+Domain Sentences" model, we consider whether adding the counts of sentences per EHR that involve each of the seven risk factor domains as identified by our topic extraction model improved the model performance. In the "Baseline+Clinical Sentiment" model, we evaluate whether adding clinical sentiment scores for each risk factor domain improved the model performance. We also experimented with combining both sets of features and found no additional improvement.

Each model configuration was trained and evaluated 100 times and the features with the highest importance for each iteration were recorded. To further fine-tune our models, we also perform three-fold cross-validated recursive feature elimination 30 times on each of the three configurations and report the performances of the models with the best performing feature sets. These can be found in Table 2.

Our baseline results show that the model trained using only the structured features produce equivalent performances as reported by prior models for readmission risk prediction across all healthcare fields (Artetxe et al., 2018). The two models that were trained using unstructured features produced better results and both outperform the baseline results. The "Baseline+Clinical Sentiment" model produced the best results, resulting in an F1 of

Model	Acc	AUC	F1
Baseline	0.63	0.63	0.63
Baseline+Domain Sentences	0.69	0.70	0.69
Baseline+Clinical Sentiment	0.72	0.72	0.72

Table 2: Results (in ascending order)

0.72, an improvement of 14.3% over the baseline.

In order to establish what features were not relevant in the classification task, we performed recursive feature elimination. We identified 13 feature values as being not predictive of readmission (they were eliminated from at least two of the three feature sets without producing a drop in performance) including: all values for marital status (Single, Married, Other, and Unknown), missing values for GAF at admission, GAF score difference between admission & discharge, GAF at discharge, Veteran status, Race, and Insight & Mode of Past Insight values reflecting a clinically positive change (Good and Improving). Poor Insight values, however, are predictive of readmission.

5 Conclusions

We have introduced and assessed the efficacy of adding NLP-based features like topic extraction and clinical sentiment features to traditional structured-feature based classification models for early readmission prediction in psychiatry patients. The approach we have introduced is a hybrid machine learning approach that combines deep learning techniques with linear methods to ensure clinical interpretability of the prediction model.

Results show not only that both the number of sentences per risk domain and the clinical sentiment analysis scores outperform the structured-feature baseline and contribute significantly to better classification results, but also that the clinical sentiment features produce the highest results in all evaluation metrics (F1 = 0.72).

These results suggest that clinical sentiment features for each of seven risk domains extracted from free-text narratives further enhance early readmission prediction. In addition, combining state-of-art MLP methods has a potential utility in generating clinical meaningful features that can be be used in downstream linear models with interpretable and transparent results. In future work, we intend to increase the size of the EHR corpus, increase the demographic spread of patients, and extract new features based on clinical expertise to increase our model performances. Additionally, we intend to continue our clinical sentiment annotation project from (Holderness et al., 2019) to increase the accuracy of that portion of our NLP pipeline.

6 Acknowledgments

This work was supported by a grant from the National Institute of Mental Health (grant no. 5R01MH109687 to Mei-Hua Hall). We would also like to thank the LOUHI 2019 Workshop reviewers for their constructive and helpful comments.

References

IH Monrad AAS. 2011. Guidelines for rating global assessment of functioning (gaf). *Annals of general psychiatry*, 10(1):2.

Arkaitz Artetxe, Andoni Beristain, and Manuel Grana. 2018. Predictive models for hospital readmission risk: A systematic review of methods. *Computer methods and programs in biomedicine*, 164:49–64.

Mats Bogren, Cecilia Mattisson, Per-Erik Isberg, and Per Nettelbladt. 2009. How common are psychotic and bipolar disorders? a 50-year follow-up of the lundby population. *Nordic journal of psychiatry*, 63(4):336–346.

Jeffrey L Greenwald, Patrick R Cronin, Victoria Carballo, Goodarz Danaei, and Garry Choy. 2017. A novel model for predicting rehospitalization risk incorporating physical function, cognitive status, and psychosocial support using natural language processing. *Medical care*, 55(3):261–266.

Eben Holderness, Philip Cawkwell, Kirsten Bolton, James Pustejovsky, and Mei-Hua Hall. 2019. Distinguishing clinical sentiment: The importance of domain adaptation in psychiatric patient health records. In *Proceedings of the 2nd Clinical Natural Language Processing Workshop*, pages 117–123.

Eben Holderness, Nicholas Miller, Kirsten Bolton, Philip Cawkwell, Marie Meteer, James Pustejovsky, and Mei Hua-Hall. 2018. Analysis of risk factor domains in psychosis patient health records. In *Proceedings of the Ninth International Workshop on Health Text Mining and Information Analysis*, pages 129–138.

Devan Kansagara, Honora Englander, Amanda Salanitro, David Kagen, Cecelia Theobald, Michele Freeman, and Sunil Kripalani. 2011. Risk prediction models for hospital readmission: a systematic review. *Jama*, 306(15):1688–1698.

Kim Lorine, Haig Goenjian, Soeun Kim, Alan M Steinberg, Kendall Schmidt, and Armen K Goenjian. 2015. Risk factors associated with psychiatric readmission. *The Journal of nervous and mental disease*, 203(6):425–430.

Roshni Mangalore and Martin Knapp. 2007. Cost of schizophrenia in England. *The journal of mental health policy and economics*, 10(1):23–41.

Harvey J Murff, Fern FitzHenry, Michael E Matheny, Nancy Gentry, Kristen L Kotter, Kimberly Crimin, Robert S Dittus, Amy K Rosen, Peter L Elkin, Steven H Brown, et al. 2011. Automated identification of postoperative complications within an electronic medical record using natural language processing. *Jama*, 306(8):848–855.

Mark Olfson, David Mechanic, Carol A Boyer, Stephen Hansell, James Walkup, and Peter J Weiden. 1999. Assessing clinical predictions of early rehospitalization in schizophrenia. *The Journal of nervous and mental disease*, 187(12):721–729.

Fabian Pedregosa, Gaël Varoquaux, Alexandre Gramfort, Vincent Michel, Bertrand Thirion, Olivier Grisel, Mathieu Blondel, Peter Prettenhofer, Ron Weiss, Vincent Dubourg, et al. 2011. Scikit-learn: Machine learning in python. *Journal of machine learning research*, 12(Oct):2825–2830.

Jonna Perälä, Jaana Suvisaari, Samuli I Saarni, Kimmo Kuoppasalmi, Erkki Isometsä, Sami Pirkola, Timo Partonen, Annamari Tuulio-Henriksson, Jukka Hintikka, Tuula Kieseppä, et al. 2007. Lifetime prevalence of psychotic and bipolar i disorders in a general population. *Archives of general psychiatry*, 64(1):19–28.

Nitish Srivastava, Geoffrey Hinton, Alex Krizhevsky, Ilya Sutskever, and Ruslan Salakhutdinov. 2014. Dropout: a simple way to prevent neural networks from overfitting. *The Journal of Machine Learning Research*, 15(1):1929–1958.

AD Tulloch, AS David, and G Thornicroft. 2016. Exploring the predictors of early readmission to psychiatric hospital. *Epidemiology and psychiatric sciences*, 25(2):181–193.

Simone N Vigod, Paul A Kurdyak, Dallas Seitz, Nathan Herrmann, Kinwah Fung, Elizabeth Lin, Christopher Perlman, Valerie H Taylor, Paula A Rochon, and Andrea Gruneir. 2015. Readmit: a clinical risk index to predict 30-day readmission after discharge from acute psychiatric units. *Journal of psychiatric research*, 61:205–213.

Theo Vos, Ryan M Barber, Brad Bell, Amelia Bertozzi-Villa, Stan Biryukov, Ian Bolliger, Fiona Charlson, Adrian Davis, Louisa Degenhardt, Daniel Dicker, et al. 2015. Global, regional, and national incidence, prevalence, and years lived with disability for 301 acute and chronic diseases and injuries in 188 countries, 1990–2013: a systematic analysis for the global burden of disease study 2013. *The Lancet*, 386(9995):743–800.

Durk Wiersma, Fokko J Nienhuis, Cees J Slooff, and Robert Giel. 1998. Natural course of schizophrenic disorders: a 15-year followup of a Dutch incidence cohort. *Schizophrenia bulletin*, 24(1):75–85.

Eric Q Wu, Howard G Birnbaum, Lizheng Shi, Daniel E Ball, Ronald C Kessler, Matthew Moulis, and Jyoti Aggarwal. 2005. The economic burden of schizophrenia in the United States in 2002. *Journal of Clinical Psychiatry*, 66(9):1122–1129.

What does the language of foods say about us?

Hoang Van,[*] Ahmad Musa,[*] Hang Chen, Mihai Surdeanu, Stephen Kobourov
Department of Computer Science, University of Arizona
{vnhh,ahmadmusa,hangchen,msurdeanu,kobourov}@email.arizona.edu

Abstract

In this work we investigate the signal contained in the language of food on social media. We experiment with a dataset of 24 million food-related tweets, and make several observations. First, the language of food has predictive power. We are able to predict if states in the United States (US) are above the median rates for type 2 diabetes mellitus (T2DM), income, poverty, and education – outperforming previous work by 4–18%. Second, we investigate the effect of socioeconomic factors (income, poverty, and education) on predicting state-level T2DM rates. Socioeconomic factors do improve T2DM prediction, with the greatest improvement coming from poverty information (6%), but, importantly, the language of food adds distinct information that is not captured by socioeconomics. Third, we analyze how the language of food has changed over a five-year period (2013 – 2017), which is indicative of the shift in eating habits in the US during that period. We find several food trends, and that the language of food is used differently by different groups such as different genders. Last, we provide an online visualization tool for real-time queries and semantic analysis.

1 Introduction

With an average of 6,000 new tweets posted every second, Twitter[1] has become a digital footprint of everyday life for a representative sample of the United States (US) population (Mislove et al., 2011). Previously, Fried et al. (2014) demonstrated that the language of food on Twitter can be used to predict health risks, political orientation, and geographic location. Here, we use predictive models to extend this analysis – exploring the ways in which the language of food can shed insight on health and the changing trends in both food culture and language use in different communities over time. We apply this methodology to the particular use case of predicting communities which are risk for type 2 diabetes mellitus (T2DM), a serious medical condition which affects over 30 million Americans and whose *diagnosis alone* costs \$327 billion each year[2]. We refer to T2DM as diabetes in the rest of the paper. We show that by combining knowledge from tweets with other social characteristics (e.g., average income, level of education) we can better predict risk of T2DM. The contributions of this work are four-fold:

1. We use the same methods proposed by Fried et al. (2014) with a much larger (7 times) tweet corpus gathered from 2013 – 2017 to predict the risk of T2DM. We collected over 24 million tweets with meal-related hashtags (e.g., *#breakfast*, *#lunch*) and localized 5 million of them to states within the US. We show that more data helps, and that by training on this larger dataset the state-level T2DM risk prediction accuracy is improved by 4–18%, compared to the results in Fried et al. (2014). We also apply the same models to predict additional state-level indicators: income, poverty, and education levels in order to further investigate the predictive power of the language of food. On these prediction tasks, our model outperforms the majority baseline by 12–34%. We believe that this work may drive immediate policy decisions for the communities deemed at risk without awaiting for similar results from major health organizations, which take months or years to be generated and disseminated.[3] Equally as important, we believe that this state-level T2DM risk prediction task may improve predicting risks

---

[*]Equal contribution.
[1]https://twitter.com/

[2]http://www.diabetes.org/advocacy/
news-events/cost-of-diabetes.html
[3]https://www.cdc.gov/nchs/nhis/about_
nhis.htm

Proceedings of the 10th International Workshop on Health Text Mining and Information Analysis (LOUHI 2019), pages 87–96
Hong Kong, November 3, 2019. ©2019 Association for Computational Linguistics
https://doi.org/10.18653/v1/D19-62

for *individuals* from their social media activity, a task which often suffers from sparsity (Bell et al., 2018).

2. Unlike (Fried et al., 2014), we also investigate the effect of socioeconomic factors on the diabetes prediction task itself. We observe that aggregated US social demographic information from average income[4], poverty[5], and education[6] is complementary to the information gained from tweet language used for predicting diabetes risk. We add the correlation between each of these socioeconomic factors and the diabetes[7] rate in US states as additional features in the models in (1). We demonstrate that the T2DM prediction model strongly benefits from the additional information, as prediction accuracy further increases by 2–6%. However, importantly, the model that relies solely on these indicators performs considerably worse than the model that includes features from the language of food, which demonstrates that the language of food provides distinct signal from these indicators.

3. Furthermore, with a dataset that spans nearly five years, we also analyze language trends over time. Specifically, using pointwise mutual information (PMI) and a custom-built collection of healthy/unhealthy food words, we investigate the strength of healthy/unhealthy food references on Twitter, and observe a downward trend for healthy food references and an upward trend for unhealthy food words in the US.

4. Lastly, we provide a visualization tool to help understand and visualize semantic relations between words and various categories such as how different genders refer to vegetarian vs. low-carb diets.[8] Our tool is based on semantic axes plots (Heimerl and Gleicher, 2018).

2   Related Work

Many previous efforts have shown that social media can serve as a source of data to detect possible health risks. For example, Akbari et al. (2016) proposed a supervised learning approach that automatically extracts public wellness events from microblogs. The proposed method addresses several problems associated with social media such as insufficient data, noisiness and variance, and interrelations among social events. A second contribution of Akbari et al. (2016) is an automatically-constructed large-scale diabetes dataset that is extended with manually handcrafted ground-truth labels (positive, negative) for wellness events such as diet, exercise and health.

Bell et al. (2016) proposed a strategy that uses a game-like quiz with data and questions acquired semi-automatically from Twitter to acquire relevant training data necessary to detect individual T2DM risk. In following work, Bell et al. (2018) predicted individual T2DM risk using a neural approach, which incorporates tweet texts with gender information and information about the recency of posts.

Sadeque et al. (2018) discussed several approaches for predicting depression status from a user's social media posts. They proposed a new latency-based F1 metric to measure the quality and speed of the model. Further, they re-implemented some of the common approaches for this task, and analyzed their results using their proposed metric. Lastly, they introduced a window-based technique that trades off between latency and precision in predicting depression status.

Our work is closest to (Fried et al., 2014). Similar to us, Fried et al. (2014) predicted latent population characteristics from Twitter data such as overweight rate or T2DM risk in US states. Our work extends (Fried et al., 2014) in several ways. First, in addition of tweets, we incorporate state-level indicators such as poverty, education, and income in our risk classifier, and demonstrate that language provides distinct signal from these indicators. Second, we use the much larger tweet dataset to infer language-of-food trends over a five-year span. Third, we provide a visualization tool to explore food trends over time, as well as semantic relations between words and categories in this context.

3   Data

We collected tweets along with their meta data with Twitter's public streaming API[9]. Tweets have been filtered by a set of seven hashtags to make the dataset more relevant to food (see distribution in Table 1). We stored the tweets and their metadata

---

[4]https://www.census.gov/topics/income-poverty/income.html

[5]https://www.census.gov/topics/income-poverty/poverty.html

[6]https://talkpoverty.org/indicator/listing/higher_ed/2017

[7]https://www.kff.org/other/state-indicator/adults-with-diabetes

[8]http://t4f.cs.arizona.edu/

[9]https://developer.twitter.com/en/docs/tweets/filter-realtime/guides/connecting.html

Term	# of tweets	# of tweets localized in US
#dinner	5,455,890	1,367,745
#breakfast	5,125,014	1,183,462
#lunch	4,969,679	1,094,681
#brunch	1,910,950	681,978
#snack	797,676	220,697
#meal	495,073	101,976
#supper	124,979	22,154
Total	24,493,223	4,362,940

Table 1: Seven meal-related hashtags and their corresponding number of tweets filtered from Twitter. The right-most column indicates the number of tweets we could localize to a US state or Washington D.C.

into a Lucene-backed Solr instance.[10] This Solr instance is used to localize the tweets in the US and annotate them with topic models afterwards.

All in all, we collected over 24 million tweets from the period between October 2, 2013 to August 28, 2018, a dataset that is seven times larger than that of Fried et al. (2014). Both datasets contain tweets filtered using the same 7 meal-related hashtags. In order to localize the tweets in the US, we use self-reported location, timezone, and geotagging information (latitude and longitude). The geolocalization is performed in two steps. First, we use regular expressions to match a user's reported location data with the names or postal abbreviations of the 50 US states (e.g., Arizona or AZ) and Washington D.C., and also with city names or known abbreviations (e.g., New York City or NYC). Second, if we cannot find a match, then we use the latitude and longitude information (if provided in the metadata) to localize a tweet. This allowed us to successfully localize approximately 5 million out of the 24 million tweets. For the remaining tweets, latitude/longitude data is converted into city, state, or country using Geopy[11], successfully localizing an additional hundred thousand tweets[12]. Each tweet is preprocessed and filtered to remove punctuation marks, usernames, URLs, and non-alphanumeric characters (but not hashtags).

4 Approach

This work aims for four main goals: predicting state-level characteristics, evaluating the effect of socioeconomic factors in these prediction

---

tasks, analyzing food trends, and using visualization tools to capture trends in the usage of the language of food by different population groups.

4.1 State-level prediction tasks

We investigate the predictive power of the language of food through four distinct prediction tasks: T2DM rate, income, poverty, and education level. We use the tweets from the above dataset as the only input for our prediction models.

T2DM rate prediction: We use the diabetes rate from the Kaiser Commission on Medicaid and Uninsured (KCMU)'s analysis of the Center for Disease Control's Behavioral Risk Factor Surveillance System (BRFSS) 2017 Survey (its most recent year)[7]. The state-level diabetes rate is defined as the percentage of adults in each state who have been told by a doctor that they have diabetes. The median diabetes rate for the US is 10.8%. For each state, we convert the diabetes rate into a binary variable with a value of 1 if the state diabetes rate is greater than or equal to the national median rate, and a value of 0 if it is below. For example, the state with highest diabetes rate, West Virginia (15.2%), is assigned a binary variable of 1 (high T2DM rates). On the other hand, states with below-national-median rate, like Arizona (10.4%), are assigned a label of 0 (low T2DM rates).

Income rate prediction: We collect income data from the United States Census Bureau (USCB)'s analysis of the American Community Survey (ACS)'s Income and Poverty in the United States: 2017[4]. The data shows that national median household income is $60,336. Similarly to above, we convert the household median income of the state into a binary variable with a value of 0 (low income) if its median household income is lower than national median, and a value of 1 (high income) if its median household income is equal or greater. For example, Alabama ($48,193) is labeled as low-income and Alaska ($74,058) is labeled as high-income.

Poverty rate prediction: To predict poverty rates, we also collect poverty data from the USCB's analysis of the ACS's Income and Poverty in the United States: 2017[5], which shows that national median poverty rate is 13.4%. Again, we assign each state a binary variable indicating whether its rate is above or below this national median.

Education rate prediction: For predicting education rate, we use the higher education attainment rate (HEAR) data from the Center of American Progress (CAP)[6]. The data shows that national median HEAR is 43.2%. Once again, the state-level HEAR is converted to a binary variable in the same manner as above.

Because each of these binary variables is at the state level, we group the tweets by state before feature extraction. We use leave-one-out cross-validation (LOOCV) as proposed by Fried et al. (2014). This approach is necessary because even though we have a large tweet corpus, we only have 51 aggregate data points (one for each state plus Washington, D.C.). For classification, we use Support Vector Machines (SVM) (Vapnik, 2013) for feature-based classification. To avoid overfitting, we tuned the classifier's hyper-parameters during training using the tweets from 2013 to 2016. We tested the tuned prediction models for each task using solely tweets from 2017.

We use two sets of features: lexical (words from tweets) and topical (sets of words appearing in similar contexts). For lexical features, we compare open (all unique tweet words or hashtags) and closed (800 food words) vocabularies, using the token counts as the tweet features. These experiments help us to determine the predictive power of the specific language of food versus the broader context in the full tweets (or socially compact hashtag). For topic model features, we use Latent Dirichlet Allocation (LDA) (Blei et al., 2003), to learn a set of topics from food tweets. Because tweets are very short in nature (up to 140 characters), this approach allows us to analyze correlations that could go beyond individual words. We chose 200 as the number of topics for LDA to learn. After LDA is trained using MALLET[13], we use it to create the set of topics for each tweet, and the topic with highest probability is then assigned to each tweet as an additional feature. Topics are counted across all tweets in a state in the same manner as the lexical features.

We also experimented with Deep Averaging Network (DAN) (Iyyer et al., 2015), a simple but robust bag-of-words model based on averaging word embeddings that has been shown to perform well in sentiment analysis and factoid question answering with little training data. In our case, we implemented DAN with embeddings generated using Word2Vec (Mikolov et al., 2013) trained over all 24 million tweets (including the ones that

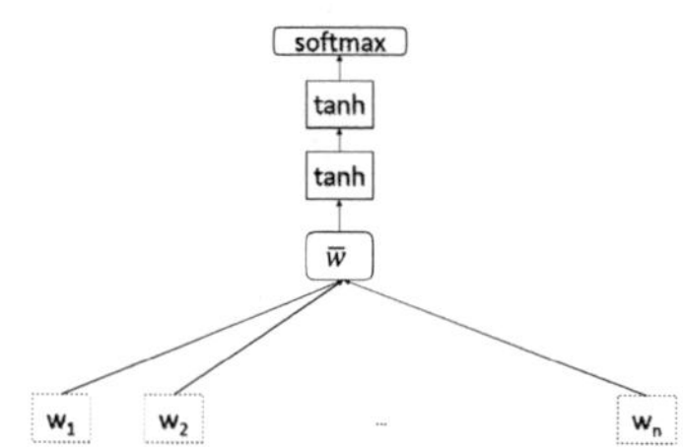

Figure 1: Deep Averaging Network for prediction tasks. The embeddings are averaged and passed to two non-linear layers (tanh).

were not localized). We compute the embedding for each token in our dataset, and pass them to the network; see Figure 1. Again using LOOCV, in each pass we leave out one state, train the network on tweets from the 50 other states and predict the T2DM rate for the left out state.

4.2 Impact of socioeconomic factors

Previous work has shown that the T2DM rate can be predicted by socioeconomic factors such as poverty (Chih-Cheng et al., 2012), income (Yelena et al., 2015), and education (Ayyagari et al., 2011). Therefore, we incorporate these factors into our prediction models (Section 4.1) to assess their contribution. We represent each socioeconomic factor and its correlation with the T2DM rate in the corresponding state as a feature, and include these new features alongside the lexical and topic-based ones. Even though in general, the correlations are relatively low (see Table 2), we will show that the model strongly benefits from the additional information leading to accuracy increases of 2–6% (see Section 5). This indicates that the language of food captures different signal and reflects distinct information from these indicators. However, because these indicators are represented as single features, as opposed to the other features (e.g., there are tens of thousands of food word features, each of which is represented as an integer count), they tended to be ignored by the classifier. To account for this, we empirically explored a series of multipliers to increase the weights of the values of these indicator features[14]. For this task, we use the same SVM classifier from Section 4.1, as well as a Random Forest (RF) classifier (Breiman, 2001)[15].

---

[13]http://mallet.cs.umass.edu/

[14]For these correlation multipliers, we experimented with powers of 10, from 10^1 to 10^6.

[15]To avoid overfitting, we do not fine-tune the RF classifier's hyperparameters.

Socioeconomic factor	Correlation with T2DM
Education	-0.37
Income	-0.14
Poverty	0.18

Table 2: Correlation between socioeconomic factors (education, income, poverty) and type 2 diabetes mellitus (T2DM) in 2017. Each correlation is calculated from the binary data described in Section 4.1. For the correlation values we used Pearson correlation (Boslaugh, 2012).

4.3 Exploring food trends

We use pointwise mutual information (PMI) between food words/hashtags and years to analyze food trends over time. We divide our corpus of tweets into four parts, each containing a complete year's set of tweets (from 2014 to 2017) and then calculate PMI for pairs (food term t, year y) using the formula:

$$PMI(t, y) = \frac{C(t, y)}{C(t) * C(y)}, \qquad (1)$$

where, $C(t, y)$ is the number of occurrences of term t in year y, $C(t)$ is the total number of occurrences of the term, and $C(y)$ is the number of tweets in year y. Intuitively, the higher the PMI value of a term in a given year, $PMI(t, y)$, the more that term is associated with tweets from that year in particular.

4.4 Semantic axes analysis

Word vector embeddings are a standard tool used to analyze text, as they capture similarity relationships between the different words. However, interpreting such embeddings and understanding the encoded grammatical and semantic relations between words can be difficult due to the high dimensionality of the embedding space (typically 50-300 dimensions).

Semantic axes visualizations allow us to view specific low dimensional projections where the new dimensions can be used to explore different semantic concepts (Heimerl and Gleicher, 2018). For our task, we generate several word embeddings from our dataset using the CBOW Word2Vec model of (Mikolov et al., 2013). Different than other visualization tools (e.g., t-SNE, PCA), when using semantic axes we need to define two semantic axes by two opposite concepts (e.g., *man* vs. *woman* and *breakfast* vs. *dinner*) and project a collection of vectors (words in embedding) based on the specific 2D space. The result is a 2D scatter plot with respect to two different concepts.

We first create a word embedding for all the tweets in our dataset. This allows us to explore the correlations between different concepts.

We further augment the semantic axes tool[16] provided by Heimerl and Gleicher (2018), to allow a concept axis to be defined by two sets of words (rather than exactly two words). For example, instead of having one axis defined by the pair (*vegetables, meat*) we can now use two sets of words (*vegetables, fruit, vegetarian, vegan*, etc., and *meat, fish, chicken, beef*, etc.). This allows us to capture more complex concepts such as "meat-eaters" that are not captured by individual words.

5 Results

We present the results for all prediction tasks of state level characteristics, as well as the evaluation of the contribution of socioeconomic factors alongside food language in predicting T2DM rate. We also investigate the shifts in eating habits over time (i.e., food trends), as well as the trends in different groups through our semantic axes experiments.

5.1 State-level characteristics prediction

In Table 3, we show the results for predicting state-level socioeconomic characteristics using various sets of features. We compare the results from our dataset with the results of Fried et al. (2014) for predicting T2DM rates. However, since Fried et al. (2014) do not experiment with predicting poverty, income, and education level, for these we compare against a majority baseline. As there are 51 states (including Washington D.C.), and each binary socioeconomic factor is based on the national median, this means that for each factor there will be 26 states either above or below (resulting in a majority baseline of 50.98%).

Comparing the effects of each type of lexical features and their combination with LDA topic features on these prediction tasks, we make several observations.

Performance comparison by feature set: First and foremost, the results demonstrate that the language of food can be used to predict health and social characteristics such as diabetes risk, income, poverty, and education level. The highest overall performance is achieved by using all tweet words (both with and without LDA). This suggests that

---

[16] http://embvis.flovis.net/

#		Diabetes	Poverty	Income	Education	Average
#	Majority baseline	50.98	50.98	50.98	50.98	50.98
	All Words					
1	Fried et al. (2014)	64.71	–	–	–	–
2	Our dataset	**74.51**	64.71	80.39	74.51	73.53
	All Words + LDA					
3	Fried et al. (2014)	64.71	–	–	–	–
4	Our dataset	**70.59**	66.67	82.35	74.51	73.53
	Hashtags					
5	Fried et al. (2014)	68.63	–	–	–	–
6	Our dataset	**74.51**	64.71	80.39	66.67	71.57
	Hashtags + LDA					
7	Fried et al. (2014)	68.63	–	–	–	–
8	Our dataset	**72.55**	62.75	84.31	68.63	72.06
	Food					
9	Fried et al. (2014)	60.78	–	–	–	–
10	Our dataset	**72.55**	62.75	64.71	62.75	65.69
	Food + LDA					
11	Fried et al. (2014)	60.78	–	–	–	–
12	Our dataset	**78.43**	62.75	62.75	62.75	66.67
	Food+Hashtags					
13	Fried et al. (2014)	62.75	–	–	–	–
14	Our dataset	**72.55**	64.71	78.43	66.67	70.59
	Food+Hashtags+LDA					
15	Fried et al. (2014)	62.75	–	–	–	–
16	Our dataset	**74.51**	64.71	84.31	68.63	73.05

Table 3: Results from using various feature sets to predict state-level characteristics: whether a given state is above or below the national median for diabetes, poverty, income, and education. We also show the average performance across all characteristics. We compare against Fried et al. (2014) as well as the majority baseline. Note that Fried et al. do not predict poverty, income, or education level. The low number of data points (51 states) is responsible for the same accuracy value in multiple experiments.

we can capture significant predictive signal from tweets when capturing food words in context.

The highest prediction performance is seen when predicting the state-level income rate, demonstrating a high correlation between food-related words and income. When predicting state-level diabetes rate, we also see strong predictive power from the language of food – all models perform above 70%, up to 78.43%. This confirms our hypothesis that there is a strong correlation between food-related words (and presumably food behaviors) and diabetes rate, one indicator of public health.

Amount and recency of data: For diabetes prediction, with our larger dataset, we improve upon the results of Fried et al. (2014) (ranging from 4 to 18%). In particular, when we use the food-word features combined with LDA topics, we increase prediction accuracy by almost 18%. These results suggest that more data matters in this type of analysis, as evidenced by the learning curves shown in Figure 2, where we compare performance against amount of training data (by year).

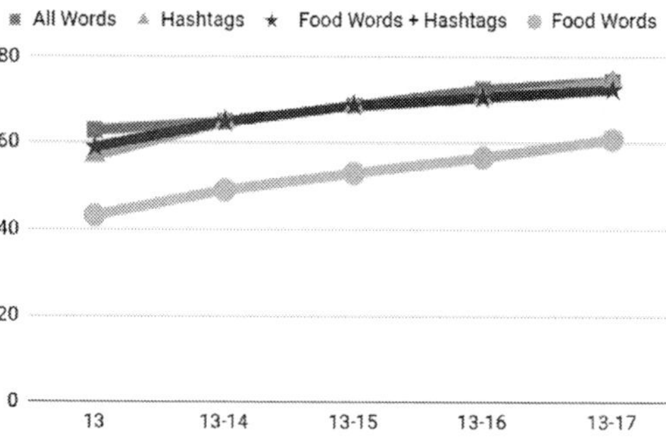

Figure 2: The learning curves for each lexical feature set in terms of predicting diabetes rate in 2017. The horizontal axis corresponds to the cumulative date range used, i.e., 13 only uses tweets from 2013, and 13-14 uses tweets from 2013 through 2014, etc. The y-axis is the state-level prediction accuracy.

We also created learning curves for prediction of T2DM, but from the opposite direction, i.e., starting from tweets from 2017 only, and then adding tweets from earlier years one year at a time. We observe that the more recent the data, the more useful it is for prediction. We hypothesize that in terms of the utility of increased data, the performance of food-word features is improved only as the amount of *relevant* data increases. For the first part of the curve (only from 17, combined 17–16, combined 17–15), the classifier's performance is improved with additional tweets. However, after this peak, additional older tweets decrease performance, suggesting that people change their eating behavior over a period spanning multiple years. The importance of recency of tweet data is also discussed in (Bell et al., 2016).

Comparison to previous work: The best performing model of Fried et al. (2014) relies on hash-tags (see Table 3, lines 5 and 7) and the worst performing model use food words (lines 9 and 11). However, with more data we find that we get the best performance with food words (line 12). We hypothesize that with smaller data, the concise semantics of hashtags are more informative, but with more data the model is able to learn the relative semantics of the food words themselves. Further, while LDA topics do not benefit any model of Fried et al. in terms of predicting diabetes, here we find that with additional data, LDA topics benefit the food words model (compare lines 10 and 12), and in fact contribute to our best performing model (line 12), perhaps because additional data leads to more representative LDA topics.

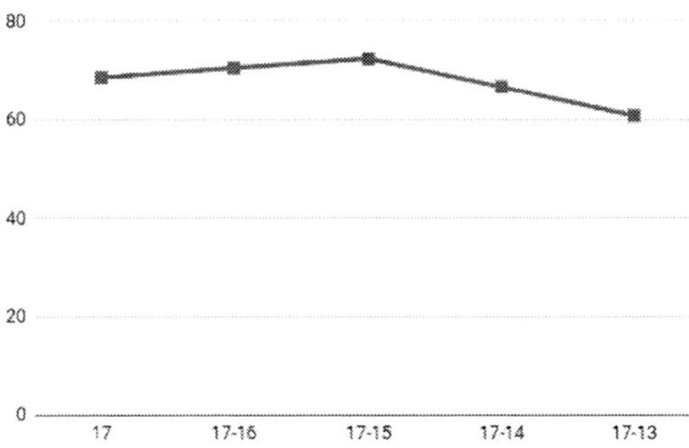

Figure 3: The learning curve using food word features, on the diabetes rate prediction task for 2017. In this figure, the data portion used for each point is in reverse order compared to Figure 2, that is, starting from most recent tweets and going back in time. The horizontal axis is labeled based on the year(s) from which the tweets used for prediction were used.

The Deep Averaging Network from Figure 1 consistently underperformed the results reported in Table 3. This approach obtained an accuracy of 60.78% for the T2DM rate prediction task, considerably lower than the 78% obtained by the best SVM configuration in Table 3. We hypothesize that the reason behind this low performance is the small number of data points (51), which is insufficient to train a neural network.

5.2 Impact of socioeconomic factors

In Table 5, we show the SVM results for predicting T2DM rate from extending the feature matrix from 5.1 with one additional feature based on the correlation between each socioeconomic factor (education, income, and poverty) and T2DM. For each factor, we compare several multipliers (see Section 4.2) to amplify the impact of the socioeconomic correlations. Consistently, we find that models with more features benefit from larger multipliers. For example, the extended food word models that have several hundred features perform best with a multiplier of 10^2, while the other extended models, which all have tens of thousands of features, perform best with a multiplier of 10^3. The best multiplier for each model, according to SVM performance, is used in our Random Forest models (Table 4).

From these extended models we see that using poverty information as an additional feature improves our SVM performance by a range of 2–8% and our RF performance by up to 6%. The other socioeconomic factors, i.e., income and education, do not help when using an SVM classi-

#	Features	Results from best performing multiplier
	All Words+LDA with RF	62.75
	Fried et al. (2014)	64.71
1	+ Education	64.71
2	+ Income	64.71
3	+ Poverty	**64.71**
	Food+LDA with RF	70.59
	Fried et al. (2014)	60.78
4	+ Education	74.51
5	+ Income	72.55
6	+ Poverty	**76.47**
	Hashtags+LDA with RF	68.63
	Fried et al. (2014)	68.63
7	+ Education	64.71
8	+ Income	64.71
9	+ Poverty	**68.63**
	Food+Hashtags+LDA with RF	66.67
	Fried et al. (2014)	62.75
10	+ Education	**72.55**
11	+ Income	**72.55**
12	+ Poverty	70.59

Table 4: Results for predicting T2DM rate using a random forest classifier with our additional socioeconomic correlation features. For each feature set, we use the best performing multiplier, as determined in the previous experiment that used a SVM classifier (Table 5). That is, the best performing multiplier for food word features is 10^2, while other features' multipliers are 10^3.

fier (Table 5), but when using a RF classifier we see up to 6% improvement (Table 4). Overall, our highest T2DM prediction performance is obtained with SVM using Food + LDA + poverty. This performance surpasses 80% accuracy and is the highest value reported for this task. Further, to the best of our knowledge, the effect of using poverty information to improve T2DM rate prediction is novel and suggests a potential avenue for improving classifiers with socioeconomic correlation information.

Importantly, predicting the T2DM below/above median labels from the poverty indicator alone has an accuracy of 58.82%. This value is considerably lower than that of the classifier that uses poverty coupled with the extended word features from tweets, which obtained 80% accuracy. This demonstrates that the language of food provides signal that is distinct from this indicator, which suggests that there is value in social media mining for the monitoring of health risks.

5.3 Food trends

Given our dataset that spans nearly five years, we are also able to investigate whether changes in food habits over time can be detected in social media language. To this end, we explored a list of 800 food words and their change in PMI values in the different years. To understand which food words

#		10^1	10^2	10^3	10^4	10^5	10^6
	All Words + LDA	70.59	–	–	–	–	–
1	+ Education	70.59	70.59	70.59	70.59	70.59	66.67
2	+ Income	70.59	70.59	70.59	66.67	66.67	66.67
3	+ Poverty	66.67	72.55	**78.43**	74.51	70.59	70.59
	Food + LDA	78.43	–	–	–	–	–
4	+ Education	70.59	74.51	68.63	70.59	68.63	68.63
5	+ Income	70.59	74.51	68.63	70.59	66.67	62.75
6	+ Poverty	78.43	**80.39**	76.47	68.63	70.59	70.59
	Hashtags+LDA	72.55	–	–	–	–	–
7	+ Education	70.59	70.59	74.51	70.59	66.67	68.63
8	+ Income	66.67	68.63	70.59	66.67	62.75	66.67
9	+ Poverty	72.55	74.51	**76.47**	64.71	68.63	68.63
	Food+Hashtags+LDA	74.51	–	–	–	–	–
10	+ Education	70.59	70.59	72.55	68.63	68.63	66.67
11	+ Income	66.67	72.55	74.51	68.63	68.63	66.67
12	+ Poverty	72.55	74.51	**78.43**	72.55	68.63	68.63

Table 5: Results for predicting T2DM rate using our SVM classifier, which is similar to that of Fried et al. (2014), but with additional socioeconomic correlation features. Columns show results under different multipliers used to boost the importance of the indicator features (see Section 4.2).

1st annotator	2nd annotator	Score
annotator 1	annotator 2	0.72
annotator 1	annotator 3	0.39
annotator 2	annotator 3	0.58

Table 6: The Cohen's kappa inter-annotator agreement scores among the three annotators.

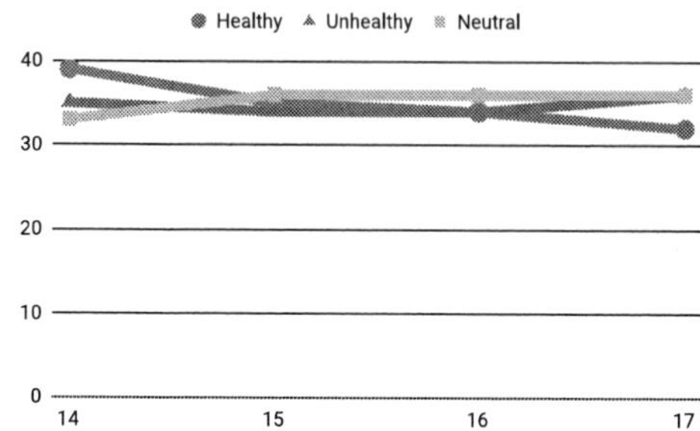

Figure 4: PMI values for each category foods annotated by the 1st annotator. The y-axis shows PMI values 10^6. The trends based on the other annotators are similar.

indicate healthy vs. unhealthy diets, we manually classified the 800 food words into three categories – healthy, unhealthy and neutral – using reliable online resources[17]. The annotations were independently performed by three annotators. The inter-annotator Kappa agreement scores[18] shown in Table 6 indicate fair to good agreement between the three annotators.

We computed PMI values for each of these 800 words and each year in our US dataset. We also computed the PMI values for the three categories and each year (here all words from each category are treated as one). The category trends in our US dataset indicate a slight increase of mentions of unhealthy food words and a slight decrease in mentions of healthy food words in US tweets; see Figure 4. These results suggest a continued decline in dietary patterns in the US, despite seemingly increased interest in health benefits from food[19].

5.4 Semantic axes visualization

As discussed in Section 4.4, visualizations can help discover correlations between different concepts, as well as look at trends over time. In Figure 5, we consider the two axes defined by *man*

vs. *woman*, and *breakfast* vs. *dinner*. Exploring the four corners we can identify particular types of foods representative of those coordinates. In the top-left we see words associated with women and breakfast (*yogurt, cupcake, pastry*), whereas in the bottom-left we see words associated with men and breakfast (*sausage, bacon, ham*). Similarly, in the top-right we see words associated with women and dinner (*mussels, halibut, eggplant*) whereas the bottom right we see words associated with men and dinner (*lasagna, lamb, teriyaki*). This data confirms common stereotypes, e.g., (1) men tend to eat more meat, whereas women often prefer fish, and (2) women are more health-conscious compared to men.

We also consider topics (defined by a collection of words) as axes, as illustrated in Table 7. The two axes now are *man* vs. *woman*, and vegetarian words vs. low-carb diets. To represent the vegetarian topic we use the words *vegan, vegetarian, tofu,* and to represent the low-carb topic we use *keto, paleo,* and *atkins*. We then average the word embedding vectors for all words in the topic to create the 2D projection.

We list the 4 corners in the projection as 4 rows in Table 7, where the left column corresponds to the concepts and the right column contains the words. Several patterns emerge: vegetarian words associated with women tend to be soups, salads,

---

[17] http://www.diabetes.org/ and https://www.healthline.com/health/diabetes/

[18] We use the scikit learn library to calculate the score. https://scikit-learn.org/stable/modules/generated/sklearn.metrics.cohen_kappa_score.html

[19] https://foodinsight.org/wp-content/uploads/2018/05/2018-FHS-Report-FINAL.pdf

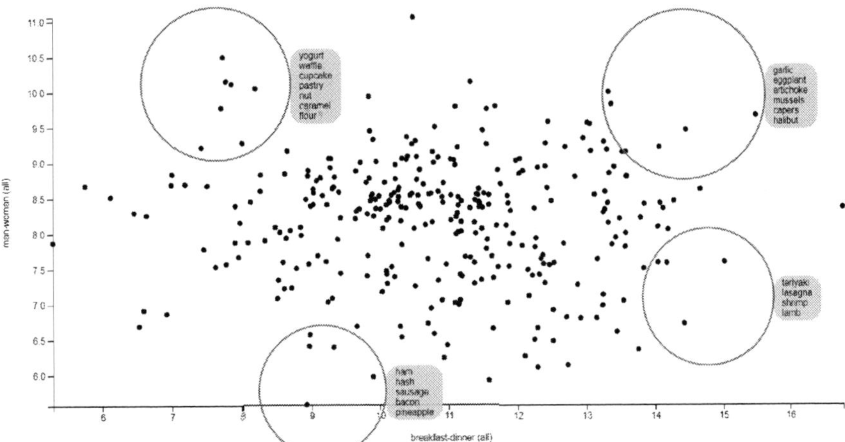

Figure 5: Semantic axes 2D plot using man vs. woman and breakfast vs. dinner as the two axes. We highlight the four corners where interesting patterns can be seen (e.g., the top-left corner is associated with with women and breakfast). Note that this image is a composite of four images, each highlighting one corner.

and gourmet types of foods (saffron, fennel). In contrast, the vegetarian words associated with men tend to be vegetables (spinach, kale, carrot). In the low-carb and women corner we find breakfast words and deserts (cupcake, pastry, caramel, wheat) whereas in the low-carb and men corner we see more hearty foods (spaghetti, hamburgers, buns).

man vs. woman, and vegetarian vs. low-carb diets

woman, vegetarian diet	mint, saffron, fennel, squash, soup, tomato, eggplant
man, vegetarian diet	beet, onion, coconut, spinach, kale, carrot
woman, low-carb diet	hazelnut, nut, cupcake, pastry, grain, caramel, wheat
man, low-carb diet	cereal, spaghetti, buns, hamburger, pepperoni, crunch

Table 7: The 4 corners in the man vs. woman and vegetarian words vs. low-carb diets plot. Each row represents one corner, The left column contains the pair of concepts; the right column contains the foods associated with those concepts.

6 Conclusion

We showed that the language of food has predictive power for non-trivial state-level health tasks such as predicting if a state has higher/lower diabetes risk than the median. When augmented with socio-economic data such as poverty indicators, performance improves further, but we demonstrate that the language of food captures different signal and reflect distinct information from these socio-economic data. We also provide visualization tools to analyze the underlying data and visualize patterns and trends. This work may have immediate use in public health, e.g., by driving rapid policy decisions for the communities deemed at health risk. Further, we hope that this work complements predicting health risk for individuals, a task that is plagued by sparsity, and which could potentially benefit from additional community-level information.

Acknowledgments

Mihai Surdeanu declares a financial interest in lum.ai. This interest has been properly disclosed to the University of Arizona Institutional Review Committee and is managed in accordance with its conflict of interest policies.

References

M. Akbari, X. Hu, N. Liqiang, and T. Chua. 2016. From tweets to wellness: Wellness event detection from twitter streams. In *Proceedings of the Thirtieth AAAI Conference on Artificial Intelligence*, AAAI'16, pages 87–93. AAAI Press.

P. Ayyagari, D. Grossman, and F. Sloan. 2011. Education and health: evidence on adults with diabetes. volume 11, pages 35–54. Springer.

D. Bell, D. Fried, L. Huangfu, M. Surdeanu, and S. Kobourov. 2016. Towards using social media to identify individuals at risk for preventable chronic illness. In *Proceedings of the Tenth International Conference on Language Resources and Evaluation (LREC 2016)*. LREC.

D. Bell, E. Laparra, A. Kousik, T. Ishihara, M. Surdeanu, and S. Kobourov. 2018. Detecting diabetes risk from social media activity. In *Ninth Interna-*

tional Workshop on Health Text Mining and Information Analysis (LOUHI).

D. M. Blei, A. Y. Ng, and M. I. Jordan. 2003. Latent dirichelet allocation. In *The Journal of Machine Learning Research*, pages 993–1022. JMLR.

S. Boslaugh. 2012. *Statistics in a Nutshell*, volume 2. O'Reilly Media, Inc, Boston, MA.

L. Breiman. 2001. Random forests. *Machine learning*, 45(1):5–32.

H. Chih-Cheng, L. Cheng-Hua, W. L Mark, H. Hsiao-Ling, C. Hsing-Yi, C. Likwang, S. Shu-Fang, S. Shyi-Jang, T. Wen-Chen, C. Ted, H. Chi-Ting, and C. Jur-Shan. 2012. Poverty increases type 2 diabetes incidence and inequality of care despite universal health coverage. In *Diabetes Care Vol 35*, pages 2286–2292. ADS.

D. Fried, M. Surdeanu, S. Kobourov, M. Hingle, and D. Bell. 2014. Analyzing the language of food on social media. In *2014 IEEE International Conference on Big Data (Big Data)*, pages 778–783. IEEE.

F. Heimerl and M. Gleicher. 2018. Interactive analysis of word vector embeddings. In *Computer Graphics Forum*, volume 37, pages 253–265. Wiley Online Library.

M. Iyyer, V. Manjunatha, J. Boyd-Graber, and H. Daumé III. 2015. Deep unordered composition rivals syntactic methods for text classification. In *Proceedings of the 53rd Annual Meeting of the Association for Computational Linguistics and the 7th International Joint Conference on Natural Language Processing (Volume 1: Long Papers)*, volume 1, pages 1681–1691.

T. Mikolov, K. Chen, G. Corrado, and J. Dean. 2013. Efficient estimation of word representations in vector space. *arXiv preprint arXiv:1301.3781*.

A. Mislove, S. Lehmann, Y. Ahn, J. Onnela, and J. N. Rosenquist. 2011. Understanding the demographics of twitter users. In *Fifth international AAAI conference on weblogs and social media*.

F. Sadeque, D. Xu, and S. Bethard. 2018. Measuring the latency of depression detection in social media. In *Proceedings of the Eleventh ACM International Conference on Web Search and Data Mining*, WSDM '18, pages 495–503, New York, NY, USA. ACM.

V. Vapnik. 2013. *The nature of statistical learning theory*. Springer science & business media.

B. Yelena, L. Mark, R. Marla, and M. John. 2015. The relationship between socioeconomic status/income and prevalence of diabetes and associated conditions: A cross-sectional population-based study in saskatchewan, canada. In *2015 International Journal for Equity in Health*, pages 93–101. BMC.

Dreaddit: A Reddit Dataset for Stress Analysis in Social Media

Elsbeth Turcan, Kathleen McKeown
Columbia University
Department of Computer Science
{eturcan, kathy}@cs.columbia.edu

Abstract

Stress is a nigh-universal human experience, particularly in the online world. While stress can be a motivator, too much stress is associated with many negative health outcomes, making its identification useful across a range of domains. However, existing computational research typically only studies stress in domains such as speech, or in short genres such as Twitter. We present Dreaddit, a new text corpus of lengthy multi-domain social media data for the identification of stress. Our dataset consists of 190K posts from five different categories of Reddit communities; we additionally label 3.5K total segments taken from 3K posts using Amazon Mechanical Turk. We present preliminary supervised learning methods for identifying stress, both neural and traditional, and analyze the complexity and diversity of the data and characteristics of each category.

1 Introduction

In our online world, social media users tweet, post, and message an incredible number of times each day, and the interconnected, information-heavy nature of our lives makes stress more prominent and easily observable than ever before. With many platforms such as Twitter, Reddit, and Facebook, the scientific community has access to a massive amount of data to study the daily worries and stresses of people across the world.[1]

Stress is a nearly universal phenomenon, and we have some evidence of its prevalence and recent increase. For example, the American Psychological Association (APA) has performed annual studies assessing stress in the United States since 2007[2] which demonstrate widespread experiences of chronic stress. Stress is a subjective experience whose effects and even definition can vary from person to person; as a baseline, the APA defines stress as a reaction to extant and future demands and pressures,[3] which can be positive in moderation. Health and psychology researchers have extensively studied the connection between too much stress and physical and mental health (Lupien et al., 2009; Calcia et al., 2016).

In this work, we present a corpus of social media text for detecting the presence of stress. We hope this corpus will facilitate the development of models for this problem, which has diverse applications in areas such as diagnosing physical and mental illness, gauging public mood and worries in politics and economics, and tracking the effects of disasters. Our contributions are as follows:

- Dreaddit, a dataset of lengthy social media posts in five categories, each including stressful and non-stressful text and different ways of expressing stress, with a subset of the data annotated by human annotators;[4]
- Supervised models, both discrete and neural, for predicting stress, providing benchmarks to stimulate further work in the area; and
- Analysis of the content of our dataset and the performance of our models, which provides insight into the problem of stress detection.

In the remainder of this paper, we will review relevant work, describe our dataset and its annotation, provide some analysis of the data and stress detection problem, present and discuss results of some supervised models on our dataset, and finally conclude with our summary and future work.

2 Related Work

Because of the subjective nature of stress, relevant research tends to focus on physical sig-

---

[1] https://www.gse.harvard.edu/news/uk/17/12/social-media-and-teen-anxiety

[2] https://www.apa.org/news/press/releases/stress/index?tab=2

[3] https://www.apa.org/helpcenter/stress-kinds

[4] Our dataset will be made available at http://www.cs.columbia.edu/~eturcan/data/dreaddit.zip.

Proceedings of the 10th International Workshop on Health Text Mining and Information Analysis (LOUHI 2019), pages 97–107
Hong Kong, November 3, 2019. ©2019 Association for Computational Linguistics
https://doi.org/10.18653/v1/D19-62

nals, such as cortisol levels in saliva (Allen et al., 2014), electroencephalogram (EEG) readings (Al-Shargie et al., 2016), or speech data (Zuo et al., 2012). This work captures important aspects of the human reaction to stress, but has the disadvantage that hardware or physical presence is required. However, because of the aforementioned proliferation of stress on social media, we believe that stress can be observed and studied purely from text.

Other threads of research have also made this observation and generally use microblog data (e.g., Twitter). The most similar work to ours includes Winata et al. (2018), who use Long Short-Term Memory Networks (LSTMs) to detect stress in speech and Twitter data; Guntuku et al. (2018), who examine the Facebook and Twitter posts of users who score highly on a diagnostic stress questionnaire; and Lin et al. (2017), who detect stress on microblogging websites using a Convolutional Neural Network (CNN) and factor graph model with a suite of discrete features. Our work is unique in that it uses data from Reddit, which is both typically longer and not typically as conducive to distant labeling as microblogs (which are labeled in the above work with hashtags or pattern matching, such as "I feel stressed"). The length of our posts will ultimately enable research into the causes of stress and will allow us to identify more implicit indicators. We also limit ourselves to text data and metadata (e.g., posting time, number of replies), whereas Winata et al. (2018) also train on speech data and Lin et al. (2017) include information from photos, neither of which is always available. Finally, we label individual parts of longer posts for acute stress using human annotators, while Guntuku et al. (2018) label users themselves for chronic stress with the users' voluntary answers to a psychological questionnaire.

Researchers have used Reddit data to examine a variety of mental health conditions such as depression (Choudhury et al., 2013) and other clinical diagnoses such as general anxiety (Cohan et al., 2018), but to our knowledge, our corpus is the first to focus on stress as a general experience, not only a clinical concept.

3 Dataset

3.1 Reddit Data

Reddit is a social media website where users post in topic-specific communities called subreddits,

> I have this feeling of dread about school right before I go to bed and I wake up with an upset stomach which lasts all day and nakes me feel like I'll throw up. This causes me to lose appetite and not wanting to drink water for fear of throwing up. I'm not sure where else to go with this, but I need help. If any of you have this, can you tell me how you deal with it? I'm tired of having this every day and feeling like I'll throw up.

Figure 1: An example of stress being expressed in social media from our dataset, from a post in r/anxiety (reproduced exactly as found). Some possible expressions of stress are highlighted.

and other users comment and vote on these posts. The lengthy nature of these posts makes Reddit an ideal source of data for studying the nuances of phenomena like stress. To collect expressions of stress, we select categories of subreddits where members are likely to discuss stressful topics:

- **Interpersonal conflict**: abuse and social domains. Posters in the abuse subreddits are largely survivors of an abusive relationship or situation sharing stories and support, while posters in the social subreddit post about any difficulty in a relationship (often but not exclusively romantic) and seek advice for how to handle the situation.
- **Mental illness**: anxiety and Post-Traumatic Stress Disorder (PTSD) domains. Posters in these subreddits seek advice about coping with mental illness and its symptoms, share support and successes, seek diagnoses, and so on.
- **Financial need**: financial domain. Posters in the financial subreddits generally seek financial or material help from other posters.

We include ten subreddits in the five domains of abuse, social, anxiety, PTSD, and financial, as detailed in Table 1, and our analysis focuses on the domain level. Using the PRAW API,[5] we scrape all available posts on these subreddits between January 1, 2017 and November 19, 2018; in total, 187,444 posts. As we will describe in subsection 3.2, we assign binary stress labels to 3,553 segments of these posts to form a supervised and semi-supervised training set. An example segment is shown in Figure 1. Highlighted phrases are in-

---

[5]https://github.com/praw-dev/praw

Domain	Subreddit Name	Total Posts	Avg Tokens/Post	Labeled Segments
abuse	r/domesticviolence	1,529	365	388
	r/survivorsofabuse	1,372	444	315
	Total	**2,901**	**402**	**703**
anxiety	r/anxiety	58,130	193	650
	r/stress	1,078	107	78
	Total	**59,208**	**191**	**728**
financial	r/almosthomeless	547	261	99
	r/assistance	9,243	209	355
	r/food_pantry	343	187	43
	r/homeless	2,384	143	220
	Total	**12,517**	**198**	**717**
PTSD	r/ptsd	4,910	265	711
social	r/relationships	107,908	578	694
	All	**187,444**	**420**	**3,553**

Table 1: **Data Statistics**. We include ten total subreddits from five domains in our dataset. Because some subreddits are more or less popular, the amount of data in each domain varies. We endeavor to label a comparable amount of data from each domain for training and testing.

dicators that the writer is stressed: the writer mentions common physical symptoms (nausea), explicitly names fear and dread, and uses language indicating helplessness and help-seeking behavior.

The average length of a post in our dataset is 420 tokens, much longer than most microblog data (e.g., Twitter's character limit as of this writing is 280 characters). While we label segments that are about 100 tokens long, we still have much additional data from the author on which to draw. We feel this is important because, while our goal in this paper is to predict stress, having longer posts will ultimately allow more detailed study of the causes and effects of stress.

In Table 2, we provide examples of labeled segments from the various domains in our dataset. The samples are fairly typical; the dataset contains mostly first-person narrative accounts of personal experiences and requests for assistance or advice. Our data displays a range of topics, language, and agreement levels among annotators, and we provide only a few examples. Lengthier examples are available in the appendix.

3.2 Data Annotation

We annotate a subset of the data using Amazon Mechanical Turk in order to begin exploring the characteristics of stress. We partition the posts into contiguous five-sentence chunks for labeling; we wish to annotate segments of the posts because we are ultimately interested in what parts of the post depict stress, but we find through manual inspection that some amount of context is important. Our posts, however, are quite long, and it would be difficult for annotators to read and annotate entire posts. This type of data will allow us in the future not only to *classify* the presence of stress, but also to *locate* its expressions in the text, even if they are diffused throughout the post.

We set up an annotation task in which English-speaking Mechanical Turk Workers are asked to label five randomly selected text segments (of five sentences each) after taking a qualification test; Workers are allowed to select "Stress", "Not Stress", or "Can't Tell" for each segment. In our instructions, we define stress as follows: "The Oxford English Dictionary defines stress as 'a state of mental or emotional strain or tension resulting from adverse or demanding circumstances'. This means that stress results from someone being uncertain that they can handle some threatening situation. We are interested in cases where that someone also feels negatively about it (sometimes we can find an event stressful, but also find it exciting and positive, like a first date or an interview).". We specifically ask Workers to decide whether the author is expressing both stress and a negative attitude about it, not whether the situation itself seems stressful. Our full instructions are available in the appendix.

We submit 4,000 segments, sampled equally from each domain and uniformly within domains,

Text	Domain	Label	Ann. Agreed
I only get it when I have a flashback or strong reaction to a trigger. I notice it sticks around even when I feel emotionally calm and can stick around for a long time after the trigger, like days or weeks. Its a new symptom I think. Also been having lots of nightmares again recently. Not sure what to do as Im not currently in therapy, but I am waiting to be seen at a mental health clinic.	PTSD	stress	6/7 (86%)
Regardless, that didn't last long, maybe half a year. I released that apartment, and most of my belongings (I kept a few boxes of my things from the military, personal effects, but little else). Looking back, there were some signs of emotional manipulation here, but it was subtle... and you know how it is, love is blind. We got engaged. It was quite the affair.	abuse	not stress	5/5 (100%)
Our dog Jett has been diagnosed with diabetes and is now in the hospital to stabilize his blood sugar. Luckily, he seems to be doing well and he will be home with us soon. Unfortunately, his bill is large enough that we just can't cover it on our own (especially with our poor financial situation). We're being evicted from our home soon and trying to find a place with this bill is just too much for us by ourselves. To help us pay the bill we've set up a GoFundMe.	financial	stress	3/5 (60%)

Table 2: **Data Examples.** Examples from our dataset with their domains, assigned labels, and number of annotators who agreed on the majority label (reproduced exactly as found, except that a link to the GoFundMe has been removed in the last example). Annotators labeled these five-sentence segments of larger posts.

to Mechanical Turk to be annotated by at least five Workers each and include in each batch one of 50 "check questions" which have been previously verified by two in-house annotators. After removing annotations which failed the check questions, and data points for which at least half of the annotators selected "Can't Tell", we are left with 3,553 labeled data points from 2,929 different posts. We take the annotators' majority vote as the label for each segment and record the percentage of annotators who agreed. The resulting dataset is nearly balanced, with 52.3% of the data (1,857 instances) labeled stressful.

Our agreement on all labeled data is $\kappa = 0.47$, using Fleiss's Kappa (Fleiss, 1971), considered "moderate agreement" by Landis and Koch (1977). We observe that annotators achieved perfect agreement on 39% of the data, and for another 32% the majority was 3/5 or less.[6] This suggests that our data displays significant variation in how stress is expressed, which we explore in the next section.

---

[6]It is possible for the majority to be less than 3/5 when more than 5 annotations were solicited.

4 Data Analysis

While all our data has the same genre and personal narrative style, we find distinctions among domains with which classification systems must contend in order to perform well, and distinctions between stressful and non-stressful data which may be useful when developing such systems. Posters in each subreddit express stress, but we expect that their different functions and stressors lead to differences in how they do so in each subreddit, domain, and broad category.

By domain. We examine the vocabulary patterns of each domain on our training data only, not including unlabeled data so that we may extend our analysis to the label level. First, we use the word categories from the Linguistic Inquiry and Word Count (LIWC) (Pennebaker et al., 2015), a lexicon-based tool that gives scores for psychologically relevant categories such as sadness or cognitive processes, as a proxy for topic prevalence and expression variety. We calculate both the percentage of tokens per domain which are included in a specific LIWC word list, and the percentage of words in a specific LIWC word list that appear

Domain	"Negemo" %	"Negemo" Coverage	"Social" %	"Anxiety" Coverage
Abuse	2.96%	39%	12.03%	58%
Anxiety	3.42%	37%	6.76%	62%
Financial	1.54%	31%	8.06%	42%
PTSD	3.29%	42%	7.95%	61%
Social	2.36%	38%	13.21%	59%
All	2.71%	62%	9.62%	81%

Table 3: **LIWC Analysis by Domain.** Results from our analysis using LIWC word lists. Each term in quotations refers to a specific word list curated by LIWC; percentage refers to the percent of words in the domain that are included in that word list, and coverage refers to the percent of words in that word list which appear in the domain.

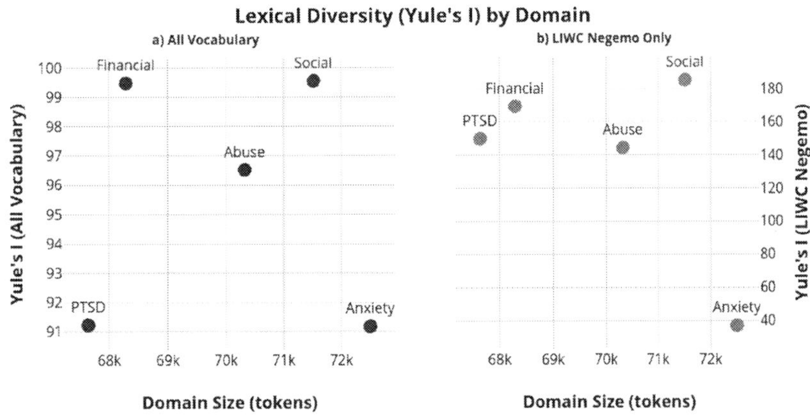

Figure 2: **Lexical Diversity by Domain.** Yule's I measure (on the y-axes) is plotted against domain size (on the x-axes) and each domain is plotted as a point on two graphics. a) measures the lexical diversity of all words in the vocabulary, while b) deletes all words that were not included in LIWC's negative emotion word list.

in each domain ("coverage" of the domain).

Results of the analysis are highlighted in Table 3. We first note that variety of expression depends on domain and topic; for example, the variety in the expression of negative emotions is particularly low in the financial domain (with 1.54% of words being negative emotion ("negemo") words and only 31% of "negemo" words used). We also see clear topic shifts among domains: the interpersonal domains contain roughly 1.5 times as many social words, proportionally, as the others; and domains are stratified by their coverage of the anxiety word list (with the most in the mental illness domains and the least in the financial domain).

We also examine the overall lexical diversity of each domain by calculating Yule's I measure (Yule, 1944). Figure 2 shows the lexical diversity of our data, both for all words in the vocabulary and for only words in LIWC's "negemo" word list. Yule's I measure reflects the repetitive-ness of the data (as opposed to the broader coverage measured by our LIWC analysis). We notice exceptionally low lexical diversity for the mental illness domains, which we believe is due to the structured, clinical language surrounding mental illnesses. For example, posters in these domains discuss topics such as symptoms, medical care, and diagnoses (Figure 1, Table 2). When we restrict our analysis to negative emotion words, this pattern persists only for anxiety; the PTSD domain has comparatively little lexical variety, but what it does have contributes to its variety of expression for negative emotions.

By label. We perform similar analyses on data labeled stressful or non-stressful by a majority of annotators. We confirm some common results in the mental health literature, including that stressful data uses more first-person pronouns (perhaps reflecting increased self-focus) and that non-stressful data uses more social words (perhaps reflecting a better social support network).

Label	1st-Person %	"Posemo" %	"Negemo" %	"Anxiety" Cover.	"Social" %
Stress	9.81%	1.77%	3.54%	78%	8.35%
Non-Stress	6.53%	2.78%	1.75%	67%	11.15%

Table 4: **LIWC Analysis by Label.** Results from our analysis using LIWC word lists, with the same definitions as in Table 3. First-person pronouns ("1st-Person") use the LIWC "I" word list.

Measure	Stress	Non-Stress
% Conjunctions	0.88%	0.74%
Tokens/Segment	100.80	93.39
Clauses/Sentence	4.86	4.33
F-K Grade	5.31	5.60
ARI	4.39	5.01

Table 5: **Complexity by Label.** Measures of syntactic complexity for stressful and non-stressful data.

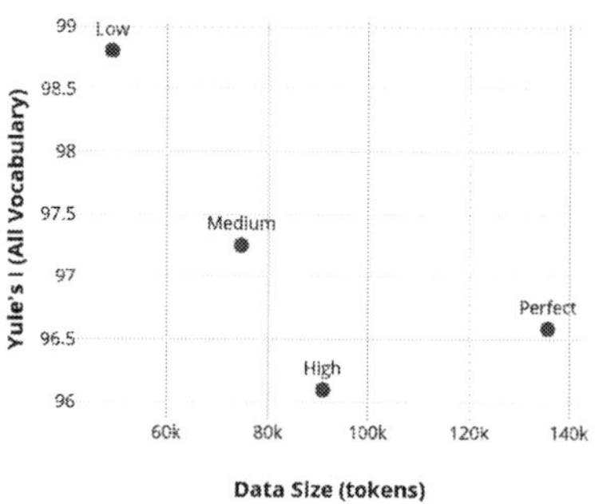

Figure 3: **Lexical Diversity by Agreement.** Yule's I measure (on the y-axis) is plotted against domain size (on the x-axis) for each level of annotator agreement. Perfect means all annotators agreed; High, 4/5 or more; Medium, 3/5 or more; and Low, everything else.

Additionally, we calculate measures of syntactic complexity, including the percentage of words that are conjunctions, average number of tokens per labeled segment, average number of clauses per sentence, Flesch-Kincaid Grade Level (Kincaid et al., 1975), and Automated Readability Index (Senter and Smith, 1967). These scores are comparable for all splits of our data; however, as shown in Table 5, we do see non-significant but persistent differences between stressful and non-stressful data, with stressful data being generally longer and more complex but also rated simpler by readability indices. These findings are intriguing and can be explored in future work.

By agreement. Finally, we examine the differences among annotator agreement levels. We find an inverse relationship between the lexical variety and the proportion of annotators who agree, as shown in Figure 3. While the amount of data and lexical variety seem to be related, Yule's I measure controls for length, so we believe that this trend reflects a difference in the type of data that encourages high or low agreement.

5 Methods

In order to train supervised models, we group the labeled segments by post and randomly select 10% of the posts ($\approx$ 10% of the labeled segments) to form a test set. This ensures that while there is a reasonable distribution of labels and domains in the train and test set, the two do not explicitly share any of the same content. This results in a total of 2,838 train data points (51.6% labeled stressful) and 715 test data points (52.4% labeled stressful). Because our data is relatively small, we train our traditional supervised models with 10-fold cross-validation; for our neural models, we break off a further random 10% of the training data for validation and average the predictions of 10 randomly-initialized trained models.

In addition to the words of the posts (both as bag-of-n-grams and distributed word embeddings), we include features in three categories:

Lexical features. Average, maximum, and minimum scores for pleasantness, activation, and imagery from the Dictionary of Affect in Language (DAL) (Whissel, 2009); the full suite of 93 LIWC features; and sentiment calculated using the `Pattern` sentiment library (Smedt and Daelemans, 2012).

Syntactic features. Part-of-speech unigrams and bigrams, the Flesch-Kincaid Grade Level, and the Automated Readability Index.

Social media features. The UTC timestamp of the post; the ratio of upvotes to downvotes on the post, where an upvote roughly corresponds to a reaction of "like" and a downvote to "dislike" (**upvote ratio**); the net score of the post (**karma**) (calculated by Reddit, $n_{\text{upvotes}} - n_{\text{downvotes}}$)[7]; and

---

[7] https://www.reddit.com/wiki/faq

the total number of comments in the entire thread under the post.

5.1 Supervised Models

We first experiment with a suite of non-neural models, including Support Vector Machines (SVMs), logistic regression, Naïve Bayes, Perceptron, and decision trees. We tune the parameters for these models using grid search and 10-fold cross-validation, and obtain results for different combinations of input and features.

For input representation, we experiment with bag-of-n-grams (for $n \in \{1..3\}$), Google News pre-trained Word2Vec embeddings (300-dimensional) (Mikolov et al., 2013), Word2Vec embeddings trained on our large unlabeled corpus (300-dimensional, to match), and BERT embeddings trained on our unlabeled corpus (768-dimensional, the top-level [CLS] embedding) (Devlin et al., 2019). We experiment with subsets of the above features, including separating the features by category (lexical, syntactic, social) and by magnitude of the Pearson correlation coefficient (r) with the training labels. Finally, we stratify the training data by annotator agreement, including separate experiments on only data for which all annotators agreed, data for which at least 4/5 annotators agreed, and so on.

We finally experiment with neural models, although our dataset is relatively small. We train both a two-layer bidirectional Gated Recurrent Neural Network (GRNN) (Cho et al., 2014) and Convolutional Neural Network (CNN) (as designed in Kim (2014)) with parallel filters of size 2 and 3, as these have been shown to be effective in the literature on emotion detection in text (e.g., Xu et al. (2018); Abdul-Mageed and Ungar (2017)). Because neural models require large amounts of data, we do not cull the data by annotator agreement for these experiments and use all the labeled data we have. We experiment with training embeddings with random initialization as well as initializing with our domain-specific Word2Vec embeddings, and we also concatenate the best feature set from our non-neural experiments onto the representations after the recurrent and convolutional/pooling layers respectively.

Finally, we apply BERT directly to our task, fine-tuning the pretrained BERT-base[8] on our clas-

---

[8] Using the implementation available at `https://github.com/huggingface/pytorch-transformers`

sification task for three epochs (as performed in Devlin et al. (2019) when applying BERT to any task). Our parameter settings for our various models are available in the appendix.

6 Results and Discussion

We present our results in Table 6. Our best model is a logistic regression classifier with Word2Vec embeddings trained on our unlabeled corpus, high-correlation features (≥ 0.4 absolute Pearson's r), and high-agreement data (at least 4/5 annotators agreed); this model achieves an F-score of 79.8 on our test set, a significant improvement over the majority baseline, the n-gram baseline, and the pre-trained embedding model, (all by the approximate randomization test, $p < 0.01$). The high-correlation features used by this model are LIWC's clout, tone, and "I" pronoun features, and we investigate the use of these features in the other model types. Particularly, we apply different architectures (GRNN and CNN) and different input representations (pretrained Word2Vec, domain-specific BERT).

We find that our logistic regression classifier described above achieves comparable performance to BERT-base (approximate randomization test, $p > 0.5$) with the added benefits of increased interpretability and less intensive training. Additionally, domain-specific word embeddings trained on our unlabeled corpus (Word2Vec, BERT) significantly outperform n-grams or pretrained embeddings, as expected, signaling the importance of domain knowledge in this problem.

We note that our basic deep learning models do not perform as well as our traditional supervised models or BERT, although they consistently, significantly outperform the majority baseline. We believe this is due to a serious lack of data; our labeled dataset is orders of magnitude smaller than neural models typically require to perform well. We expect that neural models can make good use of our large unlabeled dataset, which we plan to explore in future work. We believe that the superior performance of the pretrained BERT-base model (which uses no additional features) on our dataset supports this hypothesis as well.

In Table 7, we examine the impact of different feature sets and levels of annotator agreement on our logistic regressor with domain-specific Word2Vec embeddings and find consistent patterns supporting this model. First, we

Model	P	R	F
Majority baseline	0.5161	1.0000	0.6808
CNN + features*	0.6023	0.8455	0.7035
CNN*	0.5840	0.9322	0.7182
GRNN w/ attention + features*	0.6792	0.7859	0.7286
GRNN w/ attention*	0.7020	0.7724	0.7355
n-gram baseline*	0.7249	0.7642	0.7441
n-grams + features*	0.7474	0.7940	0.7700
LogReg w/ pretrained Word2Vec + features	0.7346	0.8103	0.7706
LogReg w/ fine-tuned BERT LM + features*	0.7704	0.8184	0.7937
LogReg w/ domain Word2Vec + features*	0.7433	0.8320	0.7980
BERT-base*	0.7518	0.8699	0.8065

Table 6: **Supervised Results.** Precision (P), recall (R), and F1-score (F) for our supervised models. Our best model achieves 79.80 F1-score on our test set, comparable to the state-of-the-art pretrained BERT-base model. In this table, "features" always refers to our best-performing feature set (≥ 0.4 absolute Pearson's r). Models marked with a * show a significant improvement over the majority baseline (approximate randomization test, $p < 0.01$).

		Agreement Threshold for Data			
		Any Majority	60% (3/5)	80% (4/5)	100% (5/5)
	None	75.40	76.31	78.48	77.69
	All	76.90	77.12	77.10	78.28
	LIWC	77.91	78.91	78.16	77.66
	DAL	75.58	77.06	78.05	77.06
	Lexical	76.42	77.92	77.54	77.88
Features	Syntactic	74.63	75.49	76.66	76.19
	Social	76.67	76.45	78.38	78.06
	$|r| \geq 0.4$	77.44	78.76	**79.80**	78.52
	$|r| \geq 0.3$	77.01	78.28	79.38	78.31
	$|r| \geq 0.2$	77.53	78.61	79.02	78.28
	$|r| \geq 0.1$	76.61	77.07	76.32	77.48

Table 7: **Feature Sets and Data Sets.** The results of our best classifier trained on different subsets of features and data. Features are grouped by type and by magnitude of their Pearson correlation with the train labels (no features had an absolute correlation greater than 0.5); data is separated by the proportion of annotators who agreed. Our best score (corresponding to our best non-neural model) is shown in bold.

see a tradeoff between data size and data quality, where lower-agreement data (which can be seen as lower-quality) results in worse performance, but the larger 80% agreement data consistently outperforms the smaller perfect agreement data. Additionally, LIWC features consistently perform well while syntactic features consistently do not, and we see a trend towards the quality of features over their quantity; those with the highest Pearson correlation with the train set (which all happen to be LIWC features) outperform sets with lower correlations, which in turn outperform the set of all features. This suggests that stress detection is a highly lexical problem, and in particular, resources developed with psychological applications in mind, like LIWC, are very helpful.

Finally, we perform an error analysis of the two best-performing models. Although the dataset is nearly balanced, both BERT-base and our best logistic regression model greatly overclassify stress, as shown in Table 8, and they broadly overlap but do differ in their predictions (disagreeing with one another on approximately 100 instances).

We note that the examples misclassified by both models are often, though not always, ones with low annotator agreement (with the average percent agreement for misclassified examples being 0.55 for BERT and 0.61 for logistic regression). Both models seem to have trouble with less explicit expressions of stress, framing negative ex-

LogReg		Gold	
		0	**1**
	0	241	105
	1	49	320

BERT		Gold	
		0	**1**
	0	240	106
	1	48	321

LogReg		BERT	
		0	**1**
	0	237	51
	1	53	374

Table 8: **Confusion Matrices.** Confusion matrices of our best models and the gold labels. 0 represents data labeled not stressed while 1 represents data labeled stressed.

Text	Gold Label	Agreement	Subreddit Name	Models Failed
Hello everyone, A very close friend of mine was in an accident a few years ago and deals with PTSD. He has horrific nightmares that wake him up and keep him in a state of fright. We live in separate provinces, so when he does have his dreams it is difficult to comfort him. Each time he calls, and I struggle with what to say on the phone.	Not Stress	60%	ptsd	Both
I asked the other day if they've set a date. He laughed in my face and said 'no' as if it were the most ridiculous thing he's ever heard. He comes home late, and showers immediately. Then, he showers every morning before he leaves. He doesn't talk to my mum and I, at all, and he's cagey and secretive about everything, to the point of hostility towards my sister.	Stress	60%	domesticviolence	BERT
If he's the textbook abuser, she is the textbook victim. She keeps giving him chances and accepting his apologies and living in this cycle of abuse. She thinks she's the one doing something wrong. I keep telling her that the only thing she is doing wrong is staying with this guy and thinking he will change. I tell her she does not deserve this treatment.	Not Stress	100%	domesticviolence	LogReg

Table 9: **Error Analysis Examples.** Examples of test samples our models failed to classify correctly. "BERT" refers to the state-of-the-art BERT-base model, while "LogReg" is our best logistic regressor described in section 6.

periences in a positive or retrospective way, and stories where another person aside from the poster is the focus; these types of errors are difficult to capture with the features we used (primarily lexical), and further work should be aware of them. We include some examples of these errors in Table 9, and further illustrative examples are available in the appendix.

7   Conclusion and Future Work

In this paper, we present a new dataset, Dreaddit, for stress classification in social media, and find the current baseline at 80% F-score on the binary stress classification problem. We believe this dataset has the potential to spur development of sophisticated, interpretable models of psychological stress. Analysis of our data and our models shows that stress detection is a highly lexical problem benefitting from domain knowledge, but we note there is still room for improvement, especially in incorporating the framing and intentions of the writer. We intend for our future work to use this dataset to contextualize stress and offer explanations using the content features of the text. Additional interesting problems applicable to this dataset include the development of effective distant labeling schemes, which is a significant first step to developing a quantitative model of stress.

Acknowledgements

We would like to thank Fei-Tzin Lee, Christopher Hidey, Diana Abagyan, and our anonymous reviewers for their insightful comments during the writing of this paper. This research was funded in part by a Presidential Fellowship from the Fu Foundation School of Engineering and Applied Science at Columbia University.

References

Muhammad Abdul-Mageed and Lyle H. Ungar. 2017. Emonet: Fine-grained emotion detection with gated recurrent neural networks. In *Proceedings of the 55th Annual Meeting of the Association for Computational Linguistics, ACL 2017, Vancouver, Canada, July 30 - August 4, Volume 1: Long Papers*, pages 718–728.

Fares Al-Shargie, Masashi Kiguchi, Nasreen Badruddin, Sarat C. Dass, and Ahmad Fadzil Mohammad Hani. 2016. Mental stress assessment using simultaneous measurement of eeg and fnirs. *Biomedical Optics Express*, 7(10):38823898.

Andrew P. Allen, Paul J. Kennedy, John F. Cryan, Timothy G. Dinan, and Gerard Clarke. 2014. Biological and psychological markers of stress in humans: Focus on the trier social stress test. *Neuroscience & Biobehavioral Reviews*, 38:94124.

Marilia A. Calcia, David R. Bonsall, Peter S. Bloomfield, Sudhakar Selvaraj, Tatiana Barichello, and Oliver D. Howes. 2016. Stress and neuroinflammation: a systematic review of the effects of stress on microglia and the implications for mental illness. *Psychopharmacology*, 233(9):1637–1650.

Kyunghyun Cho, Bart van Merrienboer, Çaglar Gülçehre, Dzmitry Bahdanau, Fethi Bougares, Holger Schwenk, and Yoshua Bengio. 2014. Learning phrase representations using RNN encoder-decoder for statistical machine translation. In *Proceedings of the 2014 Conference on Empirical Methods in Natural Language Processing, EMNLP 2014, October 25-29, 2014, Doha, Qatar, A meeting of SIGDAT, a Special Interest Group of the ACL*, pages 1724–1734.

Munmun De Choudhury, Michael Gamon, Scott Counts, and Eric Horvitz. 2013. Predicting depression via social media. In *Proceedings of the Seventh International AAAI Conference on Weblogs and Social Media*.

Arman Cohan, Bart Desmet, Andrew Yates, Luca Soldaini, Sean MacAvaney, and Nazli Goharian. 2018. SMHD: a large-scale resource for exploring online language usage for multiple mental health conditions. In *Proceedings of the 27th International Conference on Computational Linguistics*, pages 1485–1497, Santa Fe, New Mexico, USA. Association for Computational Linguistics.

Jacob Devlin, Ming-Wei Chang, Kenton Lee, and Kristina Toutanova. 2019. BERT: Pre-training of deep bidirectional transformers for language understanding. In *Proceedings of the 2019 Conference of the North American Chapter of the Association for Computational Linguistics: Human Language Technologies, Volume 1 (Long and Short Papers)*, pages 4171–4186, Minneapolis, Minnesota. Association for Computational Linguistics.

Joseph L Fleiss. 1971. Measuring nominal scale agreement among many raters. *Psychological Bulletin*, 76(5):378–382.

Sharath Chandra Guntuku, Anneke Buffone, Kokil Jaidka, Johannes C. Eichstaedt, and Lyle H. Ungar. 2018. Understanding and measuring psychological stress using social media. *CoRR*, abs/1811.07430.

Yoon Kim. 2014. Convolutional neural networks for sentence classification. *CoRR*, abs/1408.5882.

J. Peter Kincaid, Robert P. Fishburne, Richard L. Rogers, and Brad S. Chissom. 1975. Derivation of new readability formulas (automated readability index, fog count and flesch reading ease formula) for navy enlisted personnel.

Diederik P. Kingma and Jimmy Ba. 2015. Adam: A method for stochastic optimization. In *3rd International Conference on Learning Representations, ICLR 2015, San Diego, CA, USA, May 7-9, 2015, Conference Track Proceedings*.

J. Richard Landis and Gary G. Koch. 1977. The measurement of observer agreement for categorical data. *Biometrics*, 33(1):159174.

Huijie Lin, Jia Jia, Jiezhong Qiu, Yongfeng Zhang, Guangyao Shen, Lexing Xie, Jie Tang, Ling Feng, and Tat-Seng Chua. 2017. Detecting stress based on social interactions in social networks. *IEEE Transactions on Knowledge and Data Engineering*, 29(09):1820–1833.

Sonia J. Lupien, Bruce S. McEwen, Megan R. Gunnar, and Christine Heim. 2009. Effects of stress throughout the lifespan on the brain, behaviour and cognition. *Nature Reviews Neuroscience*, 10(6):434–445.

Tomas Mikolov, Ilya Sutskever, Kai Chen, Greg Corrado, and Jeffrey Dean. 2013. Distributed representations of words and phrases and their compositionality. In *Proceedings of the 26th International Conference on Neural Information Processing Systems - Volume 2*, NIPS'13, pages 3111–3119, USA. Curran Associates Inc.

F. Pedregosa, G. Varoquaux, A. Gramfort, V. Michel, B. Thirion, O. Grisel, M. Blondel, P. Prettenhofer, R. Weiss, V. Dubourg, J. Vanderplas, A. Passos, D. Cournapeau, M. Brucher, M. Perrot, and E. Duchesnay. 2011. Scikit-learn: Machine learning in Python. *Journal of Machine Learning Research*, 12:2825–2830.

James W Pennebaker, Ryan L Boyd, Kayla Jordan, and Kate Blackburn. 2015. The development and psychometric properties of liwc2015.

R.J. Senter and E.A. Smith. 1967. Automated readability index.

Tom De Smedt and Walter Daelemans. 2012. Pattern for python. *Journal of Machine Learning Research*, 13:2063–2067.

Richard Socher, Alex Perelygin, Jean Wu, Jason Chuang, Christopher D. Manning, Andrew Y. Ng, and Christopher Potts. 2013. Recursive deep models for semantic compositionality over a sentiment treebank. In *Proceedings of the 2013 Conference on Empirical Methods in Natural Language Processing, EMNLP 2013, 18-21 October 2013, Grand Hyatt Seattle, Seattle, Washington, USA, A meeting of SIGDAT, a Special Interest Group of the ACL*, pages 1631–1642.

Cynthia Whissel. 2009. Using the revised dictionary of affect in language to quantify the emotional undertones of samples of natural language. *Psychological Reports*, 105(2):509–521.

Genta Indra Winata, Onno Pepijn Kampman, and Pascale Fung. 2018. Attention-based LSTM for psychological stress detection from spoken language using distant supervision. *CoRR*, abs/1805.12307.

Peng Xu, Andrea Madotto, Chien-Sheng Wu, Ji Ho Park, and Pascale Fung. 2018. Emo2vec: Learning generalized emotion representation by multitask training. In *Proceedings of the 9th Workshop on Computational Approaches to Subjectivity, Sentiment and Social Media Analysis, WASSA@EMNLP 2018, Brussels, Belgium, October 31, 2018*, pages 292–298.

George Udny Yule. 1944. *The statistical study of literary vocabulary*. Cambridge Univ. Pr.

Xin Zuo, Tian Li, and Pascale Fung. 2012. A multilingual natural stress emotion database. In *Proceedings of the Eighth International Conference on Language Resources and Evaluation (LREC-2012)*, pages 1174–1178, Istanbul, Turkey. European Language Resources Association (ELRA).

Towards Understanding of Medical Randomized Controlled Trials by Conclusion Generation

Alexander Te-Wei Shieh* Yung-Sung Chuang* Shang-Yu Su   Yun-Nung Chen

National Taiwan University, Taipei, Taiwan

{b05401009,b05901033,f05921117}@ntu.edu.tw y.v.chen@ieee.org

Abstract

Randomized controlled trials (RCTs) represent the paramount evidence of clinical medicine. Using machines to interpret the massive amount of RCTs has the potential of aiding clinical decision-making. We propose a RCT conclusion generation task from the PubMed 200k RCT sentence classification dataset to examine the effectiveness of sequence-to-sequence models on understanding RCTs. We first build a pointer-generator baseline model for conclusion generation. Then we fine-tune the state-of-the-art GPT-2 language model, which is pre-trained with general domain data, for this new medical domain task. Both automatic and human evaluation show that our GPT-2 fine-tuned models achieve improved quality and correctness in the generated conclusions compared to the baseline pointer-generator model. Further inspection points out the limitations of this current approach and future directions to explore*.

1 Introduction

Randomized controlled trials (RCTs) are the most rigorous method to assess the effectiveness of treatments, such as surgical procedures and drugs, in clinical medicine (Sibbald and Roland, 1998). A typical RCT often constitutes of two randomized groups of patients receiving either the "intervention" (new treatment) or "control" (conventional treatment). Then, a statistical analysis is done after the experiments to determine whether the intervention has a significant effect (i.e. actually making patients better or worse). The results from various RCTs contribute to the medical decisions made by physicians every day. However, analyzing these large amounts of data could be over-whelming for clinicians (Davidoff and Miglus, 2011). With the help of machine readers, we can alleviate the burden for providing correct information that contributes to critical clinical decisions.

In this work, we aim to evaluate the capabilities of deep learning models on understanding RCTs by generating the conclusions of RCT abstracts. We achieve this by transforming the PubMed 200k RCT abstract sentence classification dataset (Dernoncourt and Lee, 2017) into a RCT conclusion generation task. Generating a *correct* and *coherent* conclusion requires the model to 1) identify the objectives of the trial, 2) understand the result and 3) generate succinct yet comprehensible texts. Therefore, this task can be a preliminary goal toward a more thorough understanding of clinical medicine literature.

To tackle this task, we first build a pointer-generator model (See et al., 2017) as the baseline. This model is widely used in abstractive summarization, which is similar to our conclusion generation task. We then leverage the high quality text generation capability of the Open AI GPT-2 (Radford et al., 2019) language model by fine-tuning the general domain GPT-2 model into a medical domain conclusion generator.

Because the correctness of RCT understanding is essential for supporting clinical decisions and neural summarization models could inaccurately present facts from the source document, we incorporate human evaluation on the correctness and quality of the generated in addition to standard ROUGE score (Lin, 2004) for automated summarization scoring. Evaluation results show the fine-tuned GPT-2 models score higher for both correctness and quality. However, there is still quite a large room for improvement both on the diversity and accuracy of the generated conclusions, providing a guidance for future research directions.

---

*These authors contribute this paper equally.

*The code is available at: https://github.com/MiuLab/RCT-Gen

Proceedings of the 10th International Workshop on Health Text Mining and Information Analysis (LOUHI 2019), pages 108–117
Hong Kong, November 3, 2019. ©2019 Association for Computational Linguistics
https://doi.org/10.18653/v1/D19-62

2 Related Work

The paper focuses on generating RCT conclusions, which is related to natural language generation. We describe the related work below and emphasize the difference between the prior work and our work. In our proposed method, we exploit the state-of-the-art language model representations for understanding the complex medical literature, and related work is then briefly described below.

2.1 Medical Natural Language Generation

Several medical domain natural language generation tasks have been studied using machine learning models, including generating radiology reports from images (Jing et al., 2018; Vaswani et al., 2017) and summarizing clinical reports (Zhang et al., 2018; Pivovarov and Elhadad, 2015) or research literature (Cohan et al., 2018). Recently, Gulden et al. (2019) studied extractive summarization on RCT descriptions.

Abstractive summarization, in which the model directly generates summaries from the source document, is closely related to our conclusion generation task. Most neural approaches for abstractive summarization were based on sequence-to-sequence recurrent neural networks (RNNs) with attention mechanisms (Devlin et al., 2019). The pointer-generator network (See et al., 2017) combined a copy mechanism that directly copies words from the source document and a coverage mechanism to avoid repetition caused by the RNN-based decoder, achieving good performance by handling unseen information. Devlin et al. (2019) further combined intra-encoder and intra-decoder attention with policy learning by using ROUGE-L score as the reward and improved the performance in terms of the evaluation metric. Hsu et al. (2018) combined an extractive model that provided attention on the sentence level and the pointer-generator architecture, and Cohan et al. (2018) also worked on abstractive summarization of long documents, including medical papers from the PubMed database, based on the pointer-generator network.

However, our goal to generate conclusions is different from abstractive summarization in that summarization is to shorten the source document while preserving most of the important information, whereas our conclusion generation model gives one or two sentences describing the main outcome of the given trial. Given the superior performance of pointer-generation networks from the above related summarization work, this paper uses the pointer-generation model as baseline and focuses on RCT conclusion generation instead of abstractive summarization.

2.2 Contextualized Representations

Recent advances of contextualized representation models, such as ELMo (Peters et al., 2018), Open AI GPT (Radford et al., 2018) and BERT (Devlin et al., 2019) achieved remarkable results across different natural language understanding tasks, such as question answering, entailment classification and named entity recognition. At the core of these models was language modeling, with either forward prediction used in GPT, bidirectional prediction used in ELMo, or masked prediction used by BERT. Variants of BERT also improved the performance of bio-medical natural language understanding tasks (Xu et al., 2019; Pugaliya et al., 2019). Peng et al. (2019) further proposed a new benchmark to evaluate the performance of contextualized models in the bio-medical domain.

Particularly, the Open AI GPT-2 model (Radford et al., 2019) has demonstrated rudimentary zero-shot summarization capabilities with only language modeling training. Its forward prediction architecture made it suitable for autoregressive generation in a sequence-to-sequence task. Most benchmarks on contextualized representation were based on sequence classification tasks such as natural language inference and multiple choice question answering (Wang et al., 2018; Peng et al., 2019). Our work, on the other hand, focuses on exploring GPT-2's capability of generating goal-directed sentences in the medical domain. Note that to our knowledge, this paper is the first attempt that investigates GPT-2 towards the medical document understanding and interpretation.

3 Task Formulation

The PubMed 200k RCT dataset was originally constructed for sequential short text classification, with each sentence labeled as "background", "objective", "methods", "results" and "conclusions". We concatenated the "background", "objective" and "results" sections of each RCT paper abstract as the model input and the goal of the model is to generate the "conclusions". Table 1 illustrates

Source: (BACKGROUND) Varenicline is believed to work , in part , by reducing craving responses to smoking cues and by reducing general levels of craving ; however , these hypotheses have never been evaluated with craving assessed in the natural environments of treatment-seeking smokers . (OBJECTIVE) Ecological momentary assessment procedures were used to assess the impact of varenicline on cue-specific and general craving in treatment-seeking smokers prior to quitting . (RESULTS) During all phases , smoking cues elicited greater craving than neutral cues ; the magnitude of this effect declined after the first week . General craving declined across each phase of the study . Relative to the placebo condition , varenicline was associated with a greater decline in general craving over the drug manipulation phase . Varenicline did not significantly attenuate cue-specific craving during any phase of the study .
Target (True Negative): Smoking cues delivered in the natural environment elicited strong craving responses in treatment-seeking smokers , but cue-specific craving was not affected by varenicline administered prior to the quit attempt . These findings suggest that the clinical efficacy of varenicline is not mediated by changes in cue-specific craving during the pre-quit period of treatment-seeking smokers .
Pointer-generator baseline model with $n = 1$ **hint word** (N/A): smoking cues are associated with a greater craving in general , and may be associated with a greater decline in general craving and
Fine-tuned GPT-2 with $n = 0$ **hint word** (False Negative): Varenicline did not reduce general craving in treatment-seeking smokers prior to quitting.
Fine-tuned GPT-2 with $n = 1$ **hint word** (True Negative): Smoking cues are associated with greater general craving than neutral cues, and varenicline does not attenuate cue-specific craving.

Table 1: An example of the GPT-2 $n = 0$ model generating a false negative conclusion (Varenicline did reduce general craving), while the GPT-2 $n = 1$ model generated a better true negative one. The "(BACKGROUND)", "(OBJECTIVE)" and "(RESULTS)" tags denote the sentence classifications according to the original PubMed RCT dataset and are not included in the actual input of our conclusion generation task.

the formulated task, where the generated conclusion needs to contain *correct* information based on the experiments and should be *concise*. After pre-processing, the number of abstracts in the training set is 189,035 and there are 2,479 conclusions used for validation. The average source paragraph length is 170.1 words (6.0 sentences), and the average target conclusion length is 41.4 words (1.8 sentences) long.

4 Models

Language model pre-training has achieved a great success among language understanding tasks with different model architectures. Because training language models requires a large amount of text data, and it is relatively difficult to acquire a lot of RCT documents, this work focuses on first pre-training language models with the transformer architecture (Vaswani et al., 2017) and then adapts the model to support the medical domain by fine-tuning. The language model pre-training from general texts is described below.

4.1 Transformer Encoder in GPT-2

We first introduce the transformer encoder (Vaswani et al., 2017) used as the backbone of the

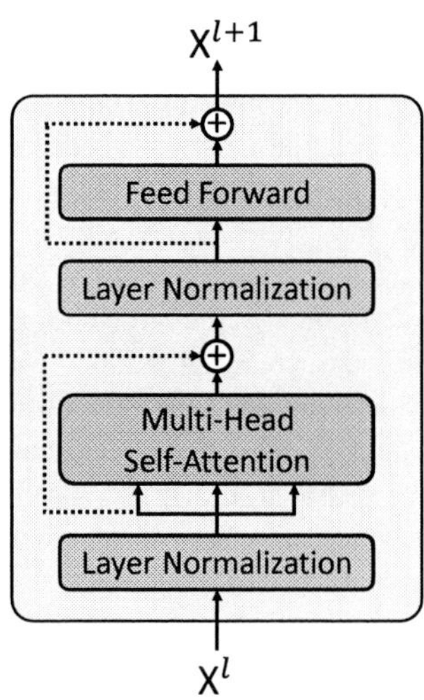

Figure 1: A modified transformer encoder block in the GPT-2 language model.

GPT-2 model. The transformer encoder is a stack of N transformer encoder blocks, where the l-th block takes a sequence of hidden representations $X^l = \{X_1^l, \cdots, X_n^l\}$ as the input and outputs an encoded sequence $X^{l+1} = \{X_1^{l+1}, \cdots, X_n^{l+1}\}$.

A transformer encoder block consists of a multi-head self-attention layer and a position-wise fully connected feed-forward layer. A residual connection (He et al., 2016) is employed around each of the two layers followed by layer normalization (Ba et al., 2016). In GPT-2, however, the layer normalization step is moved to the front of the multi-head self-attention layers and the feed-forward layers. An illustration of a GPT-2 transformer encoder block is presented in Figure 1. Each component is briefly described as follows.

Byte-Pair Encoding GPT-2 uses a special byte pair encoding (BPE) for input and output representations. It can cover essentially all Unicode strings, which is useful in processing the medical texts due to the significant out-of-vocabulary problems such as distinct nomenclature and jargon. This special BPE prevents merging characters from different categories and preserves word-level segmentation properties with a space exception.

Positional Encoding Because the transformer model relies on a self-attention mechanism with no recurrence, the model is unaware of the sequential order of inputs. To provide the model with positional information, positional encodings are applied to the input token embeddings

$$X_i^1 = \text{embed}_{\text{token}}[w_i] + \text{embed}_{\text{pos}}[i],$$

where w_i denotes the i-th input token, $\text{embed}_{\text{token}}$ and $\text{embed}_{\text{pos}}$ denote a learned token embedding matrix and a learned positional embedding matrix respectively.

Multi-Head Self-attention An attention function can be described as mapping a query to an output with a set of key-value pairs. The output is a weighted sum of values. We denote queries, keys and values as Q, K and V, respectively. Following the original implementation (Vaswani et al., 2017), a scaled dot-product attention is employed as the attention function. Hence, the output can be calculated as

$$\text{Attention}(Q, K, V) = \text{softmax}(\frac{QK^T}{\sqrt{d_k}})V,$$

where d_k denotes the dimension of key vectors.

The idea of multi-head attention is to compute multiple independent attention heads in parallel, and then concatenate the results and project again.

The multi-head self-attention in the l-th block can be calculated as

$$\text{MultiHead}(X^l) = \text{Concat}(\text{head}_1, \cdots, \text{head}_h)W^O,$$
$$\text{head}_i = \text{Attention}(X^l W_i^Q, X^l W_i^K, X^l W_i^V),$$

where X^l denotes the input sequence of the l-th block, h denotes the number of heads, W_i^Q, W_i^K, W_i^V and W^O are parameter matrices.

Position-Wise Feed-Forward Layer The second sublayer in a block is a position-wise feed-forward layer, which is applied to each position separately and independently. The output of this layer can be calculated as

$$\text{FFN}(x) = \max(0, x \cdot W_1 + b_1)W_2 + b_2,$$

where W_1 and W_2 are parameter matrices, b_1 and b_2 are parameter biases.

Residual Connection and Layer Normalization As shown in Figure 1, layer normalization is first applied on the input to the multi-head attention and feed-forward sublayers. The residual connection is then added around the two sublayers. The output of the l-th block can be calculated as

$$H^l = \text{MultiHead}(\text{LayerNorm}(X^l)) + X^l,$$
$$X^{l+1} = \text{FFN}(\text{LayerNorm}(H^l)) + H^l.$$

4.2 GPT-2 Pre-Training

The generative pre-training (GPT) via a language model objective is shown to be effective for learning representations that capture syntactic and semantic information without supervision (Peters et al., 2018; Radford et al., 2018; Devlin et al., 2019). The GPT model proposed by Radford et al. (2018) employs the transformer encoder with 12 encoder blocks. It is pre-trained on a large generic corpus that covers a wide range of topics. The training objective is to minimize the negative log-likelihood:

$$\mathcal{L} = \sum_{t=1}^{T} -\log P(w_t \mid w_{<t}, \theta),$$

where w_t denotes the t-th word in the sentence, $w_{<t}$ denotes all words prior to w_t, and θ are parameters of the transformer model.

To avoid seeing the future contexts, a masked self-attention is applied to the encoding process.

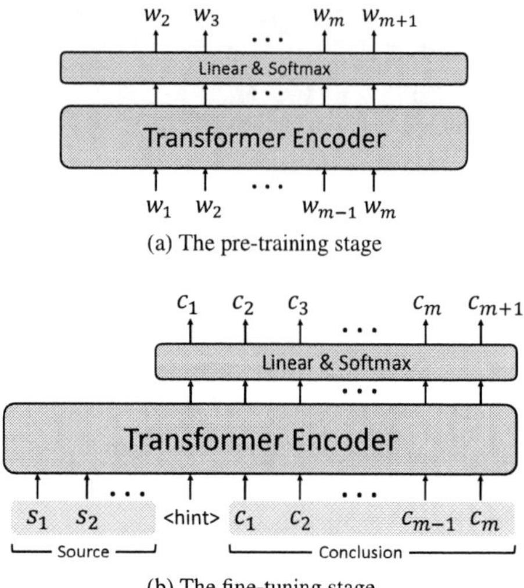

(a) The pre-training stage

(b) The fine-tuning stage.

Figure 2: Illustration of the two-stage method in the GPT model. The tag <hint> denotes where hint word tokens are introduced during fine-tuning.

In the masked self-attention, the attention function is modified into

$$\text{Attention}(Q, K, V) = \text{softmax}(\frac{QK^T}{\sqrt{d_k}} + M)V,$$

where M is a matrix representing masks. $M_{ij} = -\infty$ indicates that the j-th token has no contribution to the output of the i-th token, so it is essentially "*masked out*" when encoding the i-th token. Therefore, by setting $M_{ij} = -\infty$ for all $j > i$, we can calculate all outputs simultaneously without looking at future contexts. It was pre-trained on the WebText dataset consisting of 40 GB high quality text crawled from internet sources. We use the small version (12 layers and 117 M parameters) of the released GPT-2 models.

4.3 GPT-2 Fine-Tuning

After the model is pre-trained with a language model objective, it can be fine-tuned on downstream tasks with supervised data. In our task, we adapt the GPT-2 to the target domain by fine-tuning using RCT data. Figure 2 illustrates the learning procedure. By fine-tuning on the target data, the GPT-2 model may have the potential of understanding and generating medical texts.

In the fine-tuning stage, we modify the attention masking of the GPT-2 model so that source byte pairs are fully aware of the entire context of the source sentence, while the target byte pairs are

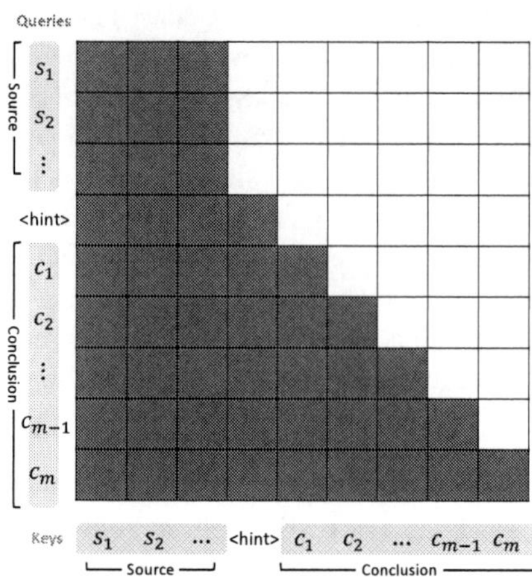

Figure 3: The attention mask used during fine-tuning. White cells denote $-\infty$ elements and grey cells denote 0 in the mask matrix.

aware of the entire source sentence plus the generated byte pairs that precede itself. That is, for context token pairs $(c_i, c_j) \in c_1, \cdots, c_m$, we set $M_{ij} = -\infty$ for all $j > i$, while for context and source token pairs (c_i, s_j), where $c_i \in c_1, \cdots, c_m$ and $s_j \in s_1, \cdots, s_n$, we set $M_{ij} = 0$. For all source token pairs $(s_i, s_j) \in s_1, \cdots, s_n$, we also set $M_{ij} = 0$. This setting is illustrated in Figure 3.

5 Experiments

Here we describe experimental details of the baseline pointer-generator model and the GPT-2 fine-tuned models.

5.1 Experimental Setup

The baseline model is a pointer-generator network (See et al., 2017) with both copy and coverage mechanisms, and is trained with a coverage loss. We adopt the implementation of Zhang et al. (2018). The vocabulary size is about 50,000, with uncased word embeddings pretrained from the PubMed RCT 200k training set and the abstracts from the PubMed dataset of long documents (Cohan et al., 2018). We concatenate $n \in \{0, 1\}$ hint words following the source sentences, where the hint words are first n words of the target conclusion. Our pointer-generator model uses beam search with beam size 5 to decode the final output conclusion.

In our GPT-2 models, we conduct conclusion generation using $n \in \{0, 1, 3, 5\}$ hint words. For $n = 0$, we append "In conclusion , " to the in-

System	ROUGE-1	ROUGE-2	ROUGE-L
PGNet $n = 0$	27.11	7.61	21.87
PGNet $n = 1$	26.88	8.19	22.63
GPT-2 $n = 0$	30.33	11.34	25.14
GPT-2 $n = 1$	**31.61**	**11.88**	**26.71**
GPT-2 $n = 3$	29.94	11.55	25.85
GPT-2 $n = 5$	29.79	11.29	25.94
GPT-2 res	24.24	6.79	20.71

Table 2: ROUGE scores of the PGNet baseline models and the GPT-2 fine-tuned models on the development set. The GPT-2 res were trained with the "results" section only. Addition of $n > 1$ hint words did not show further gains in ROUGE scores.

System	TP	TN	FP	FN	N/A	Acc.
PGnet $n = 1$	15	3	5	3	24	36%
GPT-2 $n = 0$	24	3	4	5	14	54%
GPT-2 $n = 1$	**26**	**6**	**5**	**3**	**10**	**64%**
Target	32	11	0	0	7	86%

Table 3: Human evaluation results for text understanding on the annotation questions of 50 randomly selected source documents. Note that some source documents which don't fit into the binary paradigm of positive or negative results are classified as N/A. TP: True Positive; TN: True Negative; FP: False Positive; FN: False Negative; N/A: Not Applicable.

put. Also, we perform data ablation study using only the "results" section as the model input. To address the memory constraint on our machines, we only train examples that are less than 500 byte pairs after encoding. Because GPT-2 model uses BPE for input and output, the generated conclusions are capitalized. Previous work showed that beam search did not help the generation quality of GPT-2 models (Holtzman et al., 2019), so we simply use greedy decoding to generate the conclusions . Our GPT-2 model is fine-tuned with teacher forcing, using the SGD optimizer with learning rate of 0.001, momentum of 0.9 and the decay factor of 0.0005. Our model is based on a PyTorch implementation of GPT-2 [†].

5.2 Automatic Evaluation

Table 2 shows the best validation ROUGE scores of baselines and our models. Note that the hint words are not considered in score calculation and the output of all models are lower-cased. The GPT-2 fine-tuned model significantly outperforms the pointer-generator (PGNet) baseline on all ROUGE scores, where the best performing model is GPT-2 with hint word $n = 1$, demonstrating the effectiveness of generating good conclusion in our model. However, more hint words do not bring additional gain in ROUGE scores, probably because more constraints hinder the GPT-2 model to explore potentially good conclusions. Moreover, the ablation result shows the significant drop in all ROUGE scores, indicating the importance of including the "background" and "objec-

tive" sections in the input for better content understanding.

5.3 Human Evaluation

We recruited 10 medical students with prior training in bio-statistics and epidemiology to annotate and rate the generated conclusions. Our questionnaire contains two types of questions: the annotation question and the rating question.

- **Annotation**: A annotation question contains a source document and four conclusions, namely the target conclusion written by human, the GPT-2 $n = 0$, the GPT-2 $n = 1$ and the PGnet generated conclusions. The raters are asked to classify each generated conclusions as either true positive, true negative, false positive, false negative or not applicable. We define true / false as whether the generated conclusion corresponds to the given document, and positive / negative as whether the intervention studied has a statistically significant effect, regardless of the effect being favourable or detrimental to the patients. This is to explicitly examine whether the generated conclusion is precise in terms of RCT content understanding.

- **Rating**: Rating questions use the same set-up except the question is a 5 point Likert scale for *correctness*, *quality* and *overall* impression. Each rater is given 5 annotation questions and 5 rating questions, with each source document randomly chosen from the validation set. This is to judge the generated conclusions both regarding to and regardless of the source document.

[†]https://github.com/huggingface/
pytorch-transformers

System	Correctness	Quality	Overall
PGnet $n = 1$	3.02	2.86	2.86
GPT-2 $n = 0$	**3.42**	3.66	**3.52**
GPT-2 $n = 1$	3.30	**3.94**	3.50
Target	3.92	4.08	3.98

Table 4: Human evaluation results for generation quality on the rating questions.

To mitigate bias, we do not inform which conclusion was generated or written by human, and the conclusions are lower-cased and randomly ordered in each question for fair comparison.

Table 3 presents the results from the annotation question, where the number of true positive and true negative generations from the GPT-2 fine-tuned models increase when compared to the PGNet baseline. It is clear that the proposed GPT-2 achieves better performance in terms of accuracy (the ratio of true samples). We also include the performance of human-written conclusions in the last row, which serves as the upper bound of this task. However, there is still a gap between human-written conclusions and the generated ones.

In the rating questions depicted in Table 4, the human written conclusions obtain a score nearly 4 out of 5 on all three dimensions. The GPT-2 models have comparable scores in overall impression, both scoring around 3.5 out of 5. The most significant improvement of the GPT-2 generated conclusions is the text quality, with the correctness improvement to a lesser extent. The correctness of GPT-2 $n = 1$ is slightly better than that of GPT-2 $n = 0$ in the annotation question, yet in the rating question, GPT-2 $n = 0$ has a higher averaged score. In sum, the human evaluation results demonstrate that our models significantly outperform the baseline pointer generator and tell that the proposed RCT conclusion generation task is not the same as typical summarization task, so deep text understanding is required for better performance.

6 Discussion

From the human evaluation results and our empirical inspection, we discover two major problems concerning the quality of the generated conclusions from GPT-2 models. First, there is some repetition in the generated conclusions, which impair the quality of generated text, though not as common in that of RNN-based models. We suggest additional weighted decoding or coverage mechanisms to avoid such problems. Second, the GPT-2 generated conclusions are significantly shorter than the targets. The average length generated by GPT-2 $n = 0$ and GPT-2 $n = 1$ are 19.4 and 21.0, while that of human written conclusions is 41.4. This could be caused by the limitation of greedy decoding, but the examples generated by PGnet, which applies beam search, only gives an average length of 22.6. This suggests investigation of additional measures to enrich and lengthen the generated conclusions in future work.

Another important issue is the correctness of the generation model. The GPT-2 models are able to identify simple patterns and generate conclusions with the correct relationship. However, errors occur when the study design becomes more complicated or the outcomes are complex. Therefore, future work should aim at enhancing the language understanding capabilities of generation models. Methods such as pre-training the GPT-2 models with medical domain literature or using external background knowledge might fill the missing gap in the correctness performance. This is very crucial regarding to our RCT understanding task and other tasks that require precise and reliable language generation.

Here we select 3 examples to better illustrate our evaluation methods and the discussed limitations of the current models. The example in Table 5 show two successful generations from the GPT-2 models. Table 6 shows a false positive example by the GPT-2 $n = 1$ model. On the other hand, a false negative example generated by the GPT-2 $n = 0$ can be seen in Table 7. The generated conclusions in Table 7 is also much shorter than the target conclusion written by human. Other factors that could cause this issue may include that the human authors mention information not included in the preceding source document, additional comments on the results and background knowledge and they paraphrase the same concept in different ways.

Given the above results, this paper opens a new research direction by formulating the RCT conclusion generation task and investigates the potential of language generation models towards better understanding of medical documents.

Source: Proton pump inhibitor (PPI) therapy is considered as the first choice for treatment of non-erosive reflux disease (NERD) . However , NERD is less sensitive to PPIs than erosive gastroesophageal reflux disease (GERD) and the differences between PPIs and H2 receptor antagonists are less evident in NERD than in erosive GERD . Since gastric acid secretion is lower in the Japanese population than in Western populations , we aimed to investigate whether PPI therapy is really necessary for NERD patients in Japan . Both roxatidine and omeprazole significantly improved the heartburn score at 4 and 8 weeks . The clinical response rates did not differ between roxatidine and omeprazole . Both roxatidine and omeprazole significantly relieved not only reflux but also abdominal pain and indigestion . The degrees of improvement did not differ between the two groups .
Target (True Positive)**:** Roxatidine relieved the symptoms of NERD patients with similar effectiveness to omeprazole . Therefore , roxatidine may be a good choice for NERD treatment .
GPT-2 $n = 0$ (True Positive)**:** Both roxatidine and omeprazole significantly improved the heartburn score at 4 and 8 weeks.
GPT-2 $n = 1$ (True Positive)**:** Roxatidine and omeprazole are effective in relieving symptoms of NERD in Japanese patients.

Table 5: An example of GPT-2 models generating true positive conclusions.

Source: To evaluate the efficacy of oxcarbazepine (OXC) in the treatment of agitation and aggression in patients with Alzheimer 's disease , vascular dementia or both . In total , 103 institutionalized patients at 35 sites were randomized to the trial . After 8 weeks , no statistically significant differences were found between the 2 groups for all outcomes . A trend was observed in favor of the OXC group in the reduction in the scores on the BARS (p = 0.07) .
Target (True Negative)**:** This study found no significant effect of OXC in treatment of agitation and aggression in patients with dementia .
GPT-2 $n = 0$ (True Negative)**:** OXC was not effective in the treatment of agitation and aggression in patients with Alzheimer's disease, vascular dementia or both.
GPT-2 $n = 1$ (False Positive)**:** This study suggests that OXC is effective in the treatment of agitation and aggression in patients with Alzheimer's disease.

Table 6: An example of GPT-2 $n = 0$ model generating a true negative conclusion, while the GPT-2 $n = 1$ model generated a false positive one.

Source: Atrial fibrillation (AF) is the most common complication following coronary artery bypass graft (CABG) . The mechanism of AF after CABG is not well defined ; however , it is suggested that endogenous adenosine , released in response to tissue hypoxia , may play a mechanistic role in these arrhythmias . The purpose of this study was to examine whether intravenous theophylline , via adenosine A1 receptor antagonism , would correct or modify new-onset early (<48 h post CABG) atrial fibrillation in patients post CABG , and thereby implicate endogenous adenosine as an inciting agent . Thirty patients comprised the study group . In Group A , 8 of the 15 patients (53 %) converted from AF to sinus rhythm within 15 min of theophylline administration . One patient who converted to sinus rhythm 20 min after theophylline administration was accepted as showing a negative response . In the placebo-treated group , no patient converted to sinus rhythm within 15 min (p <0.007 compared with Group A) .
Target (True Positive)**:** The mechanism of AF after CABG is not well defined and is probably multifactorial . However , this study demonstrated that antagonism of the adenosine A1 receptor can promptly convert many of these patients back to sinus rhythm , and thereby implicates endogenously released adenosine in a mechanistic role for inciting early (<48 h) post-CABG AF .
GPT-2 $n = 0$ (False Negative)**:** Intravenous theophylline, via adenosine A1 receptor antagonism, did not improve early AF in patients post CABG.
GPT-2 $n = 1$ (True Positive)**:** The results of this study suggest that intravenous theophylline, via adenosine A1 receptor antagonism, may correct or modify early AF in patients post CABG.

Table 7: An example of GPT-2 $n = 0$ model generating a false negative conclusion, while the GPT-2 $n = 1$ model generated a true positive one.

7   Conclusion and Future Work

This work introduces the RCT paper conclusion generation task as a stepping stone to the automatic understanding of clinical research literature. Our results show the general domain pre-trained GPT-2 language model can be fine-tuned to generate medical domain conclusions. The evaluation results show improvements regarding to both quality and correctness in conclusions generated by the fine-tuned GPT-2 model compared to the pointer-generator summarization model. Further study is needed to enhance the generation quality by reducing repetition errors and increasing the generation length, and to improve the correctness through better language understanding for practical clinical scenarios.

Beyond generating conclusions for RCT papers, generative language models in the medical domain with improved correctness and quality can open up new opportunities to tasks that require profound domain knowledge. For example, automatic generation of systemic review and meta-analysis articles.

Acknowledgements

We would like to thank reviewers for their insightful comments on the paper. This work was financially supported from the Young Scholar Fellowship Program by Ministry of Science and Technology (MOST) in Taiwan, under Grant 108-2636-E002-003 and 108-2634-F-002-015.

References

Jimmy Lei Ba, Jamie Ryan Kiros, and Geoffrey E Hinton. 2016. Layer normalization. *stat*, 1050:21.

Arman Cohan, Franck Dernoncourt, Doo Soon Kim, Trung Bui, Seokhwan Kim, Walter Chang, and Nazli Goharian. 2018. A discourse-aware attention model for abstractive summarization of long documents. In *Proceedings of the 2018 Conference of the North American Chapter of the Association for Computational Linguistics: Human Language Technologies, Volume 2 (Short Papers)*, pages 615–621, New Orleans, Louisiana. Association for Computational Linguistics.

Frank Davidoff and Jennifer Miglus. 2011. Delivering Clinical Evidence Where It's Needed: Building an Information System Worthy of the Profession. *JAMA*, 305(18):1906–1907.

Franck Dernoncourt and Ji Young Lee. 2017. PubMed 200k RCT: a dataset for sequential sentence classification in medical abstracts. In *Proceedings of the Eighth International Joint Conference on Natural Language Processing (Volume 2: Short Papers)*, pages 308–313, Taipei, Taiwan. Asian Federation of Natural Language Processing.

Jacob Devlin, Ming-Wei Chang, Kenton Lee, and Kristina Toutanova. 2019. BERT: Pre-training of deep bidirectional transformers for language understanding. In *Proceedings of the 2019 Conference of the North American Chapter of the Association for Computational Linguistics: Human Language Technologies, Volume 1 (Long and Short Papers)*, pages 4171–4186.

Christian Gulden, Melanie Kirchner, Christina Schttler, Marc Hinderer, Marvin Kampf, Hans-Ulrich Prokosch, and Dennis Toddenroth. 2019. Extractive summarization of clinical trial descriptions. *International Journal of Medical Informatics*, 129:114 – 121.

Kaiming He, Xiangyu Zhang, Shaoqing Ren, and Jian Sun. 2016. Deep residual learning for image recognition. In *Proceedings of the IEEE conference on computer vision and pattern recognition*, pages 770–778.

Ari Holtzman, Jan Buys, Maxwell Forbes, and Yejin Choi. 2019. The curious case of neural text degeneration. *CoRR*, abs/1904.09751.

Wan-Ting Hsu, Chieh-Kai Lin, Ming-Ying Lee, Kerui Min, Jing Tang, and Min Sun. 2018. A unified model for extractive and abstractive summarization using inconsistency loss. In *Proceedings of the 56th Annual Meeting of the Association for Computational Linguistics (Volume 1: Long Papers)*, pages 132–141, Melbourne, Australia. Association for Computational Linguistics.

Baoyu Jing, Pengtao Xie, and Eric Xing. 2018. On the automatic generation of medical imaging reports. In *Proceedings of the 56th Annual Meeting of the Association for Computational Linguistics (Volume 1: Long Papers)*, pages 2577–2586, Melbourne, Australia. Association for Computational Linguistics.

Chin-Yew Lin. 2004. ROUGE: A package for automatic evaluation of summaries. In *Text Summarization Branches Out*, pages 74–81, Barcelona, Spain. Association for Computational Linguistics.

Yifan Peng, Shankai Yan, and Zhiyong Lu. 2019. Transfer learning in biomedical natural language processing: An evaluation of BERT and elmo on ten benchmarking datasets. *CoRR*, abs/1906.05474.

Matthew E. Peters, Mark Neumann, Mohit Iyyer, Matt Gardner, Christopher Clark, Kenton Lee, and Luke Zettlemoyer. 2018. Deep contextualized word representations. In *Proc. of NAACL*.

Rimma Pivovarov and Nomie Elhadad. 2015. Automated methods for the summarization of electronic health records. *Journal of the American Medical Informatics Association*, 22(5):938–947.

Hemant Pugaliya, Karan Saxena, Shefali Garg, Sheetal Shalini, Prashant Gupta, Eric Nyberg, and Teruko Mitamura. 2019. Pentagon at mediqa 2019: Multitask learning for filtering and re-ranking answers using language inference and question entailment.

Alec Radford, Karthik Narasimhan, Tim Salimans, and Ilya Sutskever. 2018. Improving language understanding by generative pre-training.

Alec Radford, Jeff Wu, Rewon Child, David Luan, Dario Amodei, and Ilya Sutskever. 2019. Language models are unsupervised multitask learners.

Abigail See, Peter J. Liu, and Christopher D. Manning. 2017. Get to the point: Summarization with pointer-generator networks. In *Proceedings of the 55th Annual Meeting of the Association for Computational Linguistics (Volume 1: Long Papers)*, pages 1073–1083, Vancouver, Canada. Association for Computational Linguistics.

Bonnie Sibbald and Martin Roland. 1998. Understanding controlled trials: Why are randomised controlled trials important? *BMJ*, 316(7126):201.

Ashish Vaswani, Noam Shazeer, Niki Parmar, Jakob Uszkoreit, Llion Jones, Aidan N Gomez, Łukasz Kaiser, and Illia Polosukhin. 2017. Attention is all you need. In *Proceedings of the 31st International Conference on Neural Information Processing Systems*, pages 6000–6010. Curran Associates Inc.

Alex Wang, Amanpreet Singh, Julian Michael, Felix Hill, Omer Levy, and Samuel R. Bowman. 2018. GLUE: A multi-task benchmark and analysis platform for natural language understanding. *CoRR*, abs/1804.07461.

Yichong Xu, Xiaodong Liu, Chunyuan Li, Hoifung Poon, and Jianfeng Gao. 2019. Doubletransfer at MEDIQA 2019: Multi-source transfer learning for natural language understanding in the medical domain. *CoRR*, abs/1906.04382.

Yuhao Zhang, Daisy Yi Ding, Tianpei Qian, Christopher D. Manning, and Curtis P. Langlotz. 2018. Learning to summarize radiology findings. In *Proceedings of the Ninth International Workshop on Health Text Mining and Information Analysis*, pages 204–213, Brussels, Belgium. Association for Computational Linguistics.

Building a De-identification System for Real Swedish Clinical Text Using Pseudonymised Clinical Text

Hanna Berg
Department of Computer
and Systems Sciences
Stockholm University
Kista, Sweden
`hanna.berg@dsv.su.se`

Taridzo Chomutare
Norwegian Centre for E-health Research
University Hospital of North Norway
Tromsø, Norway
`Taridzo.Chomutare@ehealthresearch.no`

Hercules Dalianis[*]
Department of Computer and Systems Sciences
Stockholm University
Kista, Sweden
`hercules@dsv.su.se`

Abstract

This article presents experiments with pseudonymised Swedish clinical text used as training data to de-identify real clinical text with the future aim to transfer non-sensitive training data to other hospitals.

Conditional Random Fields (CFR) and Long Short-Term Memory (LSTM) machine learning algorithms were used to train de-identification models. The two models were trained on pseudonymised data and evaluated on real data. For benchmarking, models were also trained on real data, and evaluated on real data as well as trained on pseudonymised data and evaluated on pseudonymised data.

CRF showed better performance for some PHI information like *Date Part, First Name* and *Last Name*; consistent with some reports in the literature. In contrast, poor performances on *Location* and *Health Care Unit* information were noted, partially due to the constrained vocabulary in the pseudonymised training data.

It is concluded that it is possible to train transferable models based on pseudonymised Swedish clinical data, but even small narrative and distributional variation could negatively impact performance.

1 Introduction

Electronic health records (EHR) are produced in a steady stream, with the potential of advancing future medical care. Research on EHR data holds the potential to improve our understanding of patient care, care processes, and disease characteristics and progression. However, much of the data

---

Hercules Dalianis is also guest professor at the Norwegian Centre for E-health Research

is sensitive, containing Protected Health Information (PHI) such as personal names, addresses, phone numbers, that can identify particular individuals and thus cannot be available to the public for general scientific inquiry. Although good progress has been made in the general sub-field of de-identifying clinical text, the problem is still not fully resolved (Meystre et al., 2010; Yogarajan et al., 2018).

This study examines the use of pseudonymised health records as training data for de-identification tasks. Several ethical and scientific issues arise regarding the balance between maintaining patient confidentiality and the need for wider application of trained models. How will a de-identification system be constructed and used in a cross hospital setting without risking the privacy of patients? Is it possible to obscuring the training data by pseudonymising it and then use it for the training of a machine learning system?

De-identification and pseudonymisation are two related concepts. In this paper de-identification is used as a more general term to describe the process of finding personal health information to be able to conceal identifying information. A pseudonymised text is a text where the personal health information has been identified either manually or automatically and then replaced with realistic surrogates.

The research question in this study is whether it is possible to use de-identified and pseudonymised clinical text in Swedish as training data for de-identifying real clinical text, and hence make it possible to transfer the system cross hospital.

We highlight whether learning from the exist-

Proceedings of the 10th International Workshop on Health Text Mining and Information Analysis (LOUHI 2019), pages 118–125
Hong Kong, November 3, 2019. ©2019 Association for Computational Linguistics
https://doi.org/10.18653/v1/D19-62

ing, non-sensitive, pseudonymised Swedish clinical text can be useful in a new and different context; considering the normal variations in the distribution and nature of PHI information, and potential effects of scrubbing (Berman, 2003), that is, removing and modifying PHIs that was carried out to patient records during the de-identification process.

2 Previous research

The identification of PHI is a type of named entity recognition task where sensitive named entities specifically are identified. The first study with CRF-based de-identification for Swedish was on the gold standard Stockholm EPR PHI Corpus. The distribution of PHIs is shown in Table 1. In this instance, manual annotation with expert consensus was used to create the gold standard (Dalianis and Velupillai, 2010).

De-identification tasks based on the CRF machine learning algorithm has been carried out on this data set previously with precision scores ranging between 85% and 95%, recalls ranging between 71% and 87% and F1-scores between 0.76 and 0.91 (Dalianis and Velupillai, 2010; Berg and Dalianis, 2019).

One approach previously used for concealing the training set's sensitive data was carried out by (Dalianis and Boström, 2012), using the Stockholm EPR PHI Corpus. In the study, the textual part of the data were used to create 14 different features and part of speech tags. The textual part was then removed, and only the features and part of speech tags were used for training a Random Forest model. Fairly high precision of 89.1 % was obtained, but with a recall of 54.3 % and F1-score of 64.8.

In contrast to using only the sensitive EHR data for training, McMurry et al. (2013) integrated both publicly available scientific, medical publications and private sensitive clinical notes to develop a de-identification system. While considering the term frequencies and part of speech tags between the two data sources, they used both rule lists and decision trees for their system. This was an interesting approach since it raised the prospect of using non-sensitive data in building useful de-identification models. However, it is not clear whether medical journals have significant advantages over any other public text, like news corpora, for detecting PHI. A study similar to Mc-

Murry et al. (2013), by Berg and Dalianis (2019), showed few benefits of combining non-medical public text and sensitive clinical notes to build a de-identification system for medical records.

More recently, deep learning approaches using recurrent neural networks seem to yield significant improvements over traditional rules-based methods or statistical machine learning (Dernoncourt et al., 2017). Still, recent studies indicate that combining several approaches will yield the best results. For instance, the best system in a recent de-identification shared task was a combination of bidirectional LSTM, CRF and a rule-based subsystem (Liu et al., 2017).

Significant domain variation, such as a different language, is an important factor that was not considered in the discussed shared task. Domain differences were cited as the reason for poor performance on psychiatric notes de-identification (Stubbs et al., 2017), compared with the previous de-identification task on general clinical narratives (Stubbs et al., 2015).

Within the same language and similar clinical settings, the change of domain is likely not substantial. While in future research it may be worth considering domain adaption techniques to work towards a system meant to be used between hospitals, they were not considered in this study, beyond the use of non-sensitive dictionaries for names and location.

3 Data and methods

In this study, machine learning approaches are used since the best de-identification systems appear to be machine learning-based (Kushida et al., 2012). While rule-based methods such as using dictionaries and pattern-matching were previously more prevalent than machine learning methods for solving text-based de-identification problems (Meystre et al., 2010), today it is more typical to have both approaches used, since rule-based methods still yield better results for some PHI information (Neamatullah et al., 2008b). Dictionaries and patterns were therefore used as features within one of the models.

3.1 Data

Two different data sets for de-identification were used: Stockholm EPR PHI Psuedo Corpus (*Pseudo*) as well as the Stockholm EPR PHI Cor-

Class	Annotated	Retrieved	Relevant	Exact matches			Partial matches		
				Precision	Recall	F-score	Precision	Recall	F-score
Age	56	45	37	0.822222	0.660714	0.732673	0.904762	0.778061	0.836642
Date_Part	710	654	617	0.943425	0.869014	0.904692	0.946196	0.871730	0.907438
Full_Date	500	426	342	0.802817	0.684000	0.738661	**0.931665**	**0.802106**	**0.862045**
First_Name	923	749	713	0.951936	0.772481	0.852871	0.954606	0.773772	0.854729
Last_Name	928	816	777	**0.952206**	**0.837284**	**0.891055**	0.961653	0.845484	0.899835
Health_Care_Unit	1021	689	559	0.811321	0.547502	0.653801	0.921497	0.608116	0.732705
Location	148	73	54	0.739726	0.364865	0.488688	0.778539	0.379129	0.509933
Phone_Number	135	86	80	0.930233	0.592593	0.723982	0.954195	0.613105	0.746535
Total	4421	3538	3179	0.898530	0.719068	0.798844	0.941190	0.751441	0.835680

Additional file 5 (Table S5) - Results of the manual Consensus Gold standard using ten-fold cross-evaluation

Table 1: Results from (Dalianis and Velupillai, 2010) using the Stanford CRF.

pus (*Real*)[1].

The Stockholm EPR PHI Pseudo Corpus was produced from the Stockholm EPR PHI Corpus by automatically pseudonymising all PHIs. This process is described by Dalianis (2019). The Stockholm EPR PHI Corpus is described by Dalianis and Velupillai (2010). An example is shown in Figure 1 from the Stockholm EPR PHI Pseudo Corpus.

The Stockholm EPR PHI Corpus and the Stockholm EPR PHI Pseudo Corpus are both parts of Swedish Health Record Research Bank (HEALTH BANK). HEALTH BANK encompasses structured and unstructured data from 512 clinical units from Karolinska University Hospital collected from 2006 to 2014 (Dalianis et al., 2015).

The number of entities and types of entities in both the Stockholm EPR PHI Psuedo Corpus and the Stockholm EPR PHI Corpus is shown in Table 2. From Table 2, it can be observed that the distribution of PHI instances between the two data sets is somewhat similar, but there is a significant difference when it comes to unique instances between the two data sets. In total, the *Real* data set contains proportionally more unique instances than the *Pseudo* data set. The entities in the *Real* data set also tend to have more tokens.

3.2 Methods

Using the de-identified and pseudonymised data set, two models were trained based on two machine learning algorithms; CRF and the deep learning algorithm LSTM. The two algorithms were chosen since both have been shown to produce state of the art performance, and applying the two on Swedish clinical data sets makes for an informative comparison.

PHI classes	Pseudo	Unique	Real	Unique
First Name	885	24 %	938	79 %
Last Name	911	15 %	957	86 %
Age	51	80 %	64	97 %
Phone Number	310	78 %	327	92 %
Location	159	94 %	229	84 %
Full Date	726	25 %	770	89 %
Date Part	1897	6 %	2079	72 %
Health Care Unit	1278	13 %	2277	73 %
Total PHI instances	6217	20 %	7647	78 %

Table 2: The distribution of PHI instances between the the Stockholm EPR PHI Psuedo Corpus, *'Pseudo'*, and the Stockholm EPR PHI Corpus, *'Real'* based on the number of tokens. A PHI entity can cover from one token (one-word expression) to several tokens (multi-word-expression), for example "Karolinska" and "R54, Karolinska, Solna" respectively. The proportion of unique instances, *'Unique'*, is shown as a percentage.

The two models were evaluated on both the real data set that is annotated for PHI, but not pseudonymised, *'Pseudo-Real'*, as well as on the pseudonymised data set, *'Pseudo-Pseudo'*. For additional comparison basis models trained on the real data set were evaluated on test sets from the same data set, *'Real-Real'*.

3.2.1 CRF

In this study, the CRF algorithm implemented in CRFSuite (Okazaki, 2007) is used with the sklearn-crfsuite wrapper[2] and the LSTM architecture described by Lample et al. (2016), based on an open-source implementation with Tensorflow[3] is used.

The linear-chain Conditional Random Fields model, implemented with sklearn-CRFSuite[4],

---

[1]This research has been approved by the Regional Ethical Review Board in Stockholm (2014/1607-32).

[2]sklearn-crfsuite, `https://sklearn-crfsuite.readthedocs.io`

[3]Sequence tagging, `https://github.com/guillaumegenthial/sequence_tagging`

[4]Linear-chain CRF, `https://`

Discharge letter <u>Huddinge hospital</u>
Resp. specialist/chief physician <u>Caroline Berg</u>
Journal author <u>Marianne Lindgren</u>
Discharge Date <u>20120325</u>
Care episode <u>20120311-20120318</u>
Main diagnosis acc to ICD-10 DV073
Medical history <u>52-year-old</u> woman, familiar at the clinic. Goes to <u>Karin</u> Lundgren and to the pain clinic.

Epikris <u>Huddinge sjukhus</u>
Ansv specialist-/överläkare <u>Caroline Berg</u>
Journalförare <u>Marianne Lindgren</u>
Utskriftsdatum <u>20120325</u>
Vårdtid <u>20120311-20120318</u>
Huvuddiagnos enl. ICD-10 DV073
Anamnes <u>52-årig</u> kvinna, välkänd på kliniken. Går hos <u>Karin</u> Lundgren samt på smärtmottagningen.

Figure 1: Example of a pseudonymised record. The original Swedish pseudonymised record is to the right and the translated version is to the left. The underlined words are the surrogates, where real data has been replaced with pseudonyms.

uses lexical, orthographic, syntactic and dictionary features. The CRF is based on trial-and-error experiments with feature sets described by Berg and Dalianis (2019), and uses the same features except for section features.

3.2.2 LSTM

The long short-term memory (LSTM) needs word embeddings as features for the training. Word2vec[5] was used to produce word embeddings using shallow neural networks, based on two corpora; a clinical corpus and medical journals. For the training using real clinical data, word embeddings were produced using a clinical corpus of 200 million tokens that produced 300,824 vectors with a dimension of 300.

For the training with pseudo clinical data, word embeddings were produced using Läkartidningen corpus (The Swedish scientific medical journals from 1996 to 2005) containing 21 million tokens that produced 118,662 vectors with a dimension of 300. The reason for using Läkartidningen is that the corpus does not contain sensitive data and hence is also more easily usable for transferable cross hospital training.

4 Results

The results of the experimental work are summarised in Figure 2. As can be observed in the figure, the CRF algorithm seems to generally outperform the LSTM algorithm on all metrics; precision, recall and F1 measure.

This result is not consistent with repeated reports in the literature, where deep learning approaches such as LSTM have been shown to out-perform most other methods, including CRF. Since deep learning approaches normally require very large amounts of data, one explanation for this result could be that the word embeddings used in this study did not contain sufficient context variations required for more robust performance or an insufficient training set of annotated data.

The ability to identify date part and age entities are similar when training on pseudonymised data and real data for the CRF. In contrast, *Location*, *Health Care Unit* and *Full Date* were negatively affected when using pseudonymised training data regardless of using a CRF or LSTM model.

4.1 CRF - Results

Experimental results of the CRF algorithm are shown in Table 3. Not presented in the table is the combination of training on real data and evaluation of pseudo data (*Real-Pseudo*), but the results of this combination gave a precision of 86.37 and recall of 77.80% and an F1-score of 81.86.

4.2 LSTM - Results

The experimental results of the LSTM algorithm are shown in Table 4 and again, not presented in the table is the combination of training on real data and evaluation of pseudo data (Real-Pseudo). The result of this combination is a precision of 65.83% and recall of 74.79% and F1-score of 70.03.

5 Analysis

The training set used in this study has a substantially constrained vocabulary compared to the evaluation set, which may partially explain the overall performance achieved when evaluating on real data (Pseudo-Real). The pseudo (training)

---

sklearn-crfsuite.readthedocs.io/en/latest/

[5]word2vec, https://github.com/tmikolov/word2vec

CRF	Real-Real			Pseudo-Pseudo			Pseudo-Real		
	P %	R %	F_1-score	P %	R %	F_1-score	P %	R %	F_1-score
First Name	95.94	92.42	94.15	98.52	98.08	98.30	97.14	72.39	82.96
Last Name	97.91	93.22	95.51	98.54	97.55	98.04	96.80	38.90	55.50
Age	97.06	68.75	80.49	100.00	68.09	81.01	97.50	81.25	88.64
Phone Number	94.69	82.95	88.43	92.37	80.15	85.83	83.48	74.42	78.69
Location	80.85	58.46	67.86	93.27	74.05	82.55	57.38	53.85	55.56
Full Date	95.68	95.48	95.58	91.02	86.32	88.61	47.56	21.97	30.06
Date Part	96.27	94.94	95.60	98.29	96.05	97.16	87.04	94.79	90.75
Health Care Unit	85.40	64.00	73.17	93.75	87.50	90.52	45.29	16.30	23.97
Overall	94.03	85.30	89.45	96.31	92.22	94.22	80.44	49.83	61.54

Table 3: Entity-based evaluation for CRF with ten fold cross-validation. A comparison is made for the different combinations of training on real data and evaluation on real (Real-Real) as well as pseudo data and on training on pseudo data and evaluation on pseudo (Pseudo-Pseudo) as well as real data (Pseudo-Real).

LSTM	Real-Real			Pseudo-Pseudo			Pseudo-Real		
	P %	R %	F_1-score	P %	R %	F_1-score	P %	R %	F_1-score
First Name	91.61	86.49	88.98	81.41	78.27	79.81	73.42	72.99	73.20
Last Name	96.40	87.02	91.47	89.29	91.88	90.57	84.70	75.00	79.55
Age	87.50	58.33	70.00	80.95	36.17	50.00	83.33	31.25	45.45
Phone Number	33.53	82.22	47.64	64.83	71.21	67.87	30.87	71.32	43.09
Location	20.47	46.02	28.34	60.71	17.35	26.98	27.40	10.77	15.47
Full Date	77.67	74.06	75.82	67.76	72.20	69.91	52.58	23.00	32.00
Date Part	90.31	90.60	90.45	91.48	95.08	93.25	59.08	94.79	72.79
Health Care Unit	68.37	61.82	64.93	81.24	81.63	81.44	27.18	14.00	18.48
Overall	76.76	78.62	77.68	82.49	81.79	82.14	60.56	55.10	57.70

Table 4: Entity-based evaluation for LSTM with three fold cross-validation where 66% of the data were used for training and 33% for evaluation. 10% of the data was previously held out as a development set. A comparison is made for the different combinations of training on real data and evaluation on real as well as pseudo data and on training on pseudo data and evaluation on pseudo as well as real data.

version of the data has less PHI tokens and the entities are more often single tokens.

The Full Date structure *yyyyddmm - yyyyddmm* is commonly occurring in the pseudo data, and the dash between the dates, "-", is often incorrectly identified. For example, using the CRF algorithm on real-data training and pseudo-data testing (Real-Pseudo), of the 159 instances not identified as full dates tokens, sixty contain '-'. The pseudo data uses the structure *yyyyddmm* while the real data uses *yyddmm*, which leads to errors. For these kinds of errors on standard data formats such as dates, it is easy to see how rule-based approaches using regular expressions could significantly improve the overall performance of the system.

The weakest performance area was for location information. There is a large variety of locations in the pseudo-data. These are also fairly specific and unlikely to occur in the real data, for example, locations with very few inhabitants. These uncommon rural places have names similar to residential homes *(äldreboenden)*. There are multiple instances of the suffix *'gården'* (yard) in the location pseudo-PHI, whereas, in the real data, the same suffix is common for care units.

In the pseudo-data, the care units are more general than in the real data, often too general to be annotated in the real data set. Infirmaries are fairly common in the real data but non-existent in the pseudo data. This lack of variation in the pseudo is partially responsible for the drop in performance.

There are at least two ways to think about mitigating this poor performance. First, location and care unit could be combined as one entity type since they are conceptually very similar, and sometimes have interchangeable entity names. Secondly, using more detailed municipality street and location mapping databases as dictionaries could be considered.

6 Discussion

There is one similar study to ours but for English by Yeniterzi et al. (2010), where the authors train their de-identification system with all combinations of pseudonymised textual data (or what they call resynthesized records) and real data and their results are in line with ours. However, there are some studies on cross-domain adaptation. In cross-domain adaption there is, however,

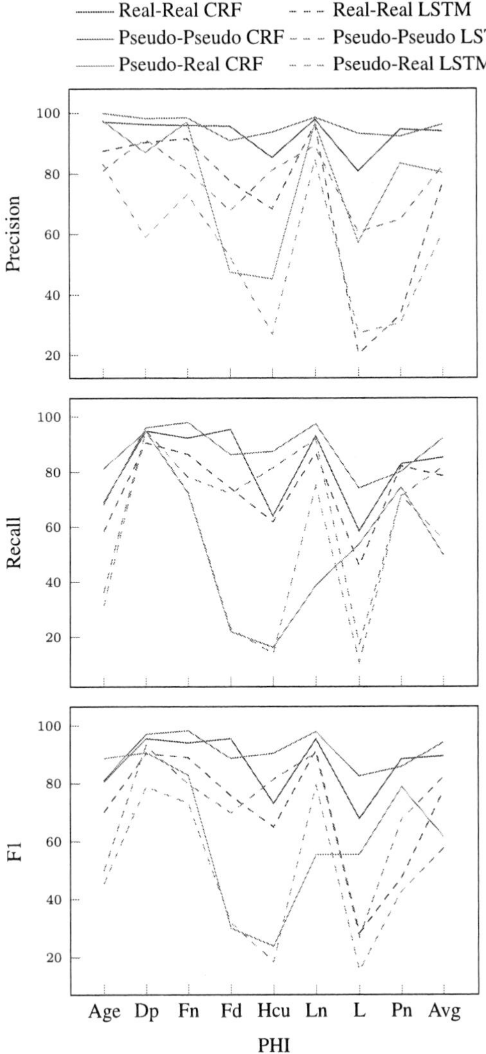

Figure 2: Line graphs visualising the results of both CRF and LSTM, and the outcomes of the evaluations. The x-axis have the PHI entities; Age, Date Part (Dp), First Name (Fn), Full Date (Fd), Health Care Unit (Hcu), Last Name (Ln), Location (L) and the average result (Avg).

a substantial domain change between the training and testing data, unlike in this study. Martinez et al. (2014) used models trained in one hospital on pathology reports in another hospital. Their system only required minor feature normalisation, and the reported results were comparable across the hospitals. Although this demonstrates feasibility, it is important to note that the pathology reports were from the same medical sub-speciality with only some narrative differences.

In this study, in addition to narrative differences between the training data and the target evaluation data, the number of care units and locations involved, as well as personal names, are widely varied. With large amounts of out of vocabulary variation, training on limited data will likely yield poor results. In practice, these data types exist in other non-sensitive sources such as city and rural location and street mapping data.

Except for location and care unit, evaluation on pseudo-data (Pseudo-Pseudo) produced better outcomes compared to performance on real-data (Pseudo-Real), which can be expected. What was a bit unexpected was the lower performance of the LSTM algorithm. The algorithm's results would potentially have been improved by larger vector data or more labelled data (Dernoncourt et al., 2017). While clinical notes have unique linguistic structures and grammatical peculiarities, non-clinical data sources could still provide important contextual information for constructing a useful vector space. Additional sources using non-sensitive data, such as public corpora in the general domain, hold a potential to improve performance on the de-identification task, therefore this line of inquiry will be followed up on in future work. In the same vein, factoring in part of speech tags from other sources of clinical data could be useful in this case. For instance, there are de-identification databases of clinical text, such as MIMIC (Neamatullah et al., 2008a; Goldberger et al., 2000), which could be used as additional information for training purposes, and using only the part of speech tags reduces security risks (Boström and Dalianis, 2012).

Current results are calculated as exact matches, and the partial match is not factored in, which may affect the result. As mentioned in the analysis the CRF algorithm rarely classifies the '-' in between dates as a part of the dates, and these are therefore not counted as matches despite the most identifying parts of the entity being identified.

To improve the general performance, a combination of both the LSTM and CRF algorithms could be performed instead of testing them independently. Combining high-performance algorithms and the use of ensemble methods seem to produce the best results as reported in the literature (Dernoncourt et al., 2017; Liu et al., 2017), and these techniques will be investigated in future work on the data sets.

7 Conclusions and future directions

The results of this study suggest that although it is possible to train models on pseudonymised data for use in different contexts, there is severe deterioration in performance for some PHI information. Even small narrative and distributional variation could negatively impact performance.

Transferring a system from one set of clinical text to a different set could result in the performance of the system deteriorating; in this study the Pseudo-Real case. This problem, what we call *The cross pseudo-real text adaptation problem*, is an issue that could happen due to the pseudonymisation/de-identification processes on the training data due to the narrative and distributional variation as well as other differences in the nature of the PHI between the training data and the target.

In the future, we will try to improve the pseudonymisation module described in Dalianis (2019) to produce a larger variation in the vocabulary as the lack of variation may affect the current result negatively.

We will also apply the learned models to other Nordic languages such as Norwegian clinical text and use the system as a pre-annotation system to assist the manual annotators in their work to create a Norwegian gold standard.

Acknowledgments

We are grateful to the DataLEASH project and Helse Nord for funding this research work.

References

Hanna Berg and Hercules Dalianis. 2019. Augmenting a De-identification System for Swedish Clinical Text Using Open Resources (and Deep learning). In *Proceedings of the Workshop on NLP and Pseudonymisation, NoDaLiDa, Turku, Finland September 30, 2019*.

Jules J Berman. 2003. Concept-match medical data scrubbing: how pathology text can be used in research. *Archives of pathology & laboratory medicine*, 127(6):680–686.

Henrik Boström and Hercules Dalianis. 2012. De-identifying health records by means of active learning. In *Proceedings of ICML 2012, The 29th International Conference on Machine Learning*, pages 1–3.

Hercules Dalianis. 2019. Pseudonymisation of Swedish Electronic Patient Records using a rule-based approach. In *Proceedings of the Workshop on NLP and Pseudonymisation, NoDaLiDa, Turku, Finland September 30, 2019*.

Hercules Dalianis and Henrik Boström. 2012. Releasing a Swedish clinical corpus after removing all words–de-identification experiments with conditional random fields and random forests. In *Proceedings of the Third Workshop on Building and Evaluating Resources for Biomedical Text Mining (BioTxtM 2012) held in conjunction with LREC*, pages 45–48.

Hercules Dalianis, Aron Henriksson, Maria Kvist, Sumithra Velupillai, and Rebecka Weegar. 2015. HEALTH BANK–A Workbench for Data Science Applications in Healthcare. In *Proceedings of the CAiSE-2015 Industry Track co-located with 27th Conference on Advanced Information Systems Engineering (CAiSE 2015), J. Krogstie, G. Juel-Skielse and V. Kabilan, (Eds.), Stockholm, Sweden, June 11, 2015, CEUR, Vol-1381*, pages 1–18.

Hercules Dalianis and Sumithra Velupillai. 2010. De-identifying Swedish clinical text - Refinement of a Gold Standard and Experiments with Conditional Random fields. *Journal of Biomedical Semantics*, 1:6.

Franck Dernoncourt, Ji Young Lee, Ozlem Uzuner, and Peter Szolovits. 2017. De-identification of patient notes with recurrent neural networks. *Journal of the American Medical Informatics Association*, 24(3):596–606.

Ary L. Goldberger, Luciani Alano Amaral, Leon Glass, Jeffrey M. Hausdorff, Plamen Ch. Ivanov, Roger G. Mark, Joseph E. Mietus, George B. Moody, Chih Kang Peng, and Harry Eugene Stanley. 2000. Physiobank, physiotoolkit, and physionet: components of a new research resource for complex physiologic signals. *Circulation*, 101 23:E215–20.

Clete A. Kushida, Deborah A. Nichols, Rik Jadrnicek, Ric Miller, James K. Walsh, and Kara Griffin. 2012. Strategies for de-identification and anonymization of electronic health record data for use in multicenter research studies. *Medical Care*, 50(7):S82–S101.

Guillaume Lample, Miguel Ballesteros, Sandeep Subramanian, Kazuya Kawakami, and Chris Dyer. 2016. Neural architectures for named entity recognition. *arXiv preprint arXiv:1603.01360*.

Zengjian Liu, Buzhou Tang, Xiaolong Wang, and Qingcai Chen. 2017. De-identification of clinical notes via recurrent neural network and conditional random field. *Journal of biomedical informatics*, 75:S34–S42.

David Martinez, Graham Pitson, Andrew MacKinlay, and Lawrence Cavedon. 2014. Cross-hospital portability of information extraction of cancer staging information. *Artificial intelligence in medicine*, 62(1):11–21.

Andrew J. McMurry, Britt Fitch, Guergana Savova, Isaac S. Kohane, and Ben Y. Reis. 2013. Improved de-identification of physician notes through integrative modeling of both public and private medical text. *BMC medical informatics and decision making*, 13:112–112. 24083569[pmid].

Stephane Meystre, Jeffrey Friedlin, Brett South, Shuying Shen, and Matthew Samore. 2010. Automatic de-identification of textual documents in the electronic health record: a review of recent research. *BMC Medical Research Methodology*, 10(1):70.

Ishna Neamatullah, Margaret M. Douglass, Li-wei H. Lehman, Andrew Reisner, Mauricio Villarroel, William J. Long, Peter Szolovits, George B. Moody, Roger G. Mark, and Gari D. Clifford. 2008a. Automated de-identification of free-text medical records. *BMC Medical Informatics and Decision Making*, 8(1):32.

Ishna Neamatullah, Margaret M Douglass, H Lehman Li-wei, Andrew Reisner, Mauricio Villarroel, William J Long, Peter Szolovits, George B. Moody, Roger G. Mark, and Gari D. Clifford. 2008b. Automated de-identification of free-text medical records. *BMC Medical Informatics and Decision Making*, 8(1):1.

Naoaki Okazaki. 2007. CRFsuite: a fast implementation of Conditional Random Fields. Accessed 2019-06-17.

Amber Stubbs, Michele Filannino, and Özlem Uzuner. 2017. De-identification of psychiatric intake records: Overview of 2016 cegs n-grid shared tasks track 1. *Journal of biomedical informatics*, 75:S4–S18.

Amber Stubbs, Christopher Kotfila, and Özlem Uzuner. 2015. Automated systems for the de-identification of longitudinal clinical narratives: Overview of 2014 i2b2/uthealth shared task track 1. *Journal of Biomedical Informatics*, 58:S11 – S19. Proceedings of the 2014 i2b2/UTHealth Shared-Tasks and Workshop on Challenges in Natural Language Processing for Clinical Data.

Reyyan Yeniterzi, John Aberdeen, Samuel Bayer, Ben Wellner, Lynette Hirschman, and Bradley Malin. 2010. Effects of personal identifier resynthesis on clinical text de-identification. *Journal of the American Medical Informatics Association*, 17(2):159–168.

Vithya Yogarajan, Michael Mayo, and Bernhard Pfahringer. 2018. A survey of automatic de-identification of longitudinal clinical narratives. *arXiv preprint arXiv:1810.06765*.

Automatic rubric-based content grading for clinical notes

Wen-wai Yim, Harold Chun, Teresa Hashiguchi,
Justin Yew, Bryan Lu
Augmedix, Inc.
{wenwai.yim,haroldchun,teresa,
justinyew,bryan}
@augmedix.com

Ashley Mills
University of Cincinnati
College of Medicine
mills2an@mail.uc.edu

Abstract

Clinical notes provide important documentation critical to medical care, as well as billing and legal needs. Too little information degrades quality of care; too much information impedes care. Training for clinical note documentation is highly variable, depending on institutions and programs. In this work, we introduce the problem of automatic evaluation of note creation through rubric-based content grading, which has the potential for accelerating and regularizing clinical note documentation training. To this end, we describe our corpus creation methods as well as provide simple feature-based and neural network baseline systems. We further provide tagset and scaling experiments to inform readers of plausible expected performances. Our baselines show promising results with content point accuracy and kappa values at 0.86 and 0.71 on the test set.

1 Introduction

Clinical notes, essential aspects of clinical care, document the principal findings of the visit, hospital stay or treatment episode, including complaints, symptoms, relevant medical history, tests performed, assessments and plans. During an encounter or soon after, notes are created based on subjective history, objective observations, as well as clinician assessment and care plans. Although this is a regular aspect of clinical care in all institutions, there is a large variability in the details taken. Clinical documentation training is often informal and institution-dependent, as systematic training of clinical documentation can be time-consuming and expensive. Training involves continued monitoring of note quality and completeness.

In this work, we present the problem of clinical note grading and provide several baseline systems for automatic content grading of a clinical note given a predefined rubric. To solve the problem, we built a simple feature-based system and a simple BERT-based system. We additionally provide training size experiments and tagset experiments for our baseline systems, as well as experiment with the relevance of using Unified Medical Language System (UMLS) and similarity features.

2 Background

Clinical notes serve as critical documentation for several purposes: medical, billing, and legal requirements. At the same time, a clinical note is written, updated, and consumed for the purpose of communication between clinicians for clinical care over a period of time. Too much irrelevant information can become an impediment to clinical care. Therefore, the identification of a base level of information is required to assess a note.

At our institution, clinical note documentation (or medical scribe) trainees are assessed via quizzes and exams in which trainees produce clinical notes based on mock patients visits (written to mimic outpatient encounters). Clinical note responses are graded against a grading rubric by training specialists, who have medical scribing experience as well as specialized training in determining note quality. The goal is to train scribes to produce a clinical note based on listening to the patient-clinician conversation during a clinical visit. Thus, the scribe expected to actively producing content, not just merely transcribe dictations from a clinician.

The purpose of a note rubric is to encapsulate the base requirements of a clinical note. Rubrics contain 40-60 rubrics items which reflect the information that needs to be captured during a medical encounter. A rubric item is typically written as a phrase and includes medically relevant attributes. For example, a rubric item discussing a

Proceedings of the 10th International Workshop on Health Text Mining and Information Analysis (LOUHI 2019), pages 126–135
Hong Kong, November 3, 2019. ©2019 Association for Computational Linguistics
https://doi.org/10.18653/v1/D19-62

symptom will typically require information about duration and severity. A rubric item discussing medication will often include dosage information. Each rubric is associated with a section of the note where it needs to be placed. For training purposes, standard note sections include: History of Present Illness (HPI), a detailed subjective record of the patient's current complaints; Review of Systems (ROS), the patient's subjective complaints grouped by organ system; Physical Exam (PE), the clinician's objective findings grouped by organ system; and Assessment and Plan (AP), the clinician's diagnosis and the next steps in treatment for each diagnosis. Figure 1 gives an example of a clinical note with these sections.

If the note contains text that satisfies a rubric item, then a content point for that rubric item is awarded. If the note contains an incorrect statement (e.g. the wrong medication dosage), then that rubric point is not awarded, regardless of a correct statement appearing elsewhere. If the note lacks the inclusion of a rubric point, then that rubric point is not awarded. At most, one content point can be awarded per rubric item. Examples of several rubric items with corresponding portions of a clinical note are shown in below.

Rubric item examples

- frequent_bm_3-4_times_per_day (HPI section), documents relevant symptom history – *"The patient complains of frequent bowel movements, 3-4 times daily."*
- pe_skin_intact_no_clubbing_cyanosis (PE section), documents physical exam performed during visit – *"Skin: Skin intact. No clubbing or cyanosis. Warm and dry."*
- plan_advise_brat_diet (AP section), documents that the provider recommended the BRAT diet to the patient – *"Recommended that the patient follow the BRAT diet for the next few days. "*

3 Related Work

Most community efforts in automatic grading have been in the context of automatic essay grading (AEG) and automatic short answer grading (ASAG), both of which harbor significant differences than our rubric-based content grading task.

AEG involves rating the quality of an essay in terms of coherence, diction, and grammar variations. Typically, an essay is given a score, e.g.

CHIEF COMPLAINT: Frequent urination

HISTORY OF PRESENT ILLNESS:
The patient is a 33 year old female who presents today complaining of frequent urination and bowel movements...

...

REVIEW OF SYSTEMS:
Constitutional: Negative for fevers, chills, sweats.

...

PHYSICAL EXAM:
General: Temperature normal. Well appearing and no acute distress

...

ASSESSMENT & PLAN:
1. Ordered urinalysis to rule out urinary tract infection
2. Put her on brat diet, counseled patient that BRAT diet is...

...

Figure 1: Abbreviated example of a clinical note. Clinical notes are typically organized by sections. The exact section types and ordering in real practice my vary by specialty and organization.

from 1-6. Applications include essay grading for standardized tests such as for the GMAT (Graduate Management Admission Test) or TOEFL (Test of English as Foreign Language) (Valenti et al., 2003). Key architects for these systems are often commercial organizations. Examples of commercial computer-assisted scoring (CAS) systems include E-rater and Intelligent Essay Assessor. Systems such as E-rater use a variety of linguistic features, including grammar, diction, and as well as including discourse level features (Attali and Burstein, 2006; Zesch et al., 2015). In another approach, the Intelligent Essay Assessor uses latent semantic analysis to abstract text to lower-rank dimension-cutting representations of documents. Scores are assigned based on similarity of new text to be graded to a corpus of previously graded text (Foltz et al., 1999). The release of the Kaggle dataset has made this type of data more available to the public (kaggle.com/c/asap aes). A key difference of AEG task from our grading task is that our efforts focus on specific content item grading and feedback, over a single holistic document level rating.

In ASAG, free text answers to a prompt are graded categorically or numerically. It is very closely related to paraphrasing, semantic similarity, and textual entailment. System reporting for this task has often been on isolated datasets with a wide range of topics and setups. Often, these systems require extensive feature engineering (Burrows et al., 2015). One example system is C-rater, produced by ETS, which grades based on the presence or absence of required content (Leacock and Chodorow, 2003; Sukkarieh and Black-more, 2009). Each required piece of content, similar to our rubric, in the text is marked as absent, present, negated, with a default of not_scored. Text goes through several processing steps, including spelling correction, concept recognition, pronoun resolution, and parsing. These features are then sent through a maximum entropy model for classification. Semantic similarity approaches apply a mixture of deep processing features, e.g. shortest path between concepts or concept similarities (Mohler and Mihalcea, 2009). In the SemEval 2013 Task 7 Challenge, the task involved classification of student answers to questions, given a reference answer, and student answers. Student answers are judged to be correct, partially_correct_incomplete, contradictory, irrelevant, or non_domain (Dzikovska et al., 2013). Although it has much in common to our rubric-based content grading setup, short answer grading has less document level issues to contend with. Moreover, specifically for our case, document-level scoring has some non-linearity with the individual classification of sub-document level text, e.g. finding one contradictory piece of evidence negates a finding of a positive piece of evidence.

The work of (Nehm et al., 2012), which attempts to award content points for specific items for college biology evolution essays, most closely resembles our task. In the task, students are awarded points based on whether they articulate key natural selection concepts, e.g. familiar plant/trait gain (mutation causing snail to produce poisonous toxin would increase fitness). The authors experimented with configuring two text analytic platforms for this task: SPSS Text Analysis 3.0 (SPSSTA) and Summarization Integrated Development Environment (SIDE). SPSSTA requires building hand-crafted vocabulary and attached rules. SIDE uses a bag-of-words representation run through a support vector machine algorithm for text classification. Key differences from our task are that rubric items are more numerous and specific; furthermore, our medium is a clinical note, which has documented linguistic and document style differences than normal essays; finally, our goal is not only to grade but to give automated in-text feedback.

In this work, we present our system for grading a clinical note given a grading rubric, which also gives subdocument level feedback. To our knowl-edge, there has been no previous attempt at clinical note automatic content grading.

4 Corpus Creation

We created a corpus by grading training notes by multiple different trainees quizzed on the same mock patient visit quiz, which included 40 rubric items. Annotation was carried out using using the ehost annotation tool (South et al., 2012). Annotators were asked to highlight sentences for if they were relevant to a rubric item. Furthermore, they were to mark whether a highlight had one of four attributes: correct, incorrect_contrary, incorrect_missingitem, and incorrect_section.

frequent_bm_3-4_times_per_day attribute examples

- correct –*"The patient reports having 3-4 bowel movements a day."* (Located in the HPI)
- incorrect_contrary –*"The patient has been having bowel movements every 30 minutes."* (Located in the HPI) Explanation: The frequency is much higher than what would be expected for 3-4 times per day. Thus the information content is considered to be inaccurate or contrary to what is given by the rubric.
- incorrect_missingitem – *"The patient reports having many bowel movements each day."* (Located in the HPI) Explanation: This statement does not give any inaccurate information but is missing a crucial element that is required to earn this rubric point, which is the frequency value is 3-4 times per day.
- incorrect_section – *"The patient reports having 3-4 bowel movements a day."* (Located in the AP) Explanation: This statement is correct, but is located in the wrong section of the note.

Parts of the note were marked for exclusion, such as short hand text (excluded because they are just notes taken by the scribe in training of the conversation) and the review of systems (ROS) part of the note (this was excluded because grading of that section was not enforced at its time of creation). Entire notes were marked for exclusion in cases where the note was intended for a different exam or in cases when the note contained all short hand and templated sections (e.g. only notes and section headers such as "Chief Complaint"). Discontinuous rubric item note contexts were linked together with a relation. The final corpus after included 338

	A1	A2	A3	A4	A5
A1	-	0.84/0.68	0.84/0.69	0.86/0.72	0.85/0.69
A2	-	-	0.86/0.72	0.87/0.73	0.87/0.73
A3	-	-	-	0.87/0.73	0.85/0.69
A4	-	-	-	-	0.88/0.76

Table 1: Content-point inter-annotator agreement (percent identity/kappa).

	A1	A2	A3	A4	A5
A1	-	0.84/0.68	0.84/0.69	0.86/0.72	0.85/0.69
A2	-	-	0.86/0.72	0.87/0.73	0.87/0.73
A3	-	-	-	0.87/0.74	0.85/0.69
A4	-	-	-	-	0.88/0.76

Table 2: Offset-level inter-annotator agreement (label/label-attribute).

notes, with a total of 10406 highlights[1]. A total of 244 rubric items required connecting discontinous highlights. The full corpus statistics are shown in Table 3.

Inter-annotator agreement was measured among 5 annotators for 9 files. We measured agreement at several levels. At a note level, we measured pairwise percent identity and Cohen's kappa (McHugh, 2012) by content points. At the text offset level, we measured precision, recall, and f1 (Hripcsak and Rothschild, 2005) for inexact overlap of highlights both at the label level (e.g. cc_frequent_urination) and at a label attribute level (e.g. cc_frequent_urination:correct). Since incorrect_section is not counted in content-based grading, for both inter-annotator agreement and for subsequent analysis, incorrect_section highlights were counted as correct unless overlapping with a incorrect_missingitem highlight, which would make it count as incorrect_missingitem. The agreement scores are shown in Table 1 and 2. Fleiss kappa (Fleiss, 1971) at the content point level was at 0.714. The rest of the corpus was divided among 5 annotators.

5 Methods

We performed classification for two types of systems: a feature-based and a simple BERT based neural network system for text classification. Since discontinuous text highlights accounted for less than 10% of items, we choose not to model this nuance. Both systems used the same pre-processing pipeline configurations shown in Figure 2.

---

[1]Includes highlights for excluding the note, as well as excluding short hand text and ROS sections

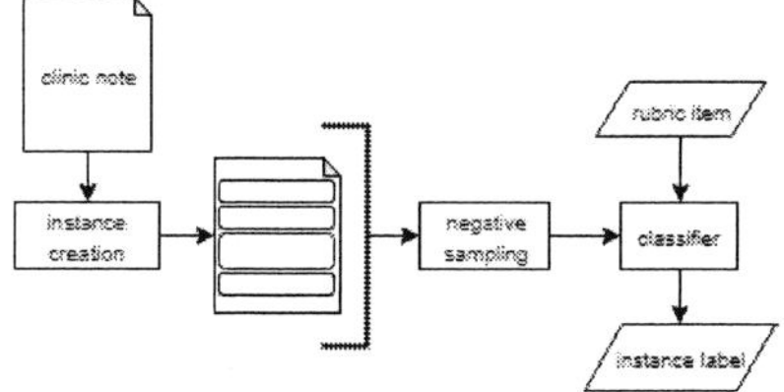

Figure 2: General pipeline

We split 338 files into 270 training, 68 test set. Tuning was performed on the training set in 5-fold cross validation.

5.1 Pipeline configurations

The text was preprocessed, where short-hand text and blank lines were removed. Sentences and words were tokenized using spaCy (spacy.io).

Instances were created by extracting minichunks (sentences or subsections in the case of the PE section) from the clinical note identified using regular expressions. We experimented with three tag-set configurations (tag_gran) with values (2lab, 3lab, 4lab), which represents {1 vs 0 points}, {correct, incorrect_contrary, vs missing} and {correct, incorrect_contrary, incorrect_missingitem, missing} per rubric item. Missing is the default value if no relevant highlight annotates the sentence.

For the minichunk level, we set a configuration negative sampling ratio (neg_samp) which specifies the factor of default class instances to non-default class instances. A similarity feature flag (simfeat_on) turns on similarity features in the feature-based system or switch the BERT-based system to one which includes matching features. For our feature based modeling, we used scikit-learn (Pedregosa et al., 2011), for our neural network pipelines we incorporated allennlp (Gardner et al.).

5.2 Feature based system

The feature based system includes an n-gram feature extraction, which then passed through a chi-squared statistical feature selection, before using a support vector machine implemented by scikit learn svc.

If the umls configuration is turned on, then Unified Medical Language System (UMLS) concept with its negation value, extracted using MetaMap, concept-grams are also added (Aronson and Lang, 2010). If turned on, similarity features for n-grams

rubric_item	total	correct	incorrect_missingitem	incorrect_contrary	incorrect_section
cc_frequent_urination	579	535	9	35	0
checks_blood_sugar_regularly_110s_before_meals	152	39	86	26	1
bp_fluctuates_150100_	365	239	91	34	1
denies_abdominal_pain_vomiting_constipation	270	98	169	3	0
denies_hematochezia	211	202	1	8	0
denies_recent_travel	257	245	5	4	3
duration_of_1_week	256	221	5	30	0
feels_loopy	125	110	8	6	1
feral_cat_in_the_house_occasionally	175	56	95	22	2
frequent_bm_3-4_times_per_day	343	249	50	44	0
frequent_urination_every_30_minutes	347	276	31	39	1
has_not_had_ua	190	177	7	6	0
healthy_diet	178	157	21	0	0
husband_was_sick_with_uti_symptoms	331	257	56	18	0
hx_of_htn	298	254	40	4	0
initially_thought_she_had_respiratory_infection	204	78	33	93	0
loose_stools_with_mucous	324	205	111	8	0
losartan_hctz_every_night_with_dinner	325	148	154	23	0
mild_dysuria	266	185	57	24	0
no_recent_antibiotics	279	263	10	5	1
pe_abdomen_hyperactive_bowel_sounds_at_llq_no_pain_with_palp	334	97	180	51	6
pe_cv_normal	315	300	10	4	1
pe_extremities_no_edema	297	282	7	2	6
pe_heent_normal_no_thyromegaly_masses_carotid_bruit	430	173	251	4	2
pe_resp_normal	331	324	6	1	0
pe_skin_intact_no_clubbing_cyanosis	276	84	173	6	13
plan_advise_brat_diet	312	249	42	15	6
plan_bp_goal__13080	155	123	11	18	3
plan_may_notice_leg_swelling_notify_if_unbearable	216	93	114	5	4
plan_prescribed_amlodipine_5mg	268	205	40	18	5
plan_recommend_30_mins_physical_activity_4-5_times_per_week	296	128	131	34	3
plan_reduce_stress_levels	119	100	13	1	5
plan_rtn_in_1_month_with_continued_bp_log	280	172	85	16	7
plan_ua_today_to_rule_out_infx	302	202	91	3	6
side_effects_of_difficulty_breathing_with_metoprolol	164	104	38	21	1
stress_work_related	223	215	7	1	0
takes_blood_pressure_every_morning	222	145	66	11	0
tried_yogurt_and_turmeric_no_improvement	176	66	106	4	0
was_seen_by_dr_reynolds_yesterday	154	44	96	13	1
weight_normal	61	52	5	3	1

Table 3: Label frequencies for full corpus

and umls concept grams (if the umls flag is on) are also added. We used jaccard similarity between 1-, 2-, and 3- grams. The full configurations include the following :

- top_n : top number of significant features to keep according to chi-squared statistic feature selection (integer)
- sec_feat : setting to determine how section information is encoded. If "embed" is set, then each feature with be concatenated with its section, e.g. "sect[hpi]=patient". If "sep" is turned on, "sect=hpi" is added as a feature.
- umls : whether or not to use umls features (binary)
- simfeat_on : whether or not to turn on similarity features (binary)
- text_off : whether or not to turn off the features not related to similarity
- umls_text_off : whether or not to turn off the umls features not related to similarity
- sent_win : window for which surrounding sentence features should be added, e.g. sent[-1]=the would be added as a feature from previous sentence unigram feature "the" if sent_win=1. (integer)

5.3 BERT based system

The neural network system made use of the previous instance creation pipeline; however in place of feature extraction, instances were transformed into BERT word vector representations. We used the output for CLS position of the embeddings to represent the whole sequence similar to that of the original paper (Devlin et al., 2018).

To mimic the feature-based system's case of simfeats_on, we include a switch to an architecture that also feeds in the CLS position output from a paired BERT classification setup. When simfeats_on is turned off, the architecture becomes that of a simple BERT sentence classification (bert). When text_off is turned on, then the architecture becomes that of a simple BERT sentence pair classification (bertpair). When simfeats_on is turned on and text_off is turned off, we have a system with both types off representations (bert+bertpair). A figure of the BERT classifier setup is show in Figure 3.

Because certain medical vocabulary may not be available with the general English trained cor-

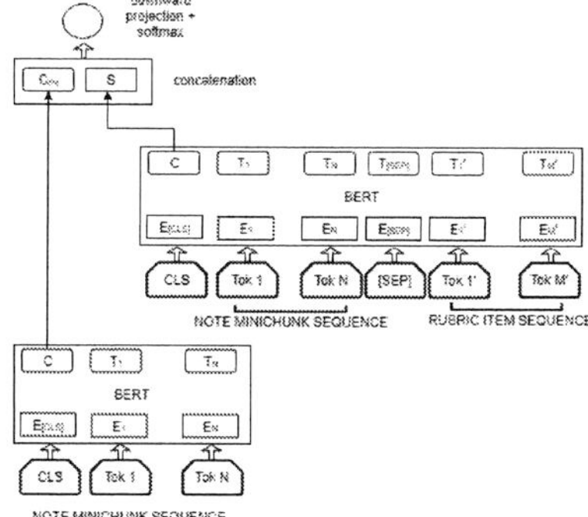

Figure 3: BERT with additional sentence pair classifier

pora, we used pre-trained BERT embedding and vocabulary from bio-bert, which is fine-tuned on pubmed data (Lee et al., 2019).

5.4 Simple baseline

For a further comparison, we have included a feature-based document-based baseline. This baseline largely follows the previously mentioned feature-based baseline though performed at the document level. Because it is performed at a document level, some attribution to a sub-document level unit becomes necessary. (Recall, we wish to be able to identify which part of the document is relevant to a rubric item as well as if we believe it is correct or otherwise.) To identify a corresponding offset labeling for this setting, we attributed a document level classification to a sentence which contained the maximum amount of important features. We defined features to be important according to their learned feature weight magnitude crossing a configurable threshold value. Thus, for a document classification, based on this logic we are able to assign a sentence related to that classification, for which we can use all our previously mentioned metrics for evaluation. We found a threshold of 10 to work well in our experiments.

5.5 Evaluation

Similar to inter-annotator agreement, we measured performance using several metrics. At the note level, we measured distance to target full document score by mean absolute error (MAE). We also measured content point accuracy and kappa to get a sense of the performance in point assignment.

At the offset level, we measured precision, recall, and f1 for rubric item label-attribute value.

For minichunk classification, offsets were translated according to the start and end of the minichunk in the document.

6 Results

Evaluations in cross-validation are reported as the average of all 5 folds. Consistent with this, graphical error bars in the learning curve figures represent the standard deviation across folds.

6.1 Experimental configurations

Feature-based parameter tuning. We started with tag_gran at 4lab, simfeats_on true, and top_n set at 1000, sent_win=0, and neg_samp=2. We then greedily searched for an optimal solution varying one parameter at a time to optimize precision, recall and f1 measure. Our final configurations were set to neg_samp=10, sect_feat set to "sep", top_n=4000, sent_win=0. We kept the same configurations for the other feature-based systems.

Neural network hyperparmeter tuning.
For the neural network models, we mainly experimented with negative sampling size, dropout, and context length. We found a neg_samp=100, dropout=0.5, epochs=2 and context_len=300 to work well.

6.2 Cross-validation and test results

Table 4 shows performances for the feature-based system using different tagsets in cross-validation. Unsurprisingly the most detailed label (4lab) at the more granualar level (minichunk) showed the best performance. Tables 4 and 5 shows the comparison of different system configurations in evaluated in cross validation. Table 6 shows results for the test set for different systems. In general, the feature-based system with the full feature-set out performed for the cross-validation and test experiments. Among the BERT systems, the simple BERT system did better than the other two configurations.

6.3 Scaling and tagset experiments

Figure 4 shows the effect of increasing training size for several metrics, under several select systems. Interestingly, 2lab and 3lab settings show similar behaviors across different metrics. For the document level baseline, tagset does not make any large difference across all three metrics. Different from other systems, 2lab and 3lab setting,

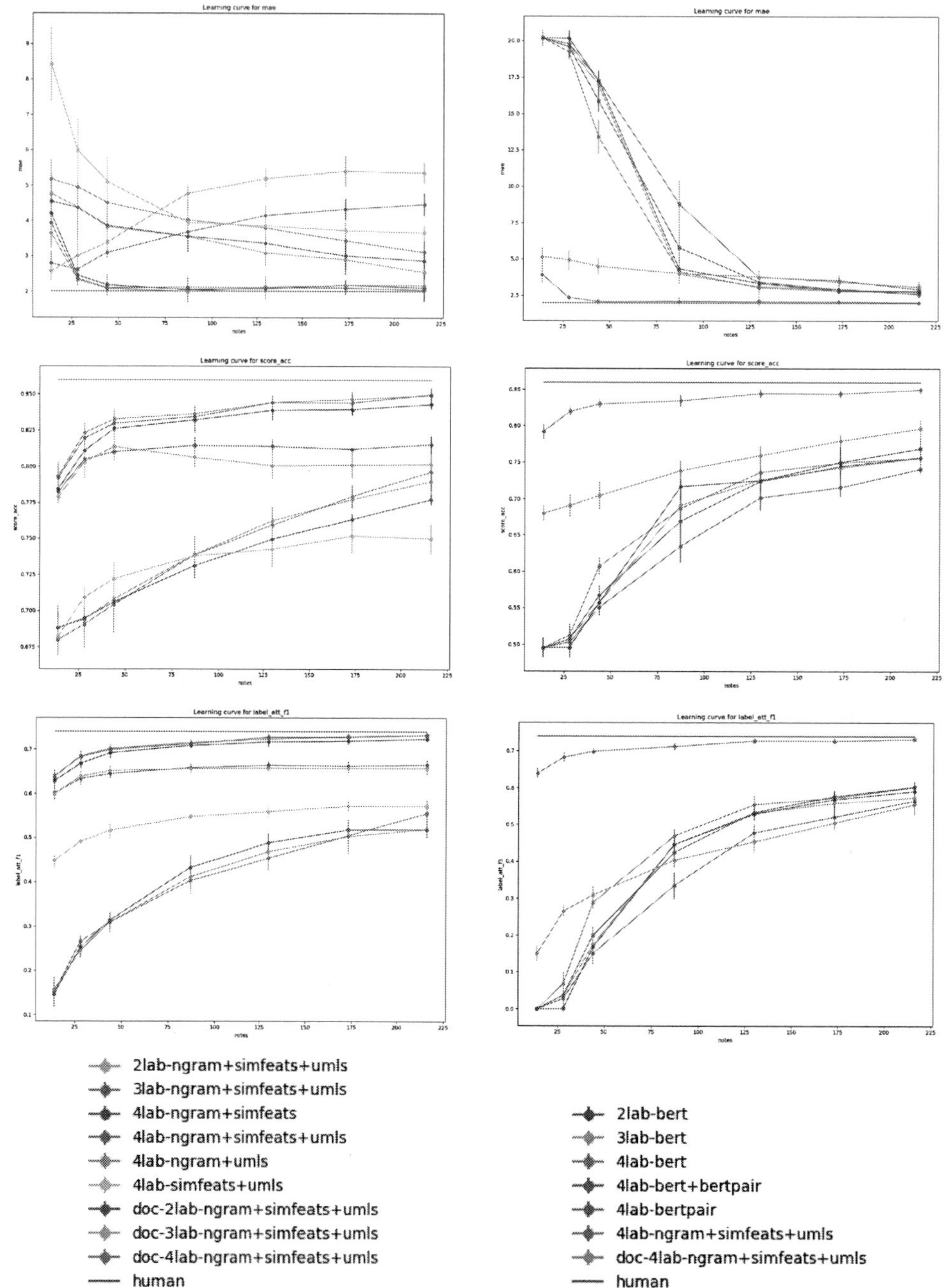

Figure 4: Learning curve experiments. 2lab, 3lab, 4lab demarks the tagset configuration. doc-* identifies the simple document classification baseline. The left column shows the performance of different n-gram configurations for 3 performance metrics. The right column shows BERT system performances for the 3 metrics along with two feature-based systems for comparison.

tagset	mae	acc	kappa	p	r	f
4lab	2.0	0.85	0.70	0.71	0.75	0.73
3lab	4.5	0.82	0.63	0.69	0.64	0.66
2lab	5.4	0.80	0.60	0.69	0.63	0.66
doc-4lab	3.1	0.80	0.59	0.58	0.53	0.55
doc-3lab	2.5	0.80	0.58	0.65	0.43	0.52
doc-2lab	2.9	0.78	0.55	0.67	0.42	0.52

Table 4: Feature-based system results for 5-fold cross-validation, varying tagsets.

system	mae	acc	kappa	p	r	f
ngram+simfeats+umls	2.0	0.85	0.70	0.71	0.75	0.73
ngram+umls	2.2	0.85	0.70	0.71	0.75	0.73
ngram+simfeats	2.1	0.84	0.69	0.70	0.74	0.72
simfeats+umls	3.7	0.75	0.50	0.63	0.53	0.57
bert+bert_pair	2.6	0.76	0.51	0.59	0.62	0.60
bert	2.8	0.76	0.51	0.60	0.61	0.60
bert_pair	2.9	0.74	0.48	0.57	0.56	0.56

Table 5: Results for 5-fold cross-validation, tag_gran=4lab.

system	mae	acc	kappa	p	r	f
ngram+simfeats+umls	2.5	0.85	0.70	0.68	0.78	0.72
ngram+umls	2.5	0.85	0.70	0.68	0.78	0.72
ngram+simfeats	2.3	0.86	0.71	0.68	0.79	0.73
simfeats+umls	3.8	0.77	0.54	0.63	0.55	0.59
bert+bert_pair	3.2	0.77	0.53	0.57	0.60	0.58
bert	2.8	0.79	0.57	0.59	0.63	0.61
bert_pair	3.1	0.76	0.51	0.58	0.60	0.59

Table 6: Detailed results for the test set, tag_gran=4lab.

6.4 UMLS and similarity features

The addition of similarity features did not provide a significant boost for the ngram feature-based system. Similarity features alone for the feature-based system underperformed at all training size levels compared to the ngram models. On the other hand, the addition of UMLS features increased performance across three metrics for all training size levels.

The BERT based system using only the simple BERT representation (without paired features), outperformed the other two settings across the three metrics at most training size levels in cross-validation. However, near higher levels of training data, BERT with BERT pair becomes comparable. The BERT pair system underperforms across all three metrics and at all training sizes.

6.5 Error Analysis

One challenging aspect of the classification task was the imbalanced categories across notes for different rubric items. Some labels were inherently less frequent, e.g. weight_normal had a total of 61 compared to cc_frequent_urination with 579 highlights. Indeed the performance amongst all rubric item scores was highly variable, with 13% f1 standard variation for the label-attribute measure. Moreover, the distribution of classes per label was also highly variable, as shown in Figure 3. For example, when no_recent_antibiotics or stress_work_related appears, in labeled data they are often correct. As a consequence, accurately predicting less populated classes becomes more difficult. For the best performing system, for example, there were instances where *Patients weight is not normal* was considered correct despite the rubric specifying the opposite. Similarly, *Patient denies feeling loopy* would be marked correct when the rubric says otherwise. When measuring at the label level for highlights instead, the performance on the test was higher by more than 10% f1, as shown in Table 7. This

when trained to maximize both precision and recall equally for label attributes, MAE will rise instead of lower such as for the other system setups. This makes sense, as both tagset settings miss crucial examples that exhibit confusing features. For example, on the 2lab setting, only positive examples are shown not those that have incorrect information or those that have missing information (i.e. partially correct information). Likewise, the 3lab setting does not have evidence for partially correct items. We found experimentally, when tuning for higher precision and lower recall, that MAE also tends to lower– suggesting that these two settings can be better maximized by tuning for MAE instead.

Like the document-based baseline, the BERT systems' performances showed that tagset did not make as big of a difference across all three metrics at different training size levels. This is possibly because there were not enough examples to even properly fine-tune for these two systems which require more training data. At higher levels of training sizes, tagsets may again come into effect. Though the BERT systems at lower training sizes start at lower performances, it quickly catches up to the document classification baseline for the MAE and f1 metrics, though never gets close to the 4lab feature-based baseline.

eval	tp	fp	fn	p	r	f
label-att	1470	700	424	0.68	0.78	0.72
label	1708	413	186	0.81	0.90	0.85

Table 7: Results for the test set, ngram+simfeats+umls tag_gran=4lab.

indicates that many errors are due to confusion between classification categories. Contradictions, labeled incorrect_contrary, for this reason was a large problem.

Manually studying errors in the test set for the best performing system, we found that rubric items frequently identified in the training, were broadly correctly classified. However, there were some rubric items that had more inconsistencies in how they were being tagged or graded. Some errors were partly due to human grading error. For example, checks_blood_sugar_regularly_110s_before_meals was a rubric item that scribes frequently missed when creating notes. Due to this, some sentences with just *"checks"* were sometimes labeled for checks_blood_sugar_regularly_110s_before_meals despite the fact that the sentences were about checking blood pressure. This leads to cases where synonymous phrases to *"blood sugar"*, *"blood glucose"*, did not get labeled as instances with *"blood glucose"* by the human graders.

7 Conclusions

In this paper we present the problem of clinical note grading and provide several baseline system evaluations at different levels of training data. We show experimentally that the choice of labeling has large effects upon the system performance. Furthermore, though neural network systems may relieve a lot of feature-engineering, this may not be plausible for smaller corpora.

Further improvements can be made by rubric-item specific pipeline specialization, as well as further augmentation of specific feature extraction modules, e.g. better negation handling. Deeper processing methods and features, including use of lemma-ization and dependency structures, would make features more generalizable. On the other hand, to maximize performance, feature-extraction can also be made more rubric-item specific, for example by hand-crafting features. For this work, we used a linear support vector machine for our classifier, but further experimentation with different classifiers for each rubric item would lead to higher performances. The BERT systems can be improved by increasing training size and adding more feed-forward layers.

Our proposed system can be used to expedite and formalize clinical note creation in a training setting. For example, instead of having a human grader view all training notes, a simple pass of the automated grading system can eliminate those that will fail with some confidence. For others, a human grader can correct the output of the system, which would speed the grading process then if the grader had to mark highlights alone. In this work, we focus on cases for which we have many examples of clinical notes generated for the same encounter with a fixed rubric. Future work will investigate grading for arbitrary notes and rubrics.

Acknowledgments

We would like to thank the Augmedix training specialist team as well as all those involved with creating the source dataset for this work.

Very special thanks to Kevin Holub and Sonny Siddhu for their efforts in initiating Augmedix's efforts in AI-assisted in-text grading for which is the motivation behind this project.

References

kaggle.com/c/asap aes. The hewlett foundation: Automated essay scoring | kaggle.

Alan R Aronson and Franois-Michel Lang. 2010. An overview of MetaMap: historical perspective and recent advances. 17(3):229–236.

Yigal Attali and Jill Burstein. 2006. Automated essay scoring with e-rater v.2 | the journal of technology, learning and assessment. 4(3).

Steven Burrows, Iryna Gurevych, and Benno Stein. 2015. The eras and trends of automatic short answer grading. 25(1):60–117.

Jacob Devlin, Ming-Wei Chang, Kenton Lee, and Kristina Toutanova. 2018. BERT: Pre-training of deep bidirectional transformers for language understanding. abs/1810.04805.

Myroslava Dzikovska, Rodney Nielsen, Chris Brew, Claudia Leacock, Danilo Giampiccolo, Luisa Bentivogli, Peter Clark, Ido Dagan, and Hoa Trang Dang. 2013. SemEval-2013 task 7: The joint student response analysis and 8th recognizing textual entailment challenge. In *Second Joint Conference on Lexical and Computational Semantics (*SEM),*

Volume 2: Proceedings of the Seventh International Workshop on Semantic Evaluation (SemEval 2013), pages 263–274, Atlanta, Georgia, USA. Association for Computational Linguistics.

JL Fleiss. 1971. Measuring nominal scale agreement among many raters. 76(5):378–382.

Peter W Foltz, Darrell Laham, and Thomas K Landauer. 1999. IMEJ article - the intelligent essay assessor: Applications to educational technology.

Matt Gardner, Joel Grus, Mark Neumann, Oyvind Tafjord, Pradeep Dasigi, Nelson F. Liu, Matthew E. Peters, Michael Schmitz, and Luke S. Zettlemoyer. AllenNLP: A deep semantic natural language processing platform. abs/1803.07640.

George Hripcsak and Adam S Rothschild. 2005. Agreement, the f-measure, and reliability in information retrieval. 12(3):296–298.

Claudia Leacock and Martin Chodorow. 2003. C-rater: Automated scoring of short-answer questions. 37(4):389–405.

Jinhyuk Lee, Wonjin Yoon, Sungdong Kim, Donghyeon Kim, Sunkyu Kim, Chan Ho So, and Jaewoo Kang. 2019. BioBERT: a pre-trained biomedical language representation model for biomedical text mining.

Mary L McHugh. 2012. Interrater reliability: the kappa statistic. 22(3):276–282.

Michael Mohler and Rada Mihalcea. 2009. Text-to-text semantic similarity for automatic short answer grading. In *EACL*.

Ross H. Nehm, Minsu Ha, and Elijah Mayfield. 2012. Transforming biology assessment with machine learning: Automated scoring of written evolutionary explanations. 21(1):183–196.

F. Pedregosa, G. Varoquaux, A. Gramfort, V. Michel, B. Thirion, O. Grisel, M. Blondel, P. Prettenhofer, R. Weiss, V. Dubourg, J. Vanderplas, A. Passos, D. Cournapeau, M. Brucher, M. Perrot, and E. Duchesnay. 2011. Scikit-learn: Machine learning in Python. *Journal of Machine Learning Research*, 12:2825–2830.

Brett R. South, Shuying Shen, Jianwei Leng, Tyler B. Forbush, Scott L. DuVall, and Wendy W. Chapman. 2012. A prototype tool set to support machine-assisted annotation. In *Proceedings of the 2012 Workshop on Biomedical Natural Language Processing*, BioNLP '12, pages 130–139, Stroudsburg, PA, USA. Association for Computational Linguistics.

Jana Z Sukkarieh and John Blackmore. 2009. c-rater: Automatic content scoring for short constructed responses.

Salvatore Valenti, Francesca Neri, and Alessandro Cucchiarelli. 2003. An overview of current research on automated essay grading. 2:319–330.

Torsten Zesch, Michael Wojatzki, and Dirk Scholten-Akoun. 2015. Task-independent features for automated essay grading. In *BEA@NAACL-HLT*.

Dilated LSTM with attention for Classification of Suicide Notes

Annika M Schoene George Lacey Alexander P Turner Nina Dethlefs
The University of Hull
Cottingham Road
Hull
HU6 7RX
amschoene@gmail.com

Abstract

In this paper we present a dilated LSTM with attention mechanism for document-level classification of suicide notes, last statements and depressed notes. We achieve an accuracy of 87.34% compared to competitive baselines of 80.35% (Logistic Model Tree) and 82.27% (Bi-directional LSTM with Attention). Furthermore, we provide an analysis of both the grammatical and thematic content of suicide notes, last statements and depressed notes. We find that the use of personal pronouns, cognitive processes and references to loved ones are most important. Finally, we show through visualisations of attention weights that the Dilated LSTM with attention is able to identify the same distinguishing features across documents as the linguistic analysis.

1 Introduction

Over recent years the use of social media platforms, such as blogging websites has become part of everyday life and there is increasing evidence emerging that social media can influence both suicide-related behaviour (Luxton et al., 2012) and other mental health conditions (Lin et al., 2016). Whilst there are efforts to tackle suicide and other mental health conditions online by social media platforms such as Facebook (Facebook, 2019), there are still concerns that there is not enough support and protection, especially for younger users (BBC, 2019). This has led to a notable increase in research of suicidal and depressed language usage (Coppersmith et al., 2015; Pestian et al., 2012) and subsequently triggered the development of new healthcare applications and methodologies that aid detection of concerning posts on social media platforms (Calvo et al., 2017). More recently, there has also been an increased use of deep learning techniques for such tasks (Schoene and Dethlefs, 2018), however there

is little evidence which features are most relevant for the accurate classification. Therefore we firstly analyse the most important linguistic features in suicide notes, depressed notes and last statements. Last Statements have been of interest to researchers in both the legal and mental health community, because an inmates last statement is written, similarly to a suicide note, closely before their death (Texas Department of Criminal Justices, 2019). However, the main difference remains that unlike in cases of suicide, inmates on death row have no choice left in regards to when, how and where they will die. Furthermore there has been extensive analysis conducted on the mental health of death row inmates where depression was one of the most common mental illnesses. Work in suicide note identification has also compared the different states of mind of depressed and suicidal people, because depression is often related to suicide (Mind, 2013). Secondly, we introduce a recurrent neural network architecture that enables us to (1) model long sequences at document level and (2) visualise the most important words to accurate classification. Finally, we evaluate the results of the linguistic analysis against the results of the neural network visualisations and demonstrate how these features align. We believe that by exploring and comparing suicide notes with last statements and depressed notes, both qualitatively and quantitatively it could help us to find further differentiating factors and aid in identifying suicidal ideation.

2 Related Work

The analysis and classification of suicide notes, depression notes and last statements has traditionally been conducted separately. Work on suicide notes has often focused on identifying suicidal ideation online (O'dea et al., 2017) or distinguish-

Proceedings of the 10th International Workshop on Health Text Mining and Information Analysis (LOUHI 2019), pages 136–145
Hong Kong, November 3, 2019. ©2019 Association for Computational Linguistics
https://doi.org/10.18653/v1/D19-62

ing genuine from forged suicide notes (Coulthard et al., 2016), whilst the main purpose of analysing last statements has been to identify psychological factors or key themes (Schuck and Ward, 2008).

Suicide Notes Recent years have seen an increase in the analysis of suicidal ideation on social media platforms, such as Twitter. Shahreen et al. (2018) searched the Twitter API for specific keywords and analysed the data using both traditional machine learning techniques as well as neural networks, achieving an accuracy of 97.6% using neural networks. Research conducted by Burnap et al. (2017) have developed a classifier to distinguish suicide-related themes such as the reports of suicides and casual references to suicide. Work by Just et al. (2017) used a dataset annotated for suicide risks by experts and a linguistic analysis tool (LIWC) to determine linguistic profiles of suicide-related twitter posts. Other work by Pestian et al. (2010) has looked into the analysis and automatic classification of sentiment in notes, where traditional machine learning algorithms were used. Another important area of suicide note research is the identification of forged suicide notes from genuine ones. Jones and Bennell (2007) have used supervised classification model and a set of linguistic features to distinguish genuine from forged suicide notes, achieving an accuracy of 82%.

Depression notes Work on identifying depression and other mental health conditions has become more prevelant over recent years, where a shared task was dedicated to distinguish depression and PTSD (Post Traumatic Stress Disorder) on Twitter using machine learning (Coppersmith et al., 2015). Morales et al. (2017) have argued that changes in cognition of people with depression can lead to different language usage, which manifests itself in the use of specific linguistic features. Research conducted byResnik et al. (2015) also used linguistic signals to detect depression with different topic modelling techniques. Work by Rude et al. (2004) used LIWC to analyse written documents by students who have experienced depression, currently depressed students as well as student who never have experienced depression, where it was found that individuals who have experienced depression used more first-person singular pronouns and negative emotion words. Nguyen et al. (2014) used LIWC to detect differences in language in online depression communities, where it was found that negative emotion words are good predictors of depressed text compared to control groups using a Lasso Model (Tibshirani, 1996). Research conducted by Morales and Levitan (2016) showed that using LIWC to identify sadness and fatigue helped to accurately classify depression.

Last statements Most work in the analysis of last statements of death row inmates has been conducted using data from The Texas Department of Criminal Justice, made available on their website (Texas Department of Criminal Justices, 2019). Recent work conducted by Foley and Kelly (2018) has primarily focused on the analysis of psychological factors, where it was found that specifically themes of *'love'* and *'spirituality'* were constant whilst requests for forgiveness declined over time. Kelly and Foley (2017) have identified that mental health conditions occur often in death row inmates with one of the most common conditions being depression. Research conducted by Heflick (2005) studied Texas last statements using qualitative methods and have found that often afterlife belief and claims on innocence are common themes in these notes. Eaton and Theuer (2009) studied qualitatively the level of apology and remorse in last statements, whilst also using logistic regression to predict the presence of apologies achieving an accuracy of 92.7%. Lester and Gunn III (2013) used the LIWC program to analyse last statements, where they have found nine main themes, including the affective and emotional processes. Also, Foley and Kelly (2018) found in a qualitative analysis that the most common themes in last statements were love (78%), spirituality (58%), regret (35%) and apology (35%).

3 Data

For our analysis and experiments we use three different datasets, which have been collected from different sources. For the experiments we use standard data preprocessing techniques and remove all identifying personal information.[1]

Last Statements Death Row This dataset has been made available by the Texas Department of Criminal Justices (2019) and contains 545 records of prisoners who have received the death penalty between 1982 and 2017 in Texas, U.S.A. A total of 431 prisoners wrote notes prior to their death.

---

[1]The authors are happy to share the datasets upon request

Due to the information available on this data we have done a basic analysis on the data available, hereafter referred to as *LS*.

Suicide Note The data for this corpus has mainly been taken from Schoene and Dethlefs (2016), but has been further extended by using notes introduced by The Kernel (2013) and Tumbler (2013). There are total of 161 suicide notes in this corpus, hereafter referred to as *GSN*.

Depression Notes We used the data collected by Schoene and Dethlefs (2016) of 142 notes written by people identifying themselves as depressed and lonely, hereafter referred to as *DL*.

4   Linguistic Analysis

To gain more insight into the content of the datasets, we performed a linguistic analysis to show differences in structure and contents of notes. For the purpose of this study we used the Linguistic Inquiry and Word Count software (LIWC) (Tausczik and Pennebaker, 2010), which has been developed to analyse textual data for psychological meaning in words. We report the average of all results across each dataset.

Dimension Analysis Firstly, we looked at the word count and different dimensions of each dataset (see Table 1). It has previously been argued by Tausczik and Pennebaker (2010) that the words people use can give insight into the emotions, thoughts and motivations of a person, where LIWC dimensions correlate with emotions as well as social relationships. The number of *words per sentences* are highest in DL writers and lowest in last statement writers. Research by Osgood and Walker (1959) has suggested that people in stressful situations break their communication down into shorter units. This may indicate alleviated stress levels in individuals writing notes prior to receiving the death sentence. *Clout* stands for the social status or confidence expressed in a person's use of language (Pennebaker et al., 2014). This dimension is highest for people writing their last statements, whereas depressed people rank lowest on this. Cohan et al. (2018) have noted that this might be due to the fact that depressed individuals often have a lower socio-economic status. The *Tone* of a note refers to the emotional tone, including both positive and negative emotions, where numbers below 50 indicate a more negative emotional tone (Cohn et al., 2004). The tone for LS is

highest overall and the lowest in DL, indicating a more overall negative tone in DL and positive tone in LS.

Type	GSN	LS	DL
Tokens per note	110.65	109.72	98.58
Word per Sent	14.87	11.42	16.88
Clout	47.73	67.68	19.94
Tone	54.83	75.43	25.51

Table 1: LIWC Dimension Analysis

Function Words and Content Words Next, we looked at selected function words and grammatical differences, which can be split into two categories called *Function Words* (see Table 2), reflecting how humans communicate and *Content words* (see Table 2), demonstrating what humans say (Tausczik and Pennebaker, 2010). Previous studies have found that whilst there is an overall lower amount of function words in a person's vocabulary, a person uses them more than 50% when communicating. Furthermore it was found that there is a difference in how human brains process function and content words (Miller, 1991). Previous research has found that function words have been connected with indicators of people's social and psychological worlds (Tausczik and Pennebaker, 2010), where it has been argued that the use of function words require basic skills. The highest amount of function words were used in DL notes, whilst both GSN and LS have a similar amount of function words. Rude et al. (2004) has found that high usage, specifically of first-person singular pronouns ("I") could indicate higher emotional and/or physical pain as the focus of their attention is towards themselves. Overall Just et al. (2017) has also identified a larger amount of personal pronouns in suicide-related social media content. Previous work by Hancock et al. (2007) has found that people use a higher amount of negations when also expressing negative emotions and used fewer words overall, compared to more positive emotions. This seem to be also true for the number of negations used in this case where amount of *Negations* were also highest in the DL corpus and lowest in the LS corpus, whilst the overall words count was lowest for DL and negative emotions highest. Furthermore, it was found that *Verbs*, *Adverb* and *Adjectives* are often used to communicate content, however previous studies have found (Jones and Bennell, 2007; Gregory,

1999) that individuals that commit suicide are under a higher drive and therefore would reference a higher amount of objects (through nouns) rather than using descriptive language such as adjectives and adverbs.

Type	GSN	LS	DL
Function	56.35	56.33	60.20
Personal pronouns	16.23	20.44	15.19
I	11.04	12.65	12.8
Negations	2.71	1.71	4.06
Verb	19.29	19.58	21.65
Adjective	4.45	2.58	4.98
Adverb	4.43	3.14	7.69

Table 2: LIWC Function and Content Words

Affect Analysis The analysis of emotions in suicide notes and last statements has often been addressed in research (Schoene and Dethlefs, 2018; Lester and Gunn III, 2013) The number of *Affect words* is highest in LS notes, whilst they are lowest in DL notes, this could be related to the emotional *Tone* of a note (see Table 1). This also applies to the amount of *Negative emotions* as they are highest in DL notes and *Positive emotions* as these are highest in LS notes. Previous research has analysed the amount of *Anger* and *Sadness* in GSN and DL notes and has shown that it is more prevalent in DL note writers as these are typical feelings expressed when people suffer from depression (Schoene and Dethlefs, 2016).

Type	GSN	LS	DL
Affect	9.1	11.58	8.44
Positive emotion	5.86	8.99	3.15
Negative emotion	3.15	2.58	5.21
Anger	0.61	0.65	1.03
Sadness	1.09	1.08	2.53

Table 3: LIWC Affect Analysis

Social and Psychological Processes *Social Processes* highlights the social relationships of note writers, where it can be seen in Table 4 that the highest amount of social processes can be found in LS and the lowest in DL. Furthermore LS notes tend to speak most about family relations and least about friends, this was also found by Kelly and Foley (2017) who found a low frequency in interpersonal relationships.

Type	GSN	LS	DL
Social processes	12.21	18.19	8.33
Family	1.17	2.17	0.47
Friends	0.77	0.38	0.73

Table 4: LIWC Social Processes

The term *Cognitive processes* encompasses a number of different aspects, where we have found that the highest amount of cognitive processes was in DL notes and the lowest in LS notes. Boals and Klein (2005) have found that people who use more cognitive mechanisms to cope with traumatic events such as break ups by using more causal words to organise and explain events and thoughts for themselves. Arguably this explains why there is a lower amount in LS notes as LS writers often have a long time to organise their thoughts, events and feelings whilst waiting for their sentence (Death Penalty Information Centre, 2019). *Insight* encompasses words such as *think* or *consider*, whilst *Cause* encompasses words that express reasoning or causation of events, e.g.: *because* or *hence*. These terms have previously been coined as *cognitive process words* by (Gregory, 1999), who argued that these words are less used in GSN notes as the writer has already finished the decision making process whilst other types of discourse would still try to justify and reason over events and choices. This can also be found in the analysis of our own data, where both GSN and LS notes show similar, but lower frequency of terms in those to categories compared to DL writers. *Tentativeness* refers to the language use that indicates a person is uncertain about a topic and uses a number of filler words. A person who use more tentative words, may have not expressed an event to another person and therefore has not processed an event yet and it has not been formed into a story (Tausczik and Pennebaker, 2010). The amount of tentative words used in DL notes is highest, whilst it is lowest in LS words. This might be due to the fact that LS writers already had to reiterate over certain events multiple times as they go through the process of prosecution.

Personal Concerns *Personal Concerns* refers to the topics most commonly brought up in the different notes, where we note that both *Money* and *Work* are most often referred to in GSN notes and lowest in LS notes. This might be due to the the fact that (Mind, 2013) lists these two topics as

Type	GSN	LS	DL
Cognitive Processes	12.19	10.85	16.77
Insight	2.37	2.3	4.07
Cause	0.95	0.8	1.94
Tentativeness	2.57	1.5	3.23

Table 5: LIWC Psychological Processes

some of the most common reasons for a person to commit suicide. *Religion* is most commonly referenced in LS notes, which confirms previous analysis of such notes (Foley and Kelly, 2018; Kelly and Foley, 2017) and lowest in DL notes. (Just et al., 2017) has found that the topic of *Death* is commonly referenced in suicide-related communication on Twitter. This was also found in this dataset, where GSN notes most commonly referenced death, whilst DL notes were least likely to reference this topic.

Type	GSN	LS	DL
Work	1.24	0.41	0.99
Money	0.68	0.18	0.31
Religion	0.82	2.7	0.09
Death	0.76	0.68	0.64

Table 6: LIWC Personal Concerns

Time Orientation and Relativity Looking at the *Time Orientation* of a note can give interesting insight into the temporal focus of attention and differences in verb tenses can show psychological distance or to which extend disclosed events have been processed (Tausczik and Pennebaker, 2010). Table 7 shows that the focus of LS letters is primarily in the past whilst GSN and DL letters focus on the present. The high focus on the past in DL notes as well as GSN notes could be, because these notes might draw on their past experiences to express the issues of their current situation or problems.The most frequent use of future tense is in LS letters which could be due to a LS notes writers common focus on afterlife (Heflick, 2005).

Type	GSN	LS	DL
Focus past	3.24	2.86	3.32
Focus present	14.39	1.43	16.11
Focus future	2.1	2.27	1.51

Table 7: LIWC Time orientation

Overall it was noted that for most analysis GSN falls between the two extremes of LS and DL.

5   Learning Model

The primary model is the Long-short-term memory (LSTM) given its suitability for language and time-series data (Hochreiter and Schmidhuber, 1997). We feed into the LSTM an input sequence $\mathbf{x} = (x_1, \ldots, x_N)$ of words in a document alongside a label $y \in Y$ denoting the class from any of the three datasets. The LSTM learns to map inputs x to outputs y via a hidden representation $\mathbf{h}_t$ which can be found recursively from an activation function.

$$f(\mathbf{h}_{t-1}, x_t), \tag{1}$$

where t denotes a time-step. During training, we minimise a loss function, in our case categorical cross-entropy as:

$$L(x, y) = -\frac{1}{N} \sum_{n \in N} x_n \log y_n. \tag{2}$$

LSTMs manage their weight updates through a number of gates that determine the amount of information that should be retained and forgotten at each time step. In particular, we distinguish an 'input gate' i that decides how much new information to add at each time-step, a 'forget gate' f that decides what information not to retain and an 'output gate' o determining the output. More formally, and following the definition by Graves (2013), this leads us to update our hidden state $\mathbf{h}$ as follows (where σ refers to the logistic sigmoid function and c is the 'cell state'):

$$i_t = \sigma(W_{xi}x_t + W_{hi}h_{t-1} + W_{ci}c_{t-1} + bi) \tag{3}$$

$$f_t = \sigma(W_{xf}x_t + W_{hf}h_{t-1} + W_{cf}c_{t-1} + bf) \tag{4}$$

$$c_t = f_t c_{t-1} + i_t \tanh(W_{xc}x_t + W_{hc}h_{t-1} + bc) \tag{5}$$

$$o_t = \sigma(W_{xo}x_t + W_{ho}h_{t-1} + W_{co}c_t + bo) \tag{6}$$

$$h_t = o_t \tanh(c_t) \tag{7}$$

A standard LSTM definition solves some of the problems of vanilla RNNs have (Hochreiter and

Schmidhuber, 1997), but it still has some short-comings when learning long dependencies. One of them is due to the cell state of an LSTM; the cell state is changed by adding some function of the inputs. When we backpropagate and take the derivative of c_t with respect to $c_t - 1$, the added term would disappear and less information would travel through the layers of a learning model.

For our implementation of a Dilated LSTM, we follow the implementation of recurrent skip connections with exponentially increasing dilations in a multi-layered learning model by Chang et al. (2017). This allows LSTMs to better learn input sequences and their dependencies and therefore temporal and complex data dependencies are learned on different layers. Whilst dilated LSTM alleviates the problem of learning long sequences, it does not contribute to identifying words in a sequence that are more important than others. Therefore we extend this network by (1) an embedding layer and (2) an attention mechanism to further improve the network's ability. A graph illustration of our learning model can be seen in Figure 2.

Dilated LSTM with Attention Each document D contains i sentences S_i, where w_i represents the words in each sentence. Firstly, we embed the words to vectors through an embedding matrix W_e, which is then used as input into the dilated LSTM.

The most important part of the dilated LSTM is the dilated recurrent skip connection, where $c_t^{(l)}$ is the cell in layer l at time t:

$$c_t^{(l)} = f(x_t^{(l)}, c_{t-s^{l-1}}^{(l)}).\qquad(8)$$

$s^{(l)}$ is the skip length; or dilation of layer l; $x_t^{(l)}$ as the input to layer l at time t; and $f(\cdot)$ denotes a LSTM cell; M and L denote dilations at different layers:

$$s^{(l)} = M^{(l-1)}, l = 1, \ldots L.\qquad(9)$$

The dilated LSTM alleviates the problem of learning long sequences, however not every word in a sequence has the same meaning or importance.

Attention layer The attention mechanism was first introduced by Bahdanau et al. (2015), but has since been used in a number of different tasks including machine translation (Luong et al., 2015), sentence pairs detection (Yin et al., 2016) , neural image captioning (Xu et al., 2015) and action recognition (Sharma et al., 2015).

Our implemenetation of the attention mechanism is inspired by Yang et al. (2016), using attention to find words that are most important to the meaning of a sentence at document level. We use the output of the dilated LSTM as direct input into the attention layer, where O denotes the output of final layer L of the Dilated LSTM at time t_{+1}.

The *attention* for each word w in a sentence s is computed as follows, where u_{it} is the hidden representation of the dilated LSTM output, α_{it} represents normalised alpha weights measuring the importance of each word and S_i is the sentence vector:

$$u_{it} = \tanh(O + b_w)\qquad(10)$$

$$\alpha_{it} = \frac{\exp\left(u_{it}^T u_w\right)}{\sum_t \exp\left(u_{it}^T u_w\right)}\qquad(11)$$

$$s_i = \sum_t \alpha_{it} o\cdot\qquad(12)$$

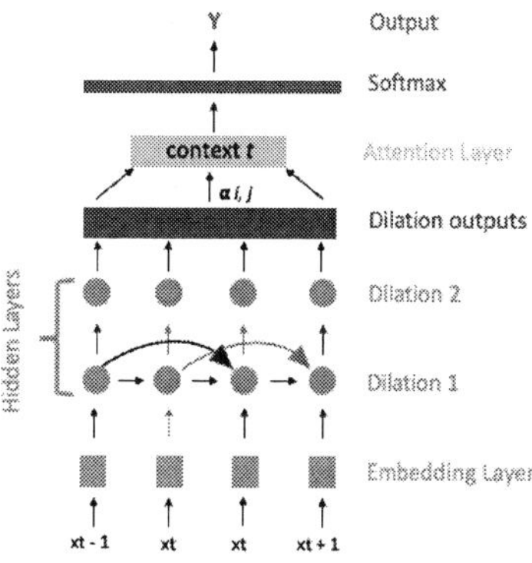

Figure 1: A 2-layer dilated LSTM with Attention.

6 Experiments and Results

For our experiments we use all three datasets, Table 8 shows the results for the experiments series. We establish three performance baselines on the datasets by using three different algorithms previously used on similar datasets. Firstly, we use the ZeroR and LMT (Logistic Model Tree) previously used by (Schoene and Dethlefs, 2016). Additionally we chose to benchmark our algorithm also against the originally proposed Bidirectional LSTM with attention proposed by Yang et al. (2016), which was also used on similar existing datasets before (Schoene and Dethlefs, 2018).

Furthermore we benchmark the Dilated Attention
LSTM against two other types of recurrent neural networks. We use *200-dimensional* word embeddings as input into each network and all neural
networks share the same hyper-parameters, where
learning rate = 0.001, batch size = 128, dropout
= 0.5, hidden size = 150 units and the *Adam* optimizer is used. For our proposed model - the
Dilated LSTM with Attention - we establish the
number of dilations empirically. There are 2 dilated layers with exponentially increasing dilations starting at *1*. Due to the size of the dataset
we have split the data into *70%* training, *15%* validation and *15%* test data. We report results based
on the test accuracy of the prediction results. It can
be seen in Table 8 that the dilated LSTM with an
attention layer outperforms the BiLSTM with Attention by 5.07%. Furthermore it was found that
both the LMT and a vanilla bi-directional LSTM
outperform a standard LSTM on this task. Previous results on similar tasks have yielded an accuracy of 69.41% using BiLSTM with Attention
(Schoene and Dethlefs, 2018) and 86 % using a
LMT (Schoene and Dethlefs, 2016).

7   Evaluation

In order to evaluate the DLSTM with attention we
look in more detail at the predicted labels and visualise examples of each note to show which features
are assigned the highest attention weights.

7.1   Label Evaluation

In Figure 2 we show the confusion matrix over the
DLSTM with attention. It can be seen that LS
notes are most often correctly predicted and DL
notes are least likely to be accurately predicted.

The same applies to results of the main compet-
ing model (Bi-directional LSTM with Attention),
Figure 3 shows that this model still misclassifies
LS notes with DL notes.

7.2   Visualisation of attention weights

In order to see which features are most important
to accurate classification we visualise examples
from the test set of each dataset, where Figures 4,
5 and 6 show the visualisation of attention weights
in the *GSN*, *LS* and *DL* datasets respectively. Furthermore, we also show three examples of the test
data with typical errors the learning model makes
in Figures 7 , 8 and 9. Words highlighted in darker
shades have a higher attention weight.

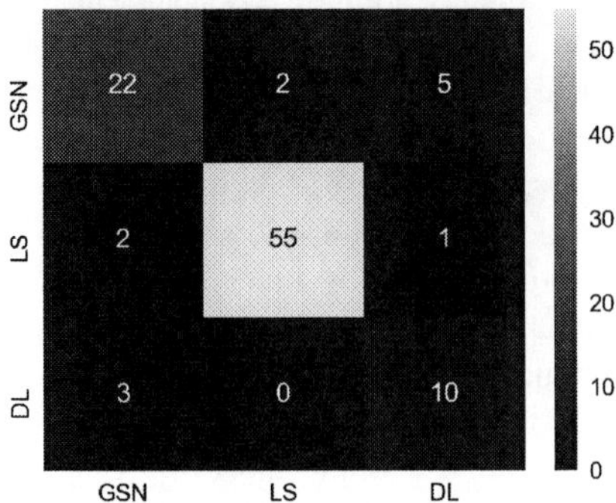

Figure 2: Confusion Matrix of test set labels - DLSTM
Attention.

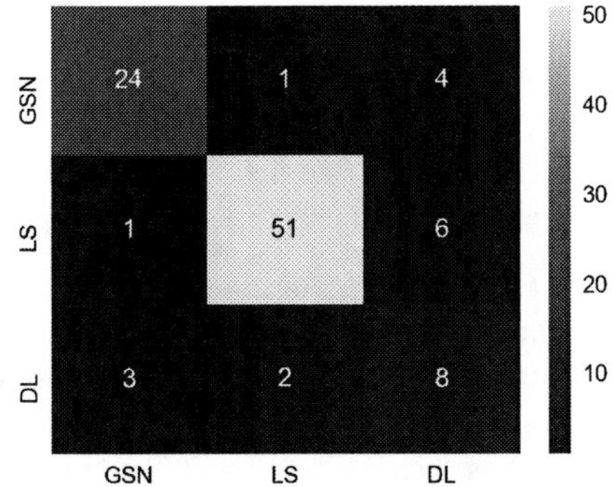

Figure 3: Confusion Matrix of test set labels - BiLSTM
Attention.

The most important words highlighted in a last
statement note (see Figure 4) are personal pronouns as well as an apology and expression of love
towards friends and family members. This corresponds with the higher amount of personal pronouns, positive emotions and references to Family
in LS notes compared to GSN and DL notes. Furthermore it can be seen that there is a low amount
of cognitive process words and more action verbs
such as *killing* or *hurt*, which could confirm that
inmates have had more time to process events and
thoughts and don't need cognitive words as a coping mechanism anymore (Boals and Klein, 2005).

Figure 5 shows a GSN note, where the most
important words are also pronouns, references to
family, requests for forgiveness and endearments.
Previous research has shown that forgiveness is an
important feature as well as the giving instructions
such as *help* or phrases like *do not follow* are key

Model	Test Accuracy	Aver. Precision	Aver. Recall	Aver. F1-score
ZeroR	42.85	0.43	0.41	0.42
LMT	80.35	0.81	0.79	0.80
LSTM	62.16	0.63	0.61	0.62
BiLSTM	65.82	0.66	0.64	0.65
BiLSTM with Attention	82.27	0.85	0.83	0.84
DLSTMAttention	**87.34**	**0.88**	**0.87**	**0.87**

Table 8: Test accuracy and F1-score of different learning models in %

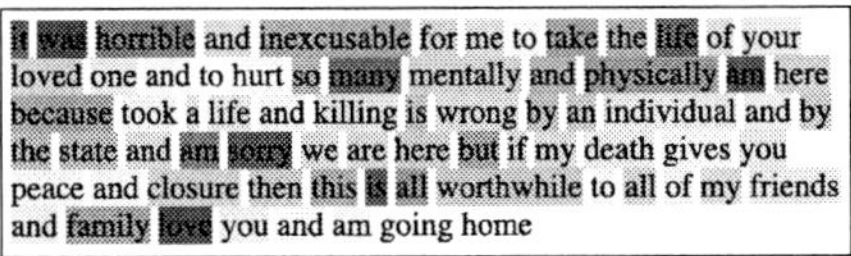

Figure 4: Example of LS note correctly classified.

to accurately classify suicide notes (Pestian et al., 2010). Terms of endearment for loved ones at the start or towards the end of a note (Gregory, 1999).

Figure 5: Example of GSN note correctly classified.

The DL note in Figure 6 shows that there is a greater amount of cognitive process verbs present, such as *feeling* or *know* as well as negations, which confirms previous analysis using LIWC.

Figure 6: Example of DL note correctly classified.

Figure 7 shows a visualisation of a LS note. In this instances the word *God* was replaced with *up*, when looking into the usage of the word *up* in other LS notes, it was found that it was commonly used in reference to religious topics such as *God*, *heaven* or *up there*.

Figure 7: LS note error analysis

In Figure 8 a visualised GSN note is shown. Whilst there is still consistency in highlighting personal pronouns (e.g.: *you*), it can be seen that the end of the note is missing and more action verbs such as *hurt* or *take* are more important.

Figure 8: GSN note error analysis

The visualisation in Figure 9 demonstrates how the personal pronoun *I* has been removed from several DL notes, where DL notes are least likely to be predicted accurately as shown in Figure 2.

Figure 9: DL note error analysis

8 Conclusion

In this paper we have presented a new learning model for classifying long sequences. We have shown that the model outperforms the baseline by 6.99 % and by 5.07 % a competitor model. Furthermore we have provided an analysis of the linguistic features on three datasets, which we have later compared in a qualitative evaluation by visualising the attention weights on examples of each dataset. We have shown that the neural network pays attention to similar linguistic features as provided by LIWC and found in human evaluated related research.

References

D. Bahdanau, K. Cho, and Y. Bengio. 2015. Neural Machine Translation by Jointly Learning to Align and Translate. In *Proc. of the International Conference on Learning Representations (ICLR)*, San Diego, CA, USA.

BBC. 2019. Facebook 'sorry' for distressing suicide posts on instagram.

Adriel Boals and Kitty Klein. 2005. Word use in emotional narratives about failed romantic relationships and subsequent mental health. *Journal of Language and Social Psychology*, 24(3):252–268.

Pete Burnap, Gualtiero Colombo, Rosie Amery, Andrei Hodorog, and Jonathan Scourfield. 2017. Multiclass machine classification of suicide-related communication on twitter. *Online social networks and media*, 2:32–44.

Rafael A Calvo, David N Milne, M Sazzad Hussain, and Helen Christensen. 2017. Natural language processing in mental health applications using non-clinical texts. *Natural Language Engineering*, 23(5):649–685.

Shiyu Chang, Yang Zhang, Wei Han, Mo Yu, Xiaoxiao Guo, Wei Tan, Xiaodong Cui, Michael Witbrock, Mark A Hasegawa-Johnson, and Thomas S Huang. 2017. Dilated recurrent neural networks. In *Advances in Neural Information Processing Systems*, pages 77–87.

Arman Cohan, Bart Desmet, Andrew Yates, Luca Soldaini, Sean MacAvaney, and Nazli Goharian. 2018. Smhd: A large-scale resource for exploring online language usage for multiple mental health conditions. *arXiv preprint arXiv:1806.05258*.

Michael A Cohn, Matthias R Mehl, and James W Pennebaker. 2004. Linguistic markers of psychological change surrounding september 11, 2001. *Psychological science*, 15(10):687–693.

Glen Coppersmith, Mark Dredze, Craig Harman, Kristy Hollingshead, and Margaret Mitchell. 2015. Clpsych 2015 shared task: Depression and ptsd on twitter. In *Proceedings of the 2nd Workshop on Computational Linguistics and Clinical Psychology: From Linguistic Signal to Clinical Reality*, pages 31–39.

Malcolm Coulthard, Alison Johnson, and David Wright. 2016. *An introduction to forensic linguistics: Language in evidence*. Routledge.

Death Penalty Information Centre. 2019. Time on death row.

Judy Eaton and Anna Theuer. 2009. Apology and remorse in the last statements of death row prisoners. *Justice Quarterly*, 26(2):327–347.

Facebook. 2019. Suicide prevention.

SR Foley and BD Kelly. 2018. Forgiveness, spirituality and love: thematic analysis of last statements from death row, texas (2002-2017). *QJM: An International Journal of Medicine*.

Alex Graves. 2013. Generating Sequences With Recurrent Neural Networks. *CoRR*, abs/1308.0850.

Adam Gregory. 1999. The decision to die: The psychology of the suicide note. *Interviewing and deception*, pages 127–156.

Jeffrey T Hancock, Christopher Landrigan, and Courtney Silver. 2007. Expressing emotion in text-based communication. In *Proceedings of the SIGCHI conference on Human factors in computing systems*, pages 929–932. ACM.

Nathan A Heflick. 2005. Sentenced to die: Last statements and dying on death row. *Omega-Journal of Death and Dying*, 51(4):323–336.

Sepp Hochreiter and Jürgen Schmidhuber. 1997. Lstm can solve hard long time lag problems. In *Advances in neural information processing systems*, pages 473–479.

Natalie J Jones and Craig Bennell. 2007. The development and validation of statistical prediction rules for discriminating between genuine and simulated suicide notes. *Archives of Suicide Research*, 11(2):219–233.

Marcel Adam Just, Lisa Pan, Vladimir L Cherkassky, Dana L McMakin, Christine Cha, Matthew K Nock, and David Brent. 2017. Machine learning of neural representations of suicide and emotion concepts identifies suicidal youth. *Nature human behaviour*, 1(12):911.

Brendan D Kelly and Sharon R Foley. 2017. Analysis of last statements prior to execution: methods, themes and future directions. *QJM: An International Journal of Medicine*, 111(1):3–6.

David Lester and John F Gunn III. 2013. Ethnic differences in the statements made by inmates about to be executed in texas. *Journal of Ethnicity in Criminal Justice*, 11(4):295–301.

Liu Yi Lin, Jaime E Sidani, Ariel Shensa, Ana Radovic, Elizabeth Miller, Jason B Colditz, Beth L Hoffman, Leila M Giles, and Brian A Primack. 2016. Association between social media use and depression among us young adults. *Depression and anxiety*, 33(4):323–331.

Minh-Thang Luong, Hieu Pham, and Christopher D Manning. 2015. Effective approaches to attention-based neural machine translation. *arXiv preprint arXiv:1508.04025*.

David D Luxton, Jennifer D June, and Jonathan M Fairall. 2012. Social media and suicide: a public health perspective. *American journal of public health*, 102(S2):S195–S200.

George Armitage Miller. 1991. The science of words.

Mind. 2013. Depression.

Michelle Morales, Stefan Scherer, and Rivka Levitan. 2017. A cross-modal review of indicators for depression detection systems. In *Proceedings of the Fourth Workshop on Computational Linguistics and Clinical PsychologyFrom Linguistic Signal to Clinical Reality*, pages 1–12.

Michelle Renee Morales and Rivka Levitan. 2016. Speech vs. text: A comparative analysis of features for depression detection systems. In *2016 IEEE Spoken Language Technology Workshop (SLT)*, pages 136–143. IEEE.

Thin Nguyen, Dinh Phung, Bo Dao, Svetha Venkatesh, and Michael Berk. 2014. Affective and content analysis of online depression communities. *IEEE Transactions on Affective Computing*, 5(3):217–226.

Bridianne O'dea, Mark E Larsen, Philip J Batterham, Alison L Calear, and Helen Christensen. 2017. A linguistic analysis of suicide-related twitter posts. *Crisis*.

Charles E Osgood and Evelyn G Walker. 1959. Motivation and language behavior: A content analysis of suicide notes. *The Journal of Abnormal and Social Psychology*, 59(1):58.

James W Pennebaker, Cindy K Chung, Joey Frazee, Gary M Lavergne, and David I Beaver. 2014. When small words foretell academic success: The case of college admissions essays. *PloS one*, 9(12):e115844.

John Pestian, Henry Nasrallah, Pawel Matykiewicz, Aurora Bennett, and Antoon Leenaars. 2010. Suicide note classification using natural language processing: A content analysis. *Biomedical informatics insights*, 3:BII–S4706.

John P Pestian, Pawel Matykiewicz, Michelle Linn-Gust, Brett South, Ozlem Uzuner, Jan Wiebe, K Bretonnel Cohen, John Hurdle, and Christopher Brew. 2012. Sentiment analysis of suicide notes: A shared task. *Biomedical informatics insights*, 5:BII–S9042.

Philip Resnik, William Armstrong, Leonardo Claudino, Thang Nguyen, Viet-An Nguyen, and Jordan Boyd-Graber. 2015. Beyond lda: exploring supervised topic modeling for depression-related language in twitter. In *Proceedings of the 2nd Workshop on Computational Linguistics and Clinical Psychology: From Linguistic Signal to Clinical Reality*, pages 99–107.

Stephanie Rude, Eva-Maria Gortner, and James Pennebaker. 2004. Language use of depressed and depression-vulnerable college students. *Cognition & Emotion*, 18(8):1121–1133.

Annika M Schoene and Nina Dethlefs. 2018. Unsupervised suicide note classification.

Annika Marie Schoene and Nina Dethlefs. 2016. Automatic identification of suicide notes from linguistic and sentiment features. In *Proceedings of the 10th SIGHUM Workshop on Language Technology for Cultural Heritage, Social Sciences, and Humanities*, pages 128–133.

Andreas RT Schuck and Janelle Ward. 2008. Dealing with the inevitable: strategies of self-presentation and meaning construction in the final statements of inmates on texas death row. *Discourse & society*, 19(1):43–62.

Nabia Shahreen, Mahfuze Subhani, and Md Mahfuzur Rahman. 2018. Suicidal trend analysis of twitter using machine learning and neural network. In *2018 International Conference on Bangla Speech and Language Processing (ICBSLP)*, pages 1–5. IEEE.

Shikhar Sharma, Ryan Kiros, and Ruslan Salakhutdinov. 2015. Action recognition using visual attention. *arXiv preprint arXiv:1511.04119*.

Yla R Tausczik and James W Pennebaker. 2010. The psychological meaning of words: Liwc and computerized text analysis methods. *Journal of language and social psychology*, 29(1):24–54.

Texas Department of Criminal Justices. 2019. Texas death row executions info and last words.

The Kernel. 2013. What suicide notes look like in the social media age.

Robert Tibshirani. 1996. Regression shrinkage and selection via the lasso. *Journal of the Royal Statistical Society: Series B (Methodological)*, 58(1):267–288.

Tumbler. 2013. Suicide notes.

Kelvin Xu, Jimmy Ba, Ryan Kiros, Kyunghyun Cho, Aaron Courville, Ruslan Salakhudinov, Rich Zemel, and Yoshua Bengio. 2015. Show, attend and tell: Neural image caption generation with visual attention. In *International conference on machine learning*, pages 2048–2057.

Zichao Yang, Diyi Yang, Chris Dyer, Xiaodong He, Alex Smola, and Eduard Hovy. 2016. Hierarchical attention networks for document classification. In *Proceedings of the 2016 Conference of the North American Chapter of the Association for Computational Linguistics: Human Language Technologies*, pages 1480–1489.

Wenpeng Yin, Hinrich Schütze, Bing Xiang, and Bowen Zhou. 2016. Abcnn: Attention-based convolutional neural network for modeling sentence pairs. *Transactions of the Association for Computational Linguistics*, 4:259–272.

Writing habits and telltale neighbors: analyzing clinical concept usage patterns with sublanguage embeddings

Denis Newman-Griffis[a,b] and **Eric Fosler-Lussier** [a]

[a]Dept of Computer Science and Engineering, The Ohio State University, Columbus, OH
[b]Rehabilitation Medicine Dept, Clinical Center, National Institutes of Health, Bethesda, MD
`{newman-griffis.1, fosler-lussier.1}` @ `osu.edu`

Abstract

Natural language processing techniques are being applied to increasingly diverse types of electronic health records, and can benefit from in-depth understanding of the distinguishing characteristics of medical document types. We present a method for characterizing the usage patterns of clinical concepts among different document types, in order to capture semantic differences beyond the lexical level. By training concept embeddings on clinical documents of different types and measuring the differences in their nearest neighborhood structures, we are able to measure divergences in concept usage while correcting for noise in embedding learning. Experiments on the MIMIC-III corpus demonstrate that our approach captures clinically-relevant differences in concept usage and provides an intuitive way to explore semantic characteristics of clinical document collections.

1 Introduction

Sublanguage analysis has played a pivotal role in natural language processing of health data, from highlighting the clear linguistic differences between biomedical literature and clinical text (Friedman et al., 2002) to supporting adaptation to multiple languages (Laippala et al., 2009). Recent studies of clinical sublanguage have extended sublanguage study to the document type level, in order to improve our understanding of the syntactic and lexical differences between highly distinct document types used in modern EHR systems (Feldman et al., 2016; Grön et al., 2019).

However, one key axis of sublanguage characterization that has not yet been explored is how domain-specific clinical *concepts* differ in their usage patterns among different document types. Established biomedical concepts may have multiple, often non-compositional surface forms

(e.g., "ALS" and "Lou Gehrig's disease"), making them difficult to analyze using lexical occurrence alone. Understanding how these concepts differ between document types can not only augment recent methods for sublanguage-based text categorization (Feldman et al., 2016), but also inform the perennial challenge of medical concept normalization (Luo et al., 2019): "depression" is much easier to disambiguate if its occurrence is known to be in a social work note or an abdominal exam.

Inspired by recent technological advances in modeling diachronic language change (Hamilton et al., 2016; Vashisth et al., 2019), we characterize concept usage differences within clinical sublanguages using nearest neighborhood structures of clinical concept embeddings. We show that overlap in nearest neighborhoods can reliably distinguish between document types while controlling for noise in the embedding process. Qualitative analysis of these nearest neighborhoods demonstrates that these distinctions are semantically relevant, highlighting sublanguage-sensitive relationships between specific concepts and between concepts and related surface forms. Our findings suggest that the structure of concept embedding spaces not only captures domain-specific semantic relationships, but can also identify a "fingerprint" of concept usage patterns within a clinical document type to inform language understanding.

2 Related Work

Sublanguage analysis historically focused on describing the characteristic grammatical structures of a particular domain (Friedman, 1986; Grishman, 2001; Friedman et al., 2002). As methods for automated analysis of large-scale data sets have improved, more studies have investigated lexical and semantic characteristics, such as usage patterns of different verbs and semantic categories

Proceedings of the 10th International Workshop on Health Text Mining and Information Analysis (LOUHI 2019), pages 146–156
Hong Kong, November 3, 2019. ©2019 Association for Computational Linguistics
https://doi.org/10.18653/v1/D19-62

Type	Docs	Lines	Tokens	Matches	Concepts	High Confidence Concepts	Consistency (%)
Case Management	967	20,106	165,608	45,306	557	111	75
Consult	98	15,514	96,515	26,109	812	0	–
Discharge Summary	59,652	14,480,154	104,027,364	30,840,589	6,381	1,599	67
ECG	209,051	1,022,023	7,307,381	2,163,682	540	14	56
Echo	45,794	2,892,069	19,752,879	6,070,772	1,233	157	65
General	8,301	307,330	2,191,618	552,789	2,559	0	–
Nursing	223,586	9,839,274	73,426,426	18,903,892	4,912	2	58
Nursing/Other	822,497	10,839,123	140,164,545	31,135,584	5,049	83	60
Nutrition	9,418	868,102	3,843,963	1,147,918	1,911	198	73
Pharmacy	103	4,887	39,163	8,935	376	0	–
Physician	141,624	26,659,749	148,306,543	39,239,425	5,538	122	57
Radiology	522,279	17,811,429	211,901,548	34,433,338	4,126	599	63
Rehab Services	5,431	585,779	2,936,022	869,485	2,239	9	62
Respiratory	31,739	1,323,495	6,358,924	2,255,725	1,039	5	63
Social Work	2,670	100,124	930,674	195,417	1,282	0	–

Table 1: Document type subcorpora in MIMIC-III. Tokenization was performed with SpaCy; Matches and Concepts refer to number of terminology string match instances and number of unique concepts embedded, respectively, using SNOMED-CT and LOINC vocabularies from UMLS 2017AB release. The number of high-confidence concepts identified for each document type is given with their mean consistency.

(Denecke, 2014), as well as more structural information such as document section patterns and syntactic features (Zeng et al., 2011; Temnikova et al., 2014). The use of terminologies to assess conceptual features of a sublanguage corpus was proposed by Walker and Amsler (1986), and Drouin (2004); Grön et al. (2019) used sublanguage features to expand existing terminologies, but large-scale characterization of concept usage in sublanguage has remained a challenging question.

Word embedding techniques have been utilized to describe diachronic language change in a number of recent studies, from evaluating broad changes over decades (Hamilton et al., 2016; Vashisth et al., 2019) to detecting fine-grained shifts in conceptualizations of psychological concepts (Vylomova et al., 2019). Embedding techniques have also been used as a mirror to analyze social biases in language data (Garg et al., 2018). Similar to our work, Ye and Fabbri (2018) investigate document type-specific embeddings from clinical data as a tool for medical language analysis. However, our approach has two significant differences: Ye and Fabbri (2018) used word embeddings only, while we utilize concept embeddings to capture concepts across multiple surface forms; more importantly, their work investigated multiple document types as a way to *control* for specific usage patterns within sublanguages in order to capture more general term similarity patterns, while our study aims to *capture* these sublanguage-specific usage patterns in order to analyze the representative differences in language

use between different expert communities.

3 Data and preprocessing

We use free text notes from the MIMIC-III critical care database (Johnson et al., 2016) for our analysis. This includes approximately 2 million text records from hospital admissions of almost 50 thousand patients to the critical care units of Beth Israel Deaconess Medical Center over a 12-year period. Each document belongs to one of 15 document types, listed in Table 1.

As sentence segmentation of clinical text is often optimized for specific document types (Griffis et al., 2016), we segmented our documents at linebreaks and tokenized using SpaCy (version 2.1.6; Honnibal and Montani 2017). All tokens were lowercased, but punctuation and deidentifier strings were retained, and no stopwords were removed.

4 Experiments

Methods for learning clinical concept representations have proliferated in recent years (Choi et al., 2016; Mencia et al., 2016; Phan et al., 2019), but often require annotations in forms such as billing codes or disambiguated concept mentions. These annotations may be supplied by human experts such as medical coders, or by adapting medical NLP tools such as MetaMap (Aronson and Lang, 2010) or cTAKES (Savova et al., 2010) to perform concept recognition (De Vine et al., 2014).

For investigating potentially divergent usage patterns of clinical concepts, these strategies face

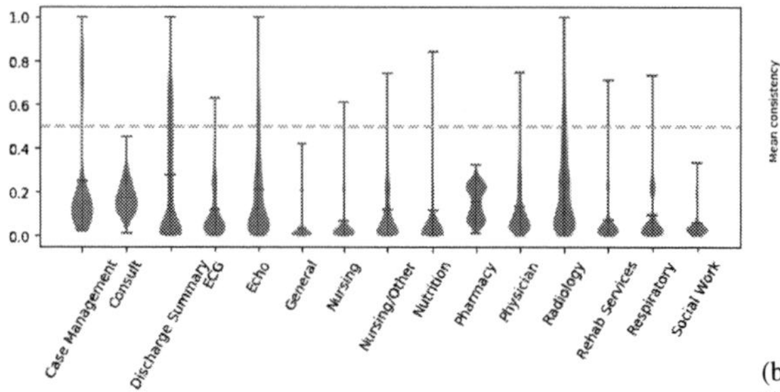

(a) Self consistency by document type; line at 50% threshold

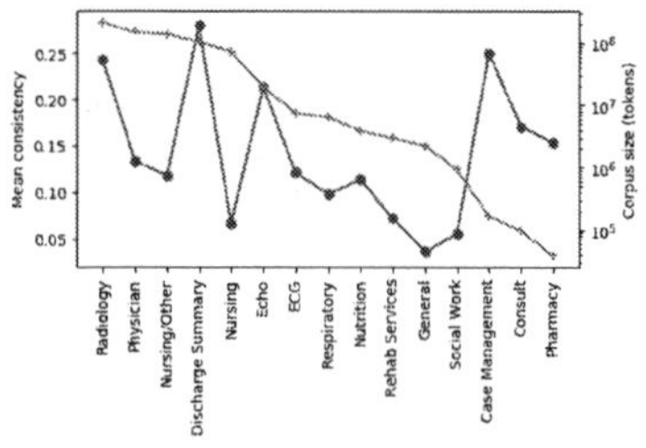

(b) Self consistency compared to corpus size (log scale), with document types sorted by decreasing corpus size.

Figure 1: Distribution of self-consistency rates (i.e., overlap in nearest neighbors between replicate embeddings of the same concept) among MIMIC document types.

serious limitations: the full diversity of MIMIC data has not been annotated for concept identifiers, and the statistical biases of trained NLP tools may suppress underlying differences in automatically-recognized concepts. We therefore take a distant supervision approach, using JET (Newman-Griffis et al., 2018). JET uses a sliding context window to jointly train embedding models for words, surface forms, and concepts, using a log-bilinear objective with negative sampling and shared embeddings for context words. It leverages known surface forms from a terminology as a source of distant supervision: each occurrence of any string in the terminology is treated as a weighted training instance for each of the concepts that string can represent. As terminologies are typically many-to-many maps between surface forms and concepts, this generally leads to a unique set of contexts being used to train the embedding of each concept, though any individual context window may be used as a sample for training multiple concepts. We constrain the scope of our analysis to only concepts and strings from SNOMED-CT and LOINC,[1] two popular high-coverage clinical vocabularies.

4.1 Identifying concepts for comparison

For each document type, we concatenate all of its documents (maintaining linebreaks), identify all occurrences of SNOMED-CT and LOINC strings in each line, and use these occurrences to train word, term, and concept embeddings with JET. Due to the size of our subcorpora, we used a window size of 5, minimum frequency of 5, embedding dimensionality of 100, initial learning rate of

0.05, and 10 iterations over each corpus.

Prior research has noted instability of nearest neighborhoods in multiple embedding methods (Wendlandt et al., 2018). We therefore train 10 sets of embeddings from each of our subcorpora, each using the same hyperparameter settings but a different random seed. We then use all 10 replicates from each subcorpus in our analyses, in order to control for variation in nearest neighborhoods introduced by random initialization and negative sampling. To evaluate the baseline reliability of concept embedding neighborhoods from each subcorpus, we calculated per-concept consistency by measuring, over all pairs of embedding sets within the 10 replicates, the average set membership overlap between the top 5 nearest neighbors by cosine similarity for each concept embedding.[2] As shown in Figure 1a, these consistency scores vary widely both within and between document types, with some document types producing no concept embeddings with consistency over 40%. Interestingly, as illustrated in Figure 1b, there is no linear relationship between log corpus size and mean concept consistency ($R^2 \approx 0.011$), suggesting that low consistency is not solely due to limited training data.

To mitigate concerns about the reliability of embeddings for comparison, a set of **high-confidence concepts** is identified for each document type by

---

[1] We used the versions distributed in the 2017AB release of the UMLS (Bodenreider, 2004).

[2] We chose five nearest neighbors for our analyses based on qualitative review of neighborhoods for concepts within different document types. We found nearest neighborhoods for concept embeddings to vary more than for word embeddings, often introducing noise beyond the top five nearest neighbors; we therefore set a conservative baseline for reliability by focusing on the closest and most stable neighbors. However, using 10 neighbors, as Wendlandt et al. (2018) did, or more could yield different qualitative patterns in document type comparisons and bears exploration.

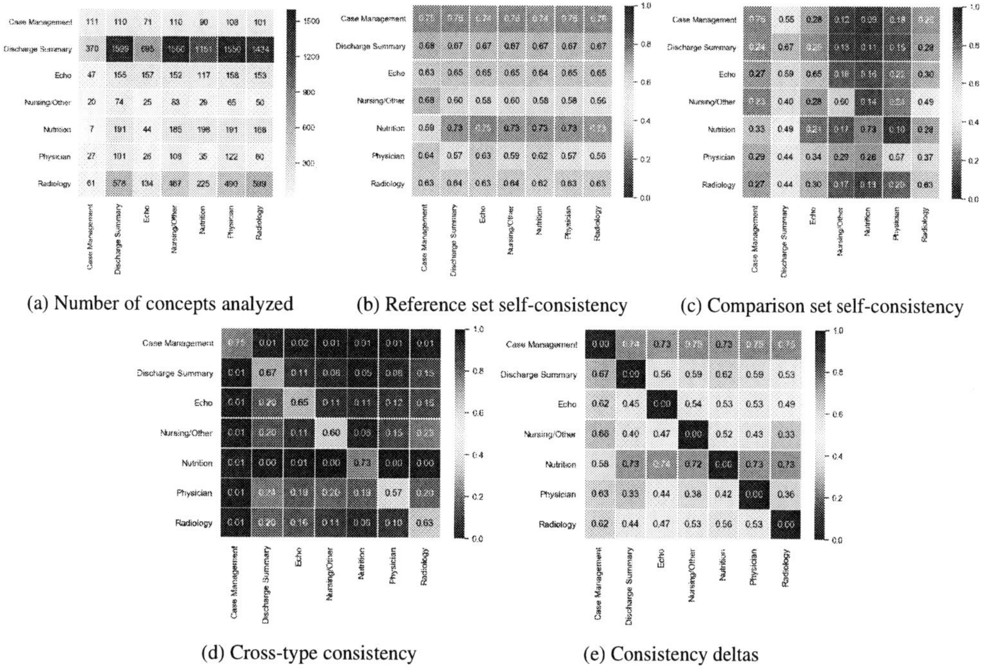

<table>
<tr><td>(a) Number of concepts analyzed</td><td>(b) Reference set self-consistency</td><td>(c) Comparison set self-consistency</td></tr>
<tr><td>(d) Cross-type consistency</td><td>(e) Consistency deltas</td></tr>
</table>

Figure 2: Comparison of concept neighborhood consistency statistics across document types, using high-confidence concepts from the reference type. Figure 2a provides the number of concepts shared between the high-confidence reference set and the comparison set. All values are the mean of the consistency distribution calculated over all concepts analyzed for the document type pair.

retaining only those with a self-consistency of at least 50%; Table 1 includes the number of high-confidence concepts identified and the mean consistency among this subset.[3] These embeddings capture reliable concept usage information for each document type, and form the basis of our comparative analysis.

4.2 Cross-corpus analysis

Our key question is what concept embeddings reveal about clinical concept usage *between* document types. To maintain a sufficient sample size, we restrict our comparison to the 7 document types with at least 50 high confidence concepts: *Case Management, Discharge Summary, Echo, Nursing/Other, Nutrition, Physician*, and *Radiology. Physician, ECG*, and *Nursing* were also used by Feldman et al. (2016) for their lexicosyn-

tactic analysis, although they combined *Nursing* documents (longer narratives) and *Nursing/Other* (which tend to be much shorter) into a single set, while we retain the distinction. Interestingly, the fourth type they analyzed, *ECG*, produced only 14 high-confidence concepts in our analysis, suggesting high semantic variability despite the large number of documents.

As learned concept sets differ between document types, the first step for comparing a document type pair is to identify the set of concepts embedded for both. For reference type A and comparison type B, we identify high-confidence concepts from A that are also present in B, and calculate four distributions using this shared set:

Reference consistency: self-consistency across each of the shared concepts, using only other shared concepts to identify nearest neighborhoods in embeddings for the reference set.

Comparison consistency: self-consistency of each shared concept in embeddings for the comparison document type, again using only shared concepts for neighbors. As the shared set is based on high-confidence concepts from the reference

---

[3] We found in our analysis that most concept consistency numbers clustered roughly bimodally, between 0-30% or 60-90%; this is reflected at a coarse level in the overall distributions in Figure 1a. Varying the threshold outside of these ranges did not have a significant impact on the number of concepts retained; the 50% threshold was chosen for simplicity. With larger corpora, yielding higher concept coverage, a higher threshold could be chosen for a stricter analysis.

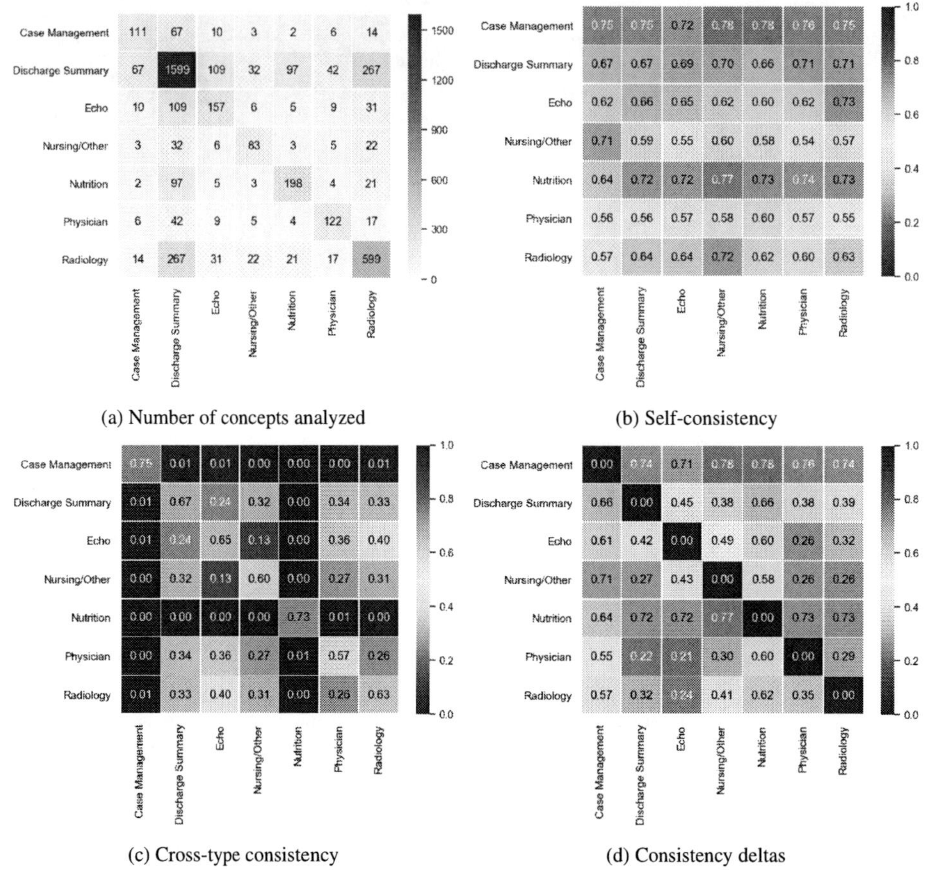

(a) Number of concepts analyzed

(b) Self-consistency

(c) Cross-type consistency

(d) Consistency deltas

Figure 3: Concept neighborhood consistency statistics, restricted to concepts that are high-confidence in both reference and comparison sets. In this case, reference self-consistency and target self-consistency are symmetric, so only reference self-consistency is presented in Figure 3b.

set, this is not symmetric with reference consistency (as the high-confidence sets may differ).

Cross-type consistency: average consistency for each shared concept calculated over all pairs of replicates (i.e., comparing the nearest neighbors of all 10 reference embedding sets to the nearest neighbors in all 10 comparison embedding sets).

Consistency deltas: the difference, for each shared concept, between its reference self-consistency and its cross-type consistency. This provides a direct evaluation of how distinct the concept usage is between two document types, where a high delta indicates highly distinct usage.

Mean values for these distributions are provided for each pair of our 7 document types in Figure 2. Comparing Figures 2b and 2c, it is clear that high-confidence concepts for one document type are typically not high-confidence for another. Most document type pairs show fairly strong diver-

gence, with deltas ranging from 30-60%. *Physician* notes have comparatively high cross-set consistency of around 20% for their high-confidence concepts, likely reflecting the all-purpose nature of these documents, which include patient history, medications, vitals, and detailed examination notes. Interestingly, *Case Management* and *Nutrition* are starkly divergent from other document types, with near-zero cross-set consistency and comparatively high self-consistency of over 70% in the compared concept sets, despite a relatively high overlap between their high-confidence sets and concepts learned for other document types.

In order to control for the low overlap between high-confidence sets in different document types, we also re-ran our consistency analysis restricted to only concepts that are high-confidence in *both* the reference and comparison sets. As shown in Figure 3, this yields considerably smaller concept

Query	Discharge Summary	Nursing/Other	Radiology
Diabetes Mellitus (C0011849)	Diabetes (C0011847)	Gestational Diabetes (C0085207)	Poorly controlled (C3853134)
	Type 2 (C0441730)	A2 immunologic symbol (C1443036)	Insulin (C0021641)
	Type 1 (C0441729)	Diabetes Mellitus, Insulin-Dependent (C0011854)	Diabetes Mellitus, Insulin-Dependent (C0011854)
	Gestational Diabetes (C0085207)	Factor V (C0015498)	Diabetes Mellitus, Non-Insulin-Dependent (C0011860)
	Diabetes Mellitus, Insulin-Dependent (C0011854)	A1 immunologic symbol (C1443035)	Stage level 5 (C0441777)

Query	Discharge Summary	Echo	Radiology
Mental state (C0278060)†	Coherent (C4068804)	Donor:Type:Point in time:^Patient:Nominal (C3263710)	Mental status changes (C0856054)
	Confusion (C0009676)	Donor person (C0013018)	Abnormal mental state (C0278061)
	Respiratory status:-:Point in time:^Patient:- (C2598168)	Respiratory arrest (C0162297)	Level of consciousness (C0234425)
	Respiratory status (C1998827)	Organ donor:Type:Point in time:^Donor:Nominal (C1716004)	Level of consciousness:Find:Pt:^Patient:Ord (C4050479)
	Abnormal mental state (C0278061)	Swallowing G-code (C4281783)	Mississippi (state) (C0026221)

Table 2: 5 nearest neighbor concepts to *Diabetes Mellitus* and *Mental state* from 3 high-confidence document types, averaging cosine similarities across all replicate embedding sets within each document type. †The two nearest neighbors to *Mental state* for all three document types were two LOINC codes using the same "mental status" string; they are omitted here for brevity.

sets for comparison, with single-digit overlap for 18/42 non-self pairings. Cross-set consistency increases somewhat, most significantly for pairings involving *Physician* or *Radiology*; however, no consistency delta falls below 20% for any non-self pair, indicating that concept neighborhoods remain distinct even within high-confidence sets.

4.3 Qualitative neighborhood analysis

Analysis of neighborhood consistency enables measuring divergence in the contextual usage patterns of clinical concepts; however, this divergence could be due to spurious or semantically uninformative correlations instead of clinically-relevant distinctions in concept similarities. To confirm that our methodology captures informative distinctions in concept usage, we qualitatively review example neighborhoods. To mitigate variability of nearest neighborhoods in embedding spaces, we identify a concept's *qualitative* nearest neighbors for a given document type by calculating its pairwise cosine distance vectors for all 10 replicates in that document type and taking the k neighbors with lowest average distance.

As with our consistency analyses, we focus on the neighborhoods of high-confidence concepts, although we do not filter the neighborhoods themselves. Of all high-confidence concepts identified in our embeddings, only two were high-confidence in 5 different document types, and these were highly generic concepts: *Interventional procedure* (C0184661) and a corresponding LOINC code (C0945766). Seven concepts were high-confidence for 4 document types; of these, two were generic procedure concepts, two were concepts for the broad gastrointestinal category, and three were versions of body weight. For a diversity of concepts, we therefore turned to the 75 concepts that were high-confidence within 3 document types. We reviewed each of these concepts, and describe our findings for three of the most broadly clinically-relevant below.

Diabetes Mellitus (C0011849) *Diabetes Mellitus* (search strings: "diabetes mellitus" and "diabetes mellitus dm") was high-confidence in *Discharge Summary*, *Nursing/Other*, and *Radiology* document types; Table 2 gives the top 5 neighbors from each type. These neighbors are semantically consistent across document types: more specific diabetes-related concepts, related biological factors; continuing down the nearest neighbors list yields related symptoms and comorbidities such as *Irritable Bowel Syndrome* (C0022104) and *Gastroesophageal reflux disease* (C0017168).

Memory loss (C0751295) *Memory loss* (search string: "memory loss") was also high-confidence in *Discharge Summary*, *Nursing/Other*, and *Radiology* documents. For brevity, its nearest neighbors are omitted from Table 2, as there is little variation among the top 5. However, the next neighbors (at only slightly greater cosine distance) vary considerably across document types, while remaining highly consistent within each individual type. In *Discharge Summary*, more high-level concepts related to overall function emerge, such as *Functional status* (C0598463), *Relationships* (C0439849), and *Rambling* (C4068735). *Radiology* yields more symptomatically-related neighbors: *Aphagia* (C0221470) is present in both, but *Radiology* includes *Disorientation* (C0233407), *Delusions* (C0011253), and *Gait, Unsteady* (C0231686). Finally, *Nursing/Other* finds concepts more related to daily life, such as *Cigars* (C0678446) and *Multifocals* (C3843228), though at a greater cosine distance than the other document types (Figure 4).

Mental state (C0278060) *Mental state* (search strings: "mental status", "mental state") was high-confidence in *Discharge Summary*, *Echo*, and *Radiology*, and highlighted an unexpected consequence of relying on the Distributional Hypothesis (Harris, 1954) for semantic characterization in sublanguage-specific corpora. The top 5 nearest neighbors (excluding two trivial LOINC codes for the same concept, also using the "mental status" search string) are given in Table 2. In *Discharge Summary* documents, "mental status" is typically referred to in detailed patient narratives, medication lists, and the like, and this yields semantically-reasonable nearest neighbors such as *Confusion* (C0009676) and *Coherent* (C4068804).

In *Echo* documents, however, "mental status" occurs most frequently within an "Indication" field of the "PATIENT/TEST INFORMATION" section. Two common patterns emerge in "Indication" texts: references to altered or reduced mental status, or patients who are vegetative and being evaluated for organ donor eligibility. Though "mental status" and "organ donor" do not cooccur, their consistent occurrence in the same contextual structures leads to extremely similar em-

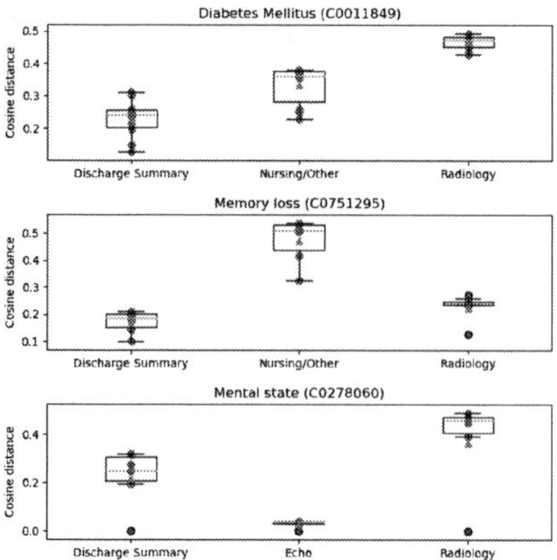

Figure 4: Cosine distance distribution of three concepts to their 10 nearest neighbors, averaged across document type replicate embeddings.

beddings (see Figure 4). A similar issue occurs in *Radiology* notes, where the "MEDICAL CONDITION" section includes several instances of elderly patients presenting with either hypothermia or altered mental status; as a result, two hypothermia concepts (C1963170 and C0020672) are in the 10 nearest neighbors to *Mental state*.

Results from *Radiology* also highlight one limitation of distant supervision for learning concept embeddings: as the word "state" is polysemous, including a geopolitical entity, geographical concepts such as *Mississippi* (C0026221) end up with similar embeddings to *Mental state*. A similar issue occurs in the neighbors for *Memory loss*; due to string polysemy, the concept *CIGAR string - sequence alignment* (C4255278) ends up with a similar embedding to *Cigars* (C0678446).

4.4 Nearest surface form embeddings

As JET learns embeddings of concepts and their surface forms jointly in a single vector space, we also analyzed the surface forms embeddings nearest to different concepts. This enabled us both to evaluate the semantic congruence of surface form and concept embeddings, and to further delve into corpus-specific contextual patterns that emerge in the vector space. As with our concept neighborhood analysis, for each of our 10 replicate embeddings in each document type, we calculated the cosine distance vector from each high-confidence concept to all of the term embeddings in the same replicate, and then averaged these distance vectors to identify neighbors robust to embedding noise. Table 3 presents surface form neighbors identified

Query	Discharge Summary	Nutrition	Case Management
	Community	Dilute	Substance
	Health center	Social work	Monitoring
Community (C0009462)	Acquired	Surgical site	Somewhat
	Residence	In situ	Hearing
	Nursing facility	Nephritis	Speech
	Discharge Summary	Echo	ECG
	ECG	ECG	ECG
	EKG	Exercise	Physician
ECG (C0013798)	Sinus tachycardia	Stress	Last
	Sinus bradycardia	Fair	No change
	Right bundle branch block	Specific	Abnormal
	Discharge Summary	Echo	Radiology
	Blood pressure	Blood pressure	Blood pressure
	Heart rate	Heart rate	Heart rate
Blood pressure (C0005823)	Pressure	Rate	Rate
	Systolic blood pressure	Exercise	Method
	Rate	Stress	Exercise

Table 3: 5 nearest neighbor surface forms to three frequent clinical concepts, across document types for which they are high-confidence.

for three high-confidence clinical concepts chosen for clinical relevance and wide usage; these concepts are discussed in the following paragraphs.

Blood pressure (C0005823) *Blood pressure* is high-confidence in *Discharge Summary*, *Echo*, and *Radiology* documents. It is a key concept that is measured frequently in various settings; intuitively, it is a sufficiently core concept that it should exhibit little variance. Its neighbor surface forms indeed indicate fairly consistent use across the three document types, referencing both related measurements ("heart rate") and related concepts ("exercise" and "stress").

Echocardiogram (C0013798) *Echocardiogram* is high-confidence in *Discharge Summary*, *Echo* (detailed summaries and interpretation written after the ECG), and *ECG* (technical notes taken during the procedure) documents. ECGs are common, and are performed for various purposes and discussed in varying detail. Interestingly, neighbor surface forms in *Discharge Summary* embeddings reflect specific pathologies, potentially capturing details determined post diagnosis and treatment. In *Echo* embeddings, the neighbors are more general surface forms evaluating the findings ("fair") and relevant history/symptoms that led to the ECG ("exercise", "stress"). *ECG* embeddings reflect their more technical nature, with surface forms such as "no change" and "abnormal" yielding high similarity.

Community (C0009462) *Community* is a very broad concept and a common word, and is discussed primarily in documents concerned with whole-person health; it is high confidence in *Discharge Summary*, *Nutrition*, and *Case Management* documents. Each of these document types reflects different usage patterns. The nearest surface forms in *Discharge Summary* embeddings reflect a focus on living conditions, referring to "health center", "residence", and "nursing facility". In *Nutrition* documents, *Community* is discussed primarily in terms of "community-acquired pneumonia", likely leading to more treatment-oriented neighbor surface forms. Finally, in *Case Management* embeddings, nearby surface forms reflect discussion of specific risk factors or resources ("substance", "monitoring") to consider in maintaining the patient's health and responding to their specific needs (e.g., "hearing", "speech").

5 Discussion

We have shown that concept embeddings learned from different clinical document type corpora reveal characteristics of how clinical concepts are used in different settings. This suggests that sublanguage-specific embeddings can help profile distinctive usage patterns for text categorization, offering greater specificity than latent topic distributions while not relying on potentially brittle lexical features. In addition, such profiles could also assist with concept normalization by providing more-informed prior probability distributions for medical vocabulary senses that are conditioned on the document or section type that they occur in.

A few limitations of our study are important to note. The embedding method we chose offers flexibility to work with arbitrary corpora and vocabularies, but its use of distant supervision introduces some undesirable noise. The example given in Section 4.3 of the similar embeddings learned for the concept *cigars* and the concept of the CIGAR string in genomic sequence editing illustrates the downside of not leveraging disambiguation techniques to filter out noisy matches. On the other hand, our restriction to strings from SNOMED-CT and LOINC provided a high-quality set of strings intended for clinical use, but also removed many potentially helpful strings from consideration. For example, the UMLS also includes the non-SNOMED/LOINC strings "diabetes" and "diabete mellitus" [*sic*] for *Diabetes Mellitus* (C0011849), both of which occur frequently in MIMIC data. Misspellings are also common in clinical data; leveraging well-developed technologies for clinical spelling correction would likely increase the coverage and confidence of sublanguage concept embeddings.

At the same time, the low volume of data analyzed in many document types introduces its own challenges for the learning process. First, though JET can in principle learn embeddings for every concept in a given terminology, this is predicated on the relevant surface forms appearing with sufficient frequency. For a small document sample, many such surface forms that would otherwise be present in a larger sample will either be missing entirely or insufficiently frequent, leading to effectively "missed" concepts. While we are not aware of another concept embedding method compatible with arbitrary unannotated corpora that could help avoid these issues, some strategies could be used to reduce the potential impact of both training noise and low sample sizes. One approach, which might also help improve concept consistency in the document types that yielded few or no high-confidence concepts, would be pretraining a shared base embedding on a large corpus such as PubMed abstracts, which could then be tuned on each document type-specific subcorpus. While this could introduce its own noise in terms of the differences between biomedical literature language and clinical language (Friedman et al., 2002), it could help control for some degree of sampling error and provide a linguistically-motivated initialization for the concept embedding models.

Finally, as we observed with *Mental state* (C0278060), relying on similarity in contextual patterns can lead to capturing more corpus-specific features with embeddigns, as opposed to (sub)language-specific features, as target corpora become smaller and more homogeneous. If a particular concept or set of concepts are always used within the same section of a document, or in the same set phrasing, the "similarity" captured by organization of an embedding space will be more informed by this writing habit endemic to the specific corpus than by clinically-informed semantic patterns that can generalize to other corpora.

6 Conclusion

Analyzing nearest neighborhoods in embedding spaces has become a powerful tool in studying diachronic language change. We have described how the same principles can be applied to sublanguage analysis, and demonstrated that the structure of concept embedding spaces captures distinctive and relevant semantic characteristics of different clinical document types. This offers a valuable tool for sublanguage characterization, and a promising avenue for developing document type "fingerprints" for text categorization and knowledge-based concept normalization.

Acknowledgments

The authors gratefully thank Guy Divita for helpful feedback on early versions of the manuscript. This research was supported by the Intramural Research Program of the National Institutes of Health and the US Social Security Administration.

References

Alan R Aronson and François-Michel Lang. 2010. An overview of metamap: historical perspective and recent advances. *Journal of the American Medical Informatics Association*, 17(3):229–36.

Olivier Bodenreider. 2004. The Unified Medical Language System (UMLS): integrating biomedical terminology. *Nucleic Acids Research*, 32(90001):D267–D270.

Edward Choi, Mohammad Taha Bahadori, Elizabeth Searles, Catherine Coffey, and Jimeng Sun. 2016. Multi-layer Representation Learning for Medical Concepts. In *Proceedings of the 22nd ACM SIGKDD International Conference on Knowledge Discovery and Data Mining*, KDD '16, pages 1495–1504, San Francisco, California, USA. ACM.

Lance De Vine, Guido Zuccon, Bevan Koopman, Laurianne Sitbon, and Peter Bruza. 2014. Medical semantic similarity with a neural language model. In

Proceedings of the 23rd ACM International Conference on Information and Knowledge Management - CIKM '14, CIKM '14, pages 1819–1822, Shanghai, China. ACM.

Kerstin Denecke. 2014. Sublanguage Analysis of Medical Weblogs. *Studies in Health Technology and Informatics*, 205:565–569.

Patrick Drouin. 2004. Detection of Domain Specific Terminology Using Corpora Comparison. In *Proceedings of the Fourth International Conference on Language Resources and Evaluation (LREC '04)*, Lisbon, Portugal. European Language Resources Association (ELRA).

K Feldman, N Hazekamp, and N V Chawla. 2016. Mining the Clinical Narrative: All Text are Not Equal. In *2016 IEEE International Conference on Healthcare Informatics (ICHI)*, pages 271–280.

Carol Friedman. 1986. Automatic structuring of sublanguage information. In Ralph Grishman and Richard Kittredge, editors, *Analyzing Language in Restricted Domains: Sublanguage Description and Processing*, pages 85–102. Lawrence Erlbaum Associates.

Carol Friedman, Pauline Kra, and Andrey Rzhetsky. 2002. Two biomedical sublanguages: A description based on the theories of Zellig Harris. *Journal of Biomedical Informatics*, 35(4):222–235.

Nikhil Garg, Londa Schiebinger, Dan Jurafsky, and James Zou. 2018. Word embeddings quantify 100 years of gender and ethnic stereotypes. *Proceedings of the National Academy of Sciences*, 115(16):E3635—E3644.

Denis Griffis, Chaitanya Shivade, Eric Fosler-Lussier, and Albert M Lai. 2016. A Quantitative and Qualitative Evaluation of Sentence Boundary Detection for the Clinical Domain. In *AMIA Summits on Translational Science Proceedings 2016*, pages 88–97. American Medical Informatics Association.

Ralph Grishman. 2001. Adaptive information extraction and sublanguage analysis. In *Proceedings of the Workshop on Adaptive Text Extraction and Mining, Seventeenth International Joint Conference on Artificial Intelligence (IJCAI-2001)*, pages 1–4, Seattle, Washington, USA.

Leonie Grön, Ann Bertels, and Kris Heylen. 2019. Leveraging Sublanguage Features for the Semantic Categorization of Clinical Terms. In *Proceedings of the 18th BioNLP Workshop and Shared Task*, pages 211–216, Florence, Italy. Association for Computational Linguistics.

William L. Hamilton, Jure Leskovec, and Dan Jurafsky. 2016. Diachronic Word Embeddings Reveal Statistical Laws of Semantic Change. In *Proceedings of the 54th Annual Meeting of the Association for Computational Linguistics (Volume 1: Long Papers)*, pages 1489–1501, Berlin, Germany. Association for Computational Linguistics.

Zellig S. Harris. 1954. Distributional Structure. *Word*, 10(2-3):146–162.

Matthew Honnibal and Ines Montani. 2017. spaCy 2: Natural language understanding with Bloom embeddings, convolutional neural networks and incremental parsing. *To appear.*

Alistair E W Johnson, Tom J Pollard, Lu Shen, Li-Wei H Lehman, Mengling Feng, Mohammad Ghassemi, Benjamin Moody, Peter Szolovits, Leo Anthony Celi, and Roger G Mark. 2016. MIMIC-III, a freely accessible critical care database. *Scientific Data*, 3:160035.

Veronika Laippala, Filip Ginter, Sampo Pyysalo, and Tapio Salakoski. 2009. Towards automated processing of clinical Finnish: Sublanguage analysis and a rule-based parser. *International Journal of Medical Informatics*, 78(12):e7 – e12.

Yen-Fu Luo, Weiyi Sun, and Anna Rumshisky. 2019. MCN: A comprehensive corpus for medical concept normalization. *Journal of Biomedical Informatics*, 92:103132.

Eneldo Loza Mencia, Gerard de Melo, and Jinseok Nam. 2016. Medical Concept Embeddings via Labeled Background Corpora. In *Proceedings of the Tenth International Conference on Language Resources and Evaluation (LREC 2016)*, pages 4629–4636. European Language Resources Association (ELRA).

Denis Newman-Griffis, Albert M Lai, and Eric Fosler-Lussier. 2018. Jointly Embedding Entities and Text with Distant Supervision. In *Proceedings of The Third Workshop on Representation Learning for NLP*, pages 195–206. Association for Computational Linguistics.

Minh C Phan, Aixin Sun, and Yi Tay. 2019. Robust Representation Learning of Biomedical Names. In *Proceedings of the 57th Annual Meeting of the Association for Computational Linguistics*, pages 3275–3285, Florence, Italy. Association for Computational Linguistics.

Guergana K Savova, James J Masanz, Philip V Ogren, Jiaping Zheng, Sunghwan Sohn, Karin C Kipper-Schuler, and Christopher G Chute. 2010. Mayo clinical text analysis and knowledge extraction system (ctakes): architecture, component evaluation and applications. *Journal of the American Medical Informatics Association*, 17(5):507–513.

Irina P Temnikova, William A Baumgartner, Negacy D Hailu, Ivelina Nikolova, Tony McEnery, Adam Kilgarriff, Galia Angelova, and K Bretonnel Cohen. 2014. Sublanguage Corpus Analysis Toolkit: A tool for assessing the representativeness and sublanguage characteristics of corpora. *Proceedings of the Ninth International Conference on Language Resources and Evaluation (LREC)*, 2014:1714–1718.

Gaurav Vashisth, Jan-Niklas Voigt-Antons, Michael Mikhailov, and Roland Roller. 2019. Exploring Diachronic Changes of Biomedical Knowledge using Distributed Concept Representations. In *Proceedings of the 18th BioNLP Workshop and Shared Task*, pages 348–358, Florence, Italy. Association for Computational Linguistics.

Ekaterina Vylomova, Sean Murphy, and Nicholas Haslam. 2019. Evaluation of Semantic Change of Harm-Related Concepts in Psychology. In *Proceedings of the 1st International Workshop on Computational Approaches to Historical Language Change*, pages 29–34, Florence, Italy. Association for Computational Linguistics.

Donald E Walker and Robert A Amsler. 1986. The use of machine-readable dictionaries in sublanguage analysis. In Ralph Grishman and Richard Kittredge, editors, *Analyzing Language in Restricted Domains: Sublanguage Description and Processing*, pages 69–83. Lawrence Erlbaum Associates, Hillsdale, NJ.

Laura Wendlandt, Jonathan K Kummerfeld, and Rada Mihalcea. 2018. Factors Influencing the Surprising Instability of Word Embeddings. In *Proceedings of the 2018 Conference of the North American Chapter of the Association for Computational Linguistics: Human Language Technologies, Volume 1 (Long Papers)*, pages 2092–2102. Association for Computational Linguistics.

Cheng Ye and Daniel Fabbri. 2018. Extracting similar terms from multiple EMR-based semantic embeddings to support chart reviews. *Journal of Biomedical Informatics*, 83:63–72.

Qing T Zeng, Doug Redd, Guy Divita, Samah Jarad, Cynthia Brandt, and Jonathan R Nebeker. 2011. Characterizing Clinical Text and Sublanguage: A Case Study of the VA Clinical Notes. *Journal of Health & Medical Informatics*, S3.

Extracting UMLS Concepts from Medical Text Using General and Domain-Specific Deep Learning Models

Kathleen C. Fraser[1]*, Isar Nejadgholi[1]*, Berry De Bruijn[1], Muqun Li[2],
Astha LaPlante[2], Khaldoun Zine El Abidine[2]
[1]National Research Council Canada
[2]Privacy Analytics Inc., Ottawa, Canada
{kathleen.fraser, isar.nejadgholi, berry.debruijn}@nrc-cnrc.gc.ca
{rachel.li, astha.agarwal, khaldoun.zineelabidine}@privacy-analytics.com

Abstract

Entity recognition is a critical first step to a number of clinical NLP applications, such as entity linking and relation extraction. We present the first attempt to apply state-of-the-art entity recognition approaches on a newly released dataset, MedMentions. This dataset contains over 4000 biomedical abstracts, annotated for UMLS semantic types. In comparison to existing datasets, MedMentions contains a far greater number of entity types, and thus represents a more challenging but realistic scenario in a real-world setting. We explore a number of relevant dimensions, including the use of contextual versus non-contextual word embeddings, general versus domain-specific unsupervised pre-training, and different deep learning architectures. We contrast our results against the well-known i2b2 2010 entity recognition dataset, and propose a new method to combine general and domain-specific information. While producing a state-of-the-art result for the i2b2 2010 task (F1 = 0.90), our results on MedMentions are significantly lower (F1 = 0.63), suggesting there is still plenty of opportunity for improvement on this new data.

1 Introduction

Entity recognition is a widely-studied task in clinical NLP, and has been the focus of a number of shared tasks, including the i2b2 2010 Shared Task (Uzuner et al., 2011), SemEval 2014 Task 7 (Pradhan et al., 2014), and SemEval 2015 Task 14 (Elhadad et al., 2015). Most previous work has focused on identifying only a few broad types of entities, such as 'problems', 'tests', and 'treatments' in the i2b2 task, and 'diseases' in the SemEval tasks. Even when corpora have been annotated for more entity types, as in the GENIA corpus of biological annotations (Ohta et al., 2002), entity

recognition tasks typically focus on only a small subset of those (Kim et al., 2004).

However, in some downstream applications it would be useful to identify *all* terms in a document which exist as concepts in the Unified Medical Language System (UMLS) Metathesaurus (Bodenreider, 2004). This resource comprises a much wider range of biomedical entity types than has previously been considered in clinical entity recognition. Additionally, the UMLS Metathesaurus defines important relationships between entity types (and the lower-level concepts associated with them) via its Semantic Network. Therefore, extracting and labelling entities with respect to their UMLS semantic type, rather than more generic types such as 'problem' or 'test', can be an important first step in many practical clinical NLP applications.

In this work, we present the first attempt to apply existing clinical entity recognition approaches to a new dataset called *MedMentions*, which is annotated for all UMLS semantic types (Mohan and Li, 2019). We compare the effectiveness of these approaches with reference to a well-known baseline dataset (i2b2 2010) and analyze the errors that occur when applying such techniques to new problems. On the basis of this error analysis, we propose an adaptation to the BERT architecture to better combine the general and clinical knowledge learned in the pre-training phase, and show that this improves over the more basic approaches.

2 Background

Early successes in clinical/biomedical entity extraction employed approaches such as conditional random fields (Jonnalagadda et al., 2012; Fu and Ananiadou, 2014; Boag et al., 2015) and semi-Markov models (De Bruijn et al., 2011), requiring numerous engineered features. In recent years,

---

*These authors contributed equally.

Proceedings of the 10th International Workshop on Health Text Mining and Information Analysis (LOUHI 2019), pages 157–167
Hong Kong, November 3, 2019. ©2019 Association for Computational Linguistics
https://doi.org/10.18653/v1/D19-62

such approaches have been surpassed in performance by deep learning models (Habibi et al., 2017). However, there is a wide range of variation possible within this set of techniques. We briefly discuss some of the parameters of interest in the following sections.

2.1 General vs. Domain-Specific Word Embeddings

Since words may have one dominant meaning in common use, and a different meaning in the medical domain, some work has explored whether word embeddings trained on medical text (e.g. clinical notes, medical journal articles) are more effective in medical entity recognition than those trained on general text sources (e.g. news, Wikipedia).

Roberts (2016) examined the effect of training word embeddings on different corpora for the task of entity extraction on the i2b2 2010 dataset. He compared six corpora: the i2b2 dataset itself, the clinical notes available in the MIMIC database (Johnson et al., 2016), MEDLINE article abstracts, WebMD forum posts, and generic text corpora from Wikipedia and Gigaword. It was found that the best F1 score was obtained by training on the MIMIC corpus, and that combining corpora also led to strong results. Si et al. (2019) also compared training embeddings on MIMIC data versus general domain data, and similarly found that pre-training on the MIMIC data led to better performance on both the i2b2 2010 and SemEval tasks. Alsentzer et al. (2019) trained embeddings only on the discharge summaries from MIMIC, and reported a marginal improvement on the i2b2 2010 task over using the entire MIMIC corpus. Peng et al. (2019) found that pre-training a BERT model on PubMed abstracts led to better performance for biomedical entity extraction, while pre-training on a combination of PubMed abstracts and MIMIC notes led to better performance when extracting entities from patient records.

2.2 Contextual vs. Non-Contextual Word Embeddings

For many years, word embeddings were non-contextual; that is, a word would have the same embedding regardless of the context in which it occurred. Popular word embeddings of this type include GloVe (Pennington et al., 2014), word2vec (Mikolov et al., 2013), and FastText (Bojanowski et al., 2017). Peters et al. (2018) popularized the idea of contextualized word embeddings, which allowed the same word to have a different representation, depending on the context. The character-based ELMo word embeddings introduced by Peters et al. (2018) can be used just as the non-contextual word embeddings were. Sheikhshabbafghi et al. (2018) trained ELMo word embeddings on a dataset of biomedical papers and achieved a new state of the art in gene mention detection on the BioCreative II gene mention shared task. This work showed that domain-specific contextual embeddings improve various types of biomedical named entities recognition. Later in 2018, BERT embeddings were also introduced (Devlin et al., 2019). The BERT architecture improved over ELMo by using a different training objective to better take into account both left and right contexts of a word, and made it possible to make use of the entire pre-trained network in the downstream task, rather than simply extracting the embedding vectors.

Si et al. (2019) compared word2vec, GloVe, FastText, ELMo, and BERT embeddings on the i2b2 2010 dataset. When using the pre-trained vectors (trained on general-domain corpora), BERT-large performed the best and word2vec performed the worst, but there was no clear advantage to the contextualized embeddings (e.g. GloVe performed better than ELMo). When the embeddings were pre-trained on MIMIC data, the contextualized embeddings did perform appreciably better than the non-contextualized embeddings.

2.3 Classifier Architecture

Much of the recent work on medical entity extraction has made use of the Long Short-Term Memory (LSTM) architecture (Hochreiter and Schmidhuber, 1997), with some variations and modifications: (1) most work uses a bi-directional LSTM (bi-LSTM), so the prediction for any word in the sequence can take into account information from both the left and right contexts, (2) some work additionally feeds the output of the bi-LSTM layer into a CRF classifier (Huang et al., 2015a; Chalapathy et al., 2016a; Lample et al., 2016; Habibi et al., 2017; Tourille et al., 2018), to predict the most likely sequence of labels, rather than just the most likely label for each word independently, and (3) some models incorporate additional information (e.g. character embeddings, or traditionally engineered features) at various points in the model (Unanue et al., 2017).

In contrast, the BERT model makes use of the Transformer architecture, an attention-based method for sequence-to-sequence modelling (Vaswani et al., 2017). Once the model has been pre-trained, in the entity extraction stage it is only necessary to add a simple classification layer on the output. However, others have also experimented with feeding the output of the BERT model to a bi-LSTM (Si et al., 2019).

3   Methods

3.1   Data

We consider three datasets in this study: the i2b2 2010 dataset, the 'full' MedMentions dataset, and the 'st21pv' MedMentions dataset. Details are shown in Table 1.

The i2b2 2010 corpus[1] consists of de-identified clinical notes (discharge summaries), annotated for three entity types: problems, tests, and treatments (Uzuner et al., 2011). The original shared task also included subtasks on assertion classification and relation extraction, but we focus here on entity extraction. In the original data release for the shared task, the training set contained 394 documents; however, the current release of the dataset contains only 170 training documents. Therefore, it is unfortunately not possible to directly compare results across the two versions of the dataset. However, a majority of the recent works are using the current release of the dataset (Zhu et al., 2018; Bhatia et al., 2019; Chalapathy et al., 2016b).

The MedMentions corpus was released earlier this year[2], and contains 4,392 abstracts from PubMed, annotated for concepts and semantic types from UMLS (2017AA release). UMLS *concepts* are fine-grained biomedical terms, with approximately 3.2 million unique concepts contained in the metathesaurus (Mohan and Li, 2019). Each concept is linked to a higher-level *semantic type*, such as 'Disease or syndrome', 'Cell component', or 'Clinical attribute'. In this work we focus on identifying the semantic type for each extracted text span, leaving the concept linking/normalization for future work. The creators of the dataset have defined an official 60%-20%-20% partitioning of the corpus into training, development, and test sets.

There are 127 semantic types in UMLS. Of these, there is only one ('Carbohydrate sequence')

which never appears in the full MedMentions dataset. Approximately 8% of the concepts in UMLS can be linked to more than one semantic type (Mohan and Li, 2019); in such cases the dataset contains a comma-separated list of all these type IDs corresponding to alphabetical order of semantic types. Where a text span has been labelled with more than one label, we select only the first one. As a result of this, there is one other type ('Enzyme') which appears in MedMentions, but only doubly-labelled with 'Amino acid, peptide, or protein', and thus does not occur in our singly-labelled training or test data. Finally, there is an extra class ('UnknownType'), for a total of 126 semantic types or classes in the 'full' training data. Of these, there are three ('Amphibian', 'Drug delivery device', and 'Vitamin') which never occur in the test data.

The full MedMentions dataset suffers from high class imbalance (e.g. there are 31,485 mentions for the semantic type 'Qualitative concept' and only two mentions for 'Fully formed anatomical structure'). Furthermore, many of the semantic types are not particularly useful in downstream clinical NLP tasks, either due to being too broad or too specialized. As a result, the creators of the MedMentions dataset also released an alternate version called 'st21pv', which stands for '21 semantic types from preferred vocabularies'. The details of how this subset was constructed are given by Mohan and Li (2019), but essentially it contains only 21 semantic types, from specific vocabularies most relevant to biomedical researchers. The raw abstracts, and partitions into training, development, and test sets are the same as in the full dataset – only the set of annotations differs.

The i2b2 and MedMentions datasets differ across a number of important dimensions: the discharge summaries in the i2b2 dataset tend to be hastily written or dictated, with short, incomplete sentences and numerous acronyms and abbreviations, compared to the academic writing style of the MedMentions abstracts. The discharge summaries also tend to be longer, averaging approximately 980 tokens per document, compared to 267 tokens per document in MedMentions. The semantic content of the documents is also different, with the discharge summaries focused exclusively on a single patient and their history, disease progression, treatment, and outcomes, while the MedMentions abstracts typically summarize

[1] www.i2b2.org/NLP/DataSets
[2] github.com/chanzuckerberg/MedMentions

	i2b2 2010		MedMentions (full)		MedMentions (st21pv)	
	Train	Test	Train	Test	Train	Test
# entity types	3	3	126	123	21	21
# documents	170	256	3513	879	3513	879
# tokens	149,743	267,837	936,247	234,910	936,247	234,910
# entities	16,520	31,161	281,719	70,305	162,908	40,101

Table 1: Properties of the datasets. For MedMentions, we combine the training and validation sets into 'Train'.

the results of a scientific study, covering a wide range of biomedical topics. Finally, there are clearly far more entity types in MedMentions than in i2b2, and greater imbalance between the different classes. Therefore, there is no guarantee that methods which perform well on the i2b2 data will also be effective on the MedMentions dataset.

3.2 Entity Recognition

Based on our review of the literature (Section 2) we experimented with two basic architectures: bi-LSTM+CRF (with pre-trained contextual and non-contextual word embeddings as input), and BERT (with a simple linear classification layer and a bi-LSTM classification layer). The details of these classifiers and their training are described below.

3.2.1 Text pre-processing

For the bi-LSTM+CRF models, input text retained casing information, but all numerical tokens were normalized to a single NUM token.

BERT uses WordPiece tokenization (Wu et al., 2016), which breaks longer words into frequently occurring sub-word units to improve handling of rare words and morphological variation. This requires additional pre- and post-processing for the entity recognition task, since the data is labelled at the word level. Following the recommendation of Devlin et al. (2019), we first re-tokenize the text using the WordPiece tokenizer, assign the given label to the first piece of each word, and assign any subsequent pieces a padding label.

In all cases, we convert the text and labels to CoNLL IOB format for input to the classifiers.

3.2.2 bi-LSTM+CRF

We use a standard bi-LSTM+CRF architecture (e.g., see (Huang et al., 2015b)), implemented in PyTorch. The bi-LSTM component has 2 bi-directional layers with hidden size of 1536 nodes. The 100-dimensional character embeddings are learned through the training process and concatenated with pre-trained GloVe embeddings (Pennington et al., 2014) as proposed by Chalapathy

et al. (2016a). We compare the performance of general GloVe embeddings, trained on Wikipedia and Gigaword, and clinical GloVe embeddings, trained on the MIMIC-III corpus (Johnson et al., 2016). In both cases the GloVe embeddings have 300 dimensions. For pre-training on MIMIC, we used a minimum frequency cut-off of 5, and a window size of 15.

We also experimented with contextual ELMo embeddings (Peters et al., 2018) and the bi-LSTM+CRF architecture, comparing general ELMo embeddings[3] with clinical ELMo embeddings.[4] The clinical ELMo embeddings were released by Zhu et al. (2018) and trained on Wikipedia pages whose titles are medical concepts in the SNOMED-CT vocabulary, as well as MIMIC-III.

The bi-LSTM+CRF models were trained using the Adam optimizer with a learning rate of 0.001 and a batch size of 32 for 10 epochs.

3.2.3 BERT

The BERT (Bidirectional Encoder Representations from Transformers) model is described by Devlin et al. (2019) and proposes to address some of the limitations observed in LSTM models. In our experiments, we use the BERT-base architecture, which has 12-layers, hidden size 768, and 12 self-attention heads. To perform the entity recognition, we added a linear layer and a softmax layer on top of the last BERT layer to determine the most probable label for each token. While this is the approach taken by Alsentzer et al. (2019), others suggest using a more complex classification model in conjunction with BERT (Si et al., 2019), and so we also experiment with a bi-LSTM layer with input and output size of 4×768 on top of the concatenation of the last four layers of BERT.

We consider four pre-trained BERT models:

- **BERT-base**[5] General domain BERT model

---

[3] github.com/allenai/allennlp/blob/master/tutorials/how_to/elmo.md
[4] github.com/noc-lab/clinical_concept_extraction
[5] github.com/google-research/bert

released by Google, pre-trained on Wikipedia and BookCorpus (Devlin et al., 2019).

- **bioBERT (v1.1)**[6] The bioBERT model is initialized with BERT-base, and then further pre-trained on biomedical abstracts from PubMed (Lee et al., 2019).
- **clinicalBERT**[7] The clinicalBERT model is initialized with bioBERT, and then further pre-trained on clinical notes from the MIMIC corpus (Alsentzer et al., 2019).
- **NCBI BERT**[8] The NCBI BERT model is initialized with BERT-base, and then further pre-trained on PubMed abstracts and MIMIC notes (Peng et al., 2019).

In the fine-tuning stage, we generally follow the recommendations in Devlin et al. (2019), and use an Adam optimizer with $\beta_1 = 0.9$, $\beta_2 = 0.999$, L2 weight decay of 0.01, and a dropout probability of 0.1 on all layers. We use a learning rate warmup over the first 10% of steps, and linear decay of the learning rate thereafter. Before training the final models, we conducted a series of hyper-parameter optimization experiments using 10-fold cross-validation on the training set. In this optimization stage we considered combinations of batch sizes in $\{16, 32\}$, learning rates in $\{0.00002, 0.00003, 0.00005, 0.0001\}$, and number of training epochs in $\{1...10\}$. We also determined that the uncased BERT-base model led to marginally better results, and so we use that in our final evaluation (bioBERT and clinicalBERT are cased, while NCBI BERT is uncased). For the BERT+bi-LSTM model, we also experimented with training only the bi-LSTM component and not fine-tuning the pre-trained layers, but found that fine-tuning led to better results in development.

3.3 Evaluation

To evaluate the systems, we use micro-averaged strict precision, recall, and F-score. This means that for any given recognized entity, it is only counted as a true positive if *both* the span and the label match exactly with the gold standard annotation. Note also that these metrics are computed on the entity-level, not the token level. For example, given the following gold and predicted label sequences:

```
GOLD: O O B-prob I-prob I-prob
PRED: O O B-prob I-prob O
```

A token-level evaluation would identify two true positives, but a strict entity-level evaluation identifies zero true positives.

4 Results

Table 2 shows the results of the entity recognition experiments for each model and dataset.

4.1 Effect of Contextual vs. Non-Contextual Word Embeddings

If we first consider the bi-LSTM+CRF results for the i2b2 dataset, we observe that the contextual ELMo embeddings lead to better results than the non-contextual GloVe embeddings, and in both cases, better results are obtained by pre-training the embeddings on domain-specific text. For MedMentions, however, for both versions of the dataset we observe that the general-domain GloVe embeddings outperform the clinical GloVe embeddings, but the clinical ELMo embeddings outperform the general ELMo embeddings. Si et al. (2019) also observed a greater benefit to using contextual embeddings when pre-training on domain-specific corpora. Here, this may be due in part to differences between the training corpora; for example, clinical GloVe was trained only on MIMIC notes, while clinical ELMo was trained on a combination of MIMIC notes and Wikipedia articles about medical concepts, which may be more similar to the biomedical abstracts contained in MedMentions.

The BERT models offer a substantial improvement in F1 over the models based on Glove or ELMo embeddings for each of the three datasets. For the i2b2 dataset, the best results are obtained using clinicalBERT (F1 = 0.88) and NCBI BERT (F1 = 0.89), each of which involved pre-training on clinical notes from MIMIC. This demonstrates the importance of pre-training on documents which are similar in nature to those seen in the labelled dataset. Consistent with this, the bioBERT model (pre-trained on biomedical abstracts) leads to the best result on both MedMentions datasets.

4.2 Effect of Classifier Structure

Finally, comparing the effectiveness of a simple linear model versus a bi-LSTM model as the top BERT layer, we observe that this change makes

---

[6]github.com/dmis-lab/biobert
[7]github.com/EmilyAlsentzer/clinicalBERT
[8]github.com/ncbi-nlp/NCBI_BERT

Model	Domain	i2b2 2010			MedMentions (full)			MedMentions (st21pv)		
		P	R	F1	P	R	F1	P	R	F1
Glove + bi-LSTM+CRF	general	0.81	0.76	0.79	0.54	0.51	0.52	0.60	0.50	0.54
Glove + bi-LSTM+CRF	clinical	0.83	0.77	0.80	0.45	0.37	0.41	0.59	0.46	0.52
ELMo + bi-LSTM+CRF	general	0.80	0.80	0.80	0.43	0.45	0.44	0.54	0.50	0.52
ELMo + bi-LSTM+CRF	clinical	0.86	0.86	0.86	0.47	0.47	0.47	0.58	0.53	0.56
BERT-base + linear	general	0.85	0.87	0.86	0.51	0.55	0.53	0.58	0.61	0.59
bioBERT + linear	biomed	0.86	0.88	0.87	0.53	0.57	0.55	0.61	0.64	0.62
clinicalBERT + linear	biomed + clinical	0.87	0.88	0.88	0.51	0.56	0.53	0.59	0.62	0.61
NCBI BERT + linear	biomed + clinical	**0.88**	**0.90**	**0.89**	0.51	0.56	0.53	0.59	0.61	0.60
bioBERT + bi-LSTM	biomed	0.86	0.88	0.87	**0.53**	**0.58**	**0.56**	**0.61**	**0.66**	**0.63**
NCBI BERT + bi-LSTM	biomed + clinical	**0.88**	**0.90**	**0.89**	0.52	0.57	0.54	0.59	0.62	0.60

Table 2: Results of entity recognition for each dataset and model.

no difference on the i2b2 dataset, but leads to the best result for both versions of MedMentions. It may be that the much larger training set in Med-Mentions is better able to effectively train the more complex classifier, and respectively that the greater complexity is necessary to properly model the large number of classes in MedMentions.

4.3   MedMentions vs i2b2 dataset

Comparing across datasets, it is clear that performance is worse on the MedMentions data than the i2b2 data. One obvious reason for this is that the MedMentions datasets have many more entity types, or classes, than the i2b2 dataset. This means that the number of examples per class seen in the training data is much lower (in some cases, only a handful), and also that the classes tend to be more easily confused. The ambiguity between classes arises at the level of the training data where, depending on context, the same text span will be associated with different labels (e.g. *neurocognitive function* is sometimes labelled as a 'biologic function' and sometimes as a 'mental process'; *PSA levels* is labelled as both a 'laboratory procedure' and a 'laboratory or test result'; and an even more highly-ambiguous term such as *probe* is variously labelled as a 'medical device', 'indicator reagent or diagnostic aid', 'nucleic acid nucleoside or nucleotide', 'functional concept', 'diagnostic procedure', 'research activity', and 'chemical viewed functionally'). Thus in MedMentions, the context becomes extremely important, and fine-grained distinctions between entity types must be learned.

5   Error Analysis

In the following section, we examine the errors made by the entity recognition systems from two different perspectives: first, we compare generally the types of errors made on the two datasets; then, we consider the role of general versus domain-specific pre-training and examine some of the specific errors that occur in each case.

5.1   Types of Errors by Dataset

Depending on the downstream application of entity recognition, different types of errors may be associated with different costs. For example, if a company is using this model in practice, the cost associated with having human annotators adjust label boundaries may be different from the cost associated with having them search for entities which have been missed altogether. Our evaluation metrics, however, do not reflect the differences among various types of errors. To further investigate the nature of the errors being made by the system, we investigated three specific types of 'partial errors'. These cases are counted as *false* in calculating the evaluation metrics, but the model actually gets at least part of the information correct:

- **Right span, wrong label:** the text span associated with an entity is correctly identified, but assigned the wrong label.
- **Right label, overlapping span:** the entity is correctly labelled, but the text span associated with the entity is not exactly the same as that indicated in the gold transcripts.
- **Wrong label, overlapping span:** the entity overlaps with one of the gold entities, but is assigned the wrong label.

In addition to these categories, errors can be complete false positives (model extracts an entity which does not overlap at all with any gold entities), or complete false negatives (model completely misses an entity from the gold transcripts).

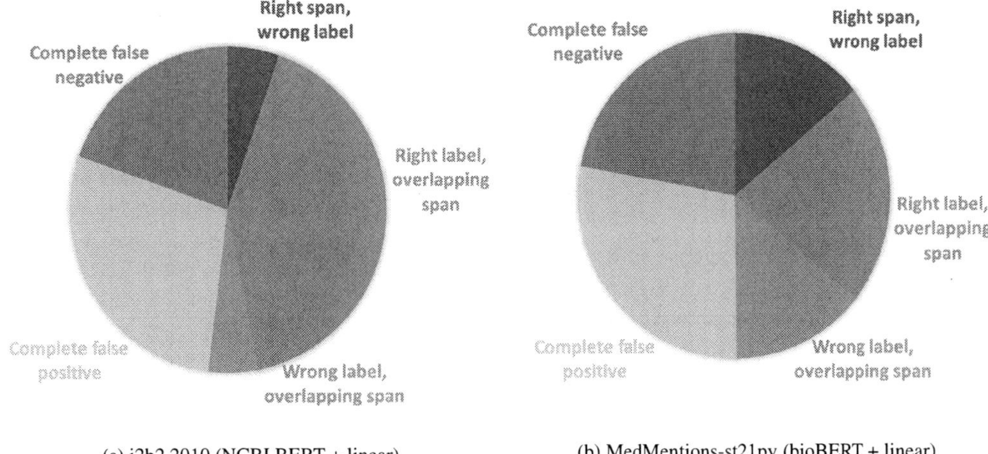

(a) i2b2 2010 (NCBI BERT + linear)   (b) MedMentions-st21pv (bioBERT + linear)

Figure 1: Types of errors made on the i2b2 and MedMentions-st21pv datasets.

In the i2b2 dataset, over half of the errors belong to one of the 'partial' error types (Figure 1a), with the biggest error category overall being **right label, overlapping span**. In many cases, the model identifies the main noun phrase associated with the entity, but not adjectives or prepositional phrases that further describe the entity: e.g., *patient feeling weaker than usual*, where the model labels *weaker* as the problem but the gold entity is *weaker than usual*, or *her glucose remained somewhat low*, where the model labels *low* as the problem, but the gold entity is *somewhat low*. One of the reasons that might lead to this kind of error is the inconsistency that exists in the human annotations. The task of identifying the span of entities can be very subjective and there is always some level of disagreement among human annotators. Also, sometimes annotation guidelines are interpreted differently by various annotators. As a result of this inconsistency, in the training stage, the model sees examples that the spans of the same entities has been labeled differently. This type of error may occur more frequently in the i2b2 dataset, due to the difficulty in annotating hurriedly written notes compared to academically written abstracts.

For MedMentions-st21pv, we again observe that roughly half the errors are 'partial' errors, but with a sizable increase in errors that involve the wrong label, with either a correct or overlapping span (Figure 1b). As discussed previously, the increase in **right span, wrong label** is likely due to the higher ambiguity between entity types in this

dataset. In many cases, errors of the type **wrong label, overlapping span** appear to be due to the inability of the annotation scheme to handle overlapping entities. For example, in *co-expression analysis*, where the model labels *co-expression* as a 'biologic function', rather than extracting the entire phrase as a 'research activity', or in *this region of the brain*, where the model simply labels *brain* as an 'anatomical structure' rather than *region of the brain* as a 'spatial concept'.

5.2 General vs Domain Knowledge

We then performed an exploratory error analysis to identify and compare the type of errors made by models using *general* and *domain-specific* embeddings. For this analysis, we considered BERT-base and NCBI BERT. BERT-base is pre-trained on Wikipedia and the Google Books database, so it transfers the general knowledge of language to the model. NCBI BERT starts its training from BERT-base, and then continues training on PubMed (biomedical abstracts) and MIMIC (clinical notes). Generally, embeddings like NCBI BERT are assumed to be more effective since they transfer both general and domain-specific knowledge. We analyze the errors of the two models to test this assumption. For simplicity, we concentrate our analysis on the i2b2 dataset, although we observe similar patterns in MedMentions.

NCBI BERT results in a higher overall F1-score than BERT-base on the i2b2 dataset, and there are 2027 entities that are correctly recognized by NCBI BERT and incorrectly recognized by BERT-base. However, there are also 1209 entities that are *correctly* recognized by BERT-base but *incorrectly* recognized by NCBI BERT. Therefore, it is not the case that NCBI BERT encodes the same knowledge as BERT-base, plus more that it has learned from PubMed and MIMIC; rather, the two systems have different strengths and weaknesses.

Qualitatively, we observed that some entities correctly recognized by NCBI BERT and missed by BERT-base involve common words that have a specialized meaning in medicine, for example in the sentence: *Suck , root , gag , grasp , and morrow were normal* . The BERT-base model does not extract any entities, while the NCBI BERT model recognizes *suck*, *root*, *gag*, and *grasp* as standard tests of infant reflexes. NCBI BERT also appears to be better at recognizing specialized acronyms and abbreviations, particularly when there is very little context, as in *Brother died 64 / MI*, where only NCBI BERT recognizes *MI* as a problem (myocardial infarction) in this brief family history, or *No CPR / No defib*, where NCBI BERT correctly labels *CPR* and *defib* as treatments, while BERT-base mis-labels them as problems.

In cases where BERT-base does better than NCBI BERT, it may be partially due to a better knowledge of well-formed text. We observed several examples where BERT-base appeared to be better at identifying the appropriate qualifiers to attach to a given entity: e.g., in *no interval development of effusion*, BERT-base correctly extracts the entire phrase *internal development of effusion*, while NCBI BERT only extracts *effusion*. Similarly, in *Right ventricular chamber size and free wall motion are normal*, BERT-base extracts *Right ventricular chamber size* as a single entity of the type 'test', while NCBI BERT splits it into *Right ventricular* and *size*.

Of course, these observations are purely anecdotal at this point, and will require future work and annotation to fully quantify the nature of the differences between the models. However, given the fact that the two models make different errors, it is at least reasonable to assume that predictions from the two models can be combined in a complementary fashion to improve the overall performance. We explore one possible architecture for doing so in the following section.

6 Concatenated Model

As a result of our error analysis, we propose a concatenated BERT model, to better combine the general knowledge from BERT-base and the clinical knowledge from the more specialized BERT models. To build such a model we concatenate the last encoding layer of a domain-specific BERT model with the last encoding layer of the general BERT model and feed this concatenation to a linear or bi-LSTM classification layer. During training we jointly fine-tune both BERT models and the classification layer. We implemented this model with both NCBI BERT and bioBERT models, since they previously led to the optimal results for i2b2 and MedMentions, respectively. NCBI BERT is concatenated with the uncased BERT-base model and bioBERT is concatenated with the cased BERT-base model.

Results for the concatenated models are given in Table 3. For all three datasets, we observe a small improvement over the best performing models in Table 2. The best result for i2b2 is achieved by concatenating the NCBI BERT and BERT-base models, with either a linear or bi-LSTM classifier on top. The resulting F1 score of 0.90 beats the previously reported state-of-the-art of 0.89 on the current release of the dataset with 170 training documents (Zhu et al., 2018). For MedMentions, concatenating bioBERT and BERT-base leads to the best results, with MedMentions-full attaining the best result using a linear classifier and MedMentions-st21pv attaining the best result with the bi-LSTM. To our knowledge, there are no prior results reported on entity (i.e. semantic type) extraction on this dataset.

Regarding the classifier layer, we observe that replacing a linear classifier layer with a bi-LSTM does not improve the results on i2b2 dataset. This is consistent with the results shown in Table 2 and indicates that a simple linear classifier is *enough* to learn the entity recognition task for i2b2 dataset. In the case of the MedMentions dataset, a bi-LSTM classifier improves the F1 score on MedMentions-st21pv but worsens it on MedMentions-full. These results show that there is room for more rigorous investigation about the classifier layer for extracting entities in MedMentions dataset. More complex neural structures with optimized hyperparameters may be needed to improve these results.

Although the improvements that we see by con-

	i2b2 2010			MedMentions (full)			MedMentions (st21pv)		
Model	P	R	F1	P	R	F1	P	R	F1
NCBI BERT concat BERT-base +linear	**0.89***	**0.90**	**0.90***	0.52	0.57	0.54	0.59	0.62	0.61
bioBERT concat BERT-base +linear	0.85	0.88	0.86	**0.54***	**0.59***	**0.56***	0.60	0.65	0.62
NCBI BERT concat BERT-base +bi-LSTM	**0.89***	**0.90**	**0.90***	0.50	0.55	0.53	0.59	0.62	0.61
bioBERT concat BERT-base +bi-LSTM	0.86	0.87	0.87	0.53	0.58	0.55	**0.63***	**0.65**	**0.64***

Table 3: Results of entity recognition using concatenated BERT models. An asterisk indicates an improvement over the best result from Table 2.

catenating the models are relatively small, they are consistent across the three datasets. This suggests that explicitly inputting both general and domain-specific information to the entity recognizer, rather than sequentially pre-training on different domains and hoping that the model 'remembers' information from each domain, can be a promising direction for future research.

7 Conclusion

We have presented the results of a set of medical entity recognition experiments on a new dataset, MedMentions. We contrasted these results with those obtained on the well-studied i2b2 2010 dataset. We explored a number of relevant dimensions, including the use of various embedding models (contextual versus non-contextual, general versus domain-specific, and LSTM versus attention-based) as well as linear versus bi-LSTM classifier layers. We also proposed a new modification to the previous BERT-based named entity recognition architectures, which allows the classifier to incorporate information from both general and domain-specific BERT embeddings. Our results on i2b2 are state-of-the-art, and our results on MedMentions set a benchmark for future work on this new dataset.

As popular public datasets become more and more studied over time, there is a chance that even if individual researchers follow good train-validate-test protocols, we eventually overfit to the datasets as a *community*, since there is so much published information available about what works well to improve performance on the test set. One goal of this work was to explore the gap in performance between a well-known clinical entity recognition dataset and a new, unstudied dataset. The same models and training procedures lead to significantly lower performance on the MedMentions dataset, for a variety of reasons: greater number of entity types, more class ambiguity, higher class imbalance, etc. Ultimately, we find that the model which performs best on i2b2 2010 is *not* the model that performs best on MedMentions, and that results on MedMentions can be improved by pre-training on more similar documents (biomedical abstracts), and by using more complex models (BERT + bi-LSTM rather than BERT + linear). We hope that other researchers will continue to advance the state-of-the-art on this new dataset.

Acknowledgement

We would like to thank Dr. Khaled Emam for his support and insightful feedback. We also thank Ms Lynn Wei for her technical support and her assistance in data collection, cleaning and preparation.

References

Emily Alsentzer, John R Murphy, Willie Boag, Wei-Hung Weng, Di Jin, Tristan Naumann, and Matthew McDermott. 2019. Publicly available clinical BERT embeddings. *arXiv preprint arXiv:1904.03323*.

Parminder Bhatia, Busra Celikkaya, and Mohammed Khalilia. 2019. Joint entity extraction and assertion detection for clinical text. In *Proceedings of the 57th Annual Meeting of the Association for Computational Linguistics*, pages 954–959.

William Boag, Kevin Wacome, Tristan Naumann, and Anna Rumshisky. 2015. CliNER: a lightweight tool for clinical named entity recognition. *AMIA Joint Summits on Clinical Research Informatics (poster)*.

Olivier Bodenreider. 2004. The Unified Medical Language System (UMLS): Integrating biomedical terminology. *Nucleic acids research*, 32(suppl_1):D267–D270.

Piotr Bojanowski, Edouard Grave, Armand Joulin, and Tomas Mikolov. 2017. Enriching word vectors with subword information. *Transactions of the Association for Computational Linguistics*, 5:135–146.

Raghavendra Chalapathy, Ehsan Zare Borzeshi, and Massimo Piccardi. 2016a. Bidirectional LSTM-CRF for clinical concept extraction. *arXiv preprint arXiv:1611.08373*.

Raghavendra Chalapathy, Ehsan Zare Borzeshi, and Massimo Piccardi. 2016b. Bidirectional lstm-crf for clinical concept extraction. In *Proceedings of the Clinical Natural Language Processing Workshop (ClinicalNLP)*, pages 7–12.

Berry De Bruijn, Colin Cherry, Svetlana Kiritchenko, Joel Martin, and Xiaodan Zhu. 2011. Machine-learned solutions for three stages of clinical information extraction: the state of the art at i2b2 2010. *Journal of the American Medical Informatics Association*, 18(5):557–562.

Jacob Devlin, Ming-Wei Chang, Kenton Lee, and Kristina Toutanova. 2019. BERT: Pre-training of deep bidirectional transformers for language understanding. In *Proceedings of NAACL-HLT*, pages 4171–4186.

Noémie Elhadad, Sameer Pradhan, Sharon Gorman, Suresh Manandhar, Wendy Chapman, and Guergana Savova. 2015. SemEval-2015 task 14: Analysis of clinical text. In *Proceedings of the 9th International Workshop on Semantic Evaluation (SemEval 2015)*, pages 303–310.

Xiao Fu and Sophia Ananiadou. 2014. Improving the extraction of clinical concepts from clinical records. *Proceedings of BioTxtM14*, pages 47–53.

Maryam Habibi, Leon Weber, Mariana Neves, David Luis Wiegandt, and Ulf Leser. 2017. Deep learning with word embeddings improves biomedical named entity recognition. *Bioinformatics*, 33(14):i37–i48.

Sepp Hochreiter and Jürgen Schmidhuber. 1997. Long short-term memory. *Neural Computation*, 9(8):1735–1780.

Zhiheng Huang, Wei Xu, and Kai Yu. 2015a. Bidirectional LSTM-CRF models for sequence tagging. *arXiv preprint arXiv:1508.01991*.

Zhiheng Huang, Wei Xu, and Kai Yu. 2015b. Bidirectional lstm-crf models for sequence tagging. *arXiv preprint arXiv:1508.01991*.

Alistair EW Johnson, Tom J Pollard, Lu Shen, H Lehman Li-wei, Mengling Feng, Mohammad Ghassemi, Benjamin Moody, Peter Szolovits, Leo Anthony Celi, and Roger G Mark. 2016. MIMIC-III, a freely accessible critical care database. *Scientific Data*, 3:160035.

Siddhartha Jonnalagadda, Trevor Cohen, Stephen Wu, and Graciela Gonzalez. 2012. Enhancing clinical concept extraction with distributional semantics. *Journal of biomedical informatics*, 45(1):129–140.

Jin-Dong Kim, Tomoko Ohta, Yoshimasa Tsuruoka, Yuka Tateisi, and Nigel Collier. 2004. Introduction to the bio-entity recognition task at JNLPBA. In *Proceedings of the International Joint Workshop on Natural Language Processing in Biomedicine and its Applications (NLPBA/BioNLP)*, pages 70–75.

Guillaume Lample, Miguel Ballesteros, Sandeep Subramanian, Kazuya Kawakami, and Chris Dyer. 2016. Neural architectures for named entity recognition. In *Proceedings of the 2016 Conference of the North American Chapter of the Association for Computational Linguistics: Human Language Technologies*, pages 260–270.

Jinhyuk Lee, Wonjin Yoon, Sungdong Kim, Donghyeon Kim, Sunkyu Kim, Chan Ho So, and Jaewoo Kang. 2019. BioBERT: pre-trained biomedical language representation model for biomedical text mining. *arXiv preprint arXiv:1901.08746*.

Tomas Mikolov, Ilya Sutskever, Kai Chen, Greg S Corrado, and Jeff Dean. 2013. Distributed representations of words and phrases and their compositionality. In *Advances in neural information processing systems*, pages 3111–3119.

Sunil Mohan and Donghui Li. 2019. MedMentions: A large biomedical corpus annotated with UMLS concepts. *arXiv preprint arXiv:1902.09476*.

Tomoko Ohta, Yuka Tateisi, and Jin-Dong Kim. 2002. The GENIA corpus: An annotated research abstract corpus in molecular biology domain. In *Proceedings of the second international conference on Human Language Technology Research*, pages 82–86. Morgan Kaufmann Publishers Inc.

Yifan Peng, Shankai Yan, and Zhiyong Lu. 2019. Transfer learning in biomedical natural language processing: An evaluation of BERT and ELMo on ten benchmarking datasets. *arXiv preprint arXiv:1906.05474*.

Jeffrey Pennington, Richard Socher, and Christopher Manning. 2014. GloVe: Global vectors for word representation. In *Proceedings of the 2014 Conference on Empirical Methods in Natural Language Processing (EMNLP)*, pages 1532–1543.

Matthew E Peters, Mark Neumann, Mohit Iyyer, Matt Gardner, Christopher Clark, Kenton Lee, and Luke Zettlemoyer. 2018. Deep contextualized word representations. *arXiv preprint arXiv:1802.05365*.

Sameer Pradhan, Noémie Elhadad, Wendy Chapman, Suresh Manandhar, and Guergana Savova. 2014. SemEval-2014 task 7: Analysis of clinical text. In *Proceedings of the 8th International Workshop on Semantic Evaluation (SemEval 2014)*, pages 54–62.

Kirk Roberts. 2016. Assessing the corpus size vs. similarity trade-off for word embeddings in clinical NLP. In *Proceedings of the Clinical Natural Language Processing Workshop (ClinicalNLP)*, pages 54–63.

Golnar Sheikhshabbafghi, Inanc Birol, and Anoop Sarkar. 2018. In-domain context-aware token embeddings improve biomedical named entity recognition. In *Proceedings of the Ninth International Workshop on Health Text Mining and Information Analysis*, pages 160–164.

Yuqi Si, Jingqi Wang, Hua Xu, and Kirk Roberts. 2019. Enhancing clinical concept extraction with contextual embedding. *arXiv preprint arXiv:1902.08691*.

Julien Tourille, Matthieu Doutreligne, Olivier Ferret, Aurélie Névéol, Nicolas Paris, and Xavier Tannier. 2018. Evaluation of a sequence tagging tool for biomedical texts. In *Proceedings of the Ninth International Workshop on Health Text Mining and Information Analysis*, pages 193–203.

Inigo Jauregi Unanue, Ehsan Zare Borzeshi, and Massimo Piccardi. 2017. Recurrent neural networks with specialized word embeddings for health-domain named-entity recognition. *Journal of biomedical informatics*, 76:102–109.

Özlem Uzuner, Brett R South, Shuying Shen, and Scott L DuVall. 2011. 2010 i2b2/VA challenge on concepts, assertions, and relations in clinical text. *Journal of the American Medical Informatics Association*, 18(5):552–556.

Ashish Vaswani, Noam Shazeer, Niki Parmar, Jakob Uszkoreit, Llion Jones, Aidan N Gomez, Łukasz Kaiser, and Illia Polosukhin. 2017. Attention is all you need. In *Advances in Neural Information Processing Systems (NeurIPS)*, pages 5998–6008.

Yonghui Wu, Mike Schuster, Zhifeng Chen, Quoc V Le, Mohammad Norouzi, Wolfgang Macherey, Maxim Krikun, Yuan Cao, Qin Gao, Klaus Macherey, et al. 2016. Google's neural machine translation system: Bridging the gap between human and machine translation. *arXiv preprint arXiv:1609.08144*.

Henghui Zhu, Ioannis Ch Paschalidis, and Amir Tahmasebi. 2018. Clinical concept extraction with contextual word embedding. *arXiv preprint arXiv:1810.10566*.

Ontological attention ensembles for capturing semantic concepts in ICD code prediction from clinical text

Matúš Falis[1,*] **Maciej Pajak**[1,*] **Aneta Lisowska**[1,*] **Patrick Schrempf**[1,3],
Lucas Deckers[1], **Shadia Mikhael**[1], **Sotirios A. Tsaftaris**[1,2] **and Alison Q. O'Neil**[1,2]

[1] Canon Medical Research Europe, Edinburgh, UK
[2] University of Edinburgh, Edinburgh, UK
[3] University of St Andrews, St Andrews, UK

{matus.falis, maciej.pajak, aneta.lisowska}@eu.medical.canon,
{patrick.schrempf, lucas.deckers, shadia.mikhael}@eu.medical.canon,
s.tsaftaris@ed.ac.uk, alison.oneil@eu.medical.canon

Abstract

We present a semantically interpretable system for automated ICD coding of clinical text documents. Our contribution is an ontological attention mechanism which matches the structure of the ICD ontology, in which shared attention vectors are learned at each level of the hierarchy, and combined into label-dependent ensembles. Analysis of the attention heads shows that shared concepts are learned by the lowest common denominator node. This allows child nodes to focus on the differentiating concepts, leading to efficient learning and memory usage. Visualisation of the multi-level attention on the original text allows explanation of the code predictions according to the semantics of the ICD ontology. On the MIMIC-III dataset we achieve a 2.7% absolute (11% relative) improvement from 0.218 to 0.245 macro-F1 score compared to the previous state of the art across 3,912 codes. Finally, we analyse the labelling inconsistencies arising from different coding practices which limit performance on this task.

1 Introduction

Classification of clinical free-text documents poses some difficult technical challenges. One task of active research is the assignment of diagnostic and procedural International Classification of Diseases (ICD) codes. These codes are assigned retrospectively to hospital admissions based on the medical record, for population disease statistics and for reimbursements for hospitals in countries such as the United States. As manual coding is both time-consuming and error-prone, automation of the coding process is desirable. Coding errors may result in unpaid claims and loss of revenue (Adams et al., 2002).

Automated matching of unstructured text to medical codes is difficult because of the large number of possible codes, the high class imbalance in the data, and the ambiguous language and frequent lack of exposition in clinical text. However, the release of large datasets such as MIMIC-III (Johnson et al., 2016) has paved the way for progress, enabling rule-based systems (Farkas and Szarvas, 2008) and classical machine learning methods such as support vector machines (Suominen et al., 2008), to be superseded by neural network-based approaches (Baumel et al., 2017; Karimi et al., 2017; Shi et al., 2018; Duarte et al., 2018; Rios and Kavuluru, 2018). The most successful reported model on the ICD coding task is a shallow convolutional neural network (CNN) model with label-dependent attention introduced by Mullenbach et al. (2018) and extended by Sadoughi et al. (2018) with multi-view convolution and a modified label regularisation module.

One of the common features of the aforementioned neural network models is the use of attention mechanisms (Vaswani et al., 2017). This mirrors advances in general representation learning. In the text domain, use of multi-headed attention has been core to the development of *Transformer*-based language models (Devlin et al., 2018; Radford et al., 2019). In the imaging domain, authors have had success with combining attention vectors learned at the global and local levels with *Double Attention* networks (Chen et al., 2018). In the domain of structured (coded) medical data, Choi et al. (2017) leveraged the ontological structure of the ICD and SNOMED CT coding systems in their *GRAM* model, to combine the attention vectors of a code and its ancestors in order to predict the codes for the next patient visit based on the codes assigned in the previous visit.

Our contributions are:

1. A structured ontological attention ensemble mechanism which provides improved accuracy, efficiency, and interpretability.

---

*equal contribution

Proceedings of the 10th International Workshop on Health Text Mining and Information Analysis (LOUHI 2019), pages 168–177
Hong Kong, November 3, 2019. ©2019 Association for Computational Linguistics
https://doi.org/10.18653/v1/D19-62

Dataset	# Documents	# Unique patients	# ICD-9 Codes	# Unique ICD-9 codes
Training	47,719	36,997	758,212	8,692
Development	1,631	1,374	28,896	3,012
Test	3,372	2,755	61,578	4,085
Total	**52,722**	**41,126**	**848,686**	**8,929**

Table 1: Distribution of documents and codes in the MIMIC-III dataset.

2. An analysis of the multi-level attention weights with respect to the text input, which allows us to interpret the code predictions according to the semantics of the ICD ontology.

3. An analysis of the limitations of the MIMIC-III dataset, in particular the labelling inconsistencies arising from variable coding practices between coders and between timepoints.

2   Dataset

We used the MIMIC-III dataset (Johnson et al., 2016) ("Medical Information Mart for Intensive Care") which comes from the intensive care unit of the Beth Israel Deaconess Medical Center in Boston. We concatenated the hospital discharge summaries associated with each admission to form a single document and combined the corresponding ICD-9 codes. The data was split into training, development, and test patient sets according to the split of Mullenbach et al. (2018) (see Table 1).

3   Methods

We formulate the problem as a multi-label binary classification task, for which each hospital discharge summary is labelled with the presence or absence of the complete set of ICD-9 codes for the associated admission. Our model is a CNN similar to those of (Mullenbach et al., 2018; Sadoughi et al., 2018). Inspired by the graph-based attention model of (Choi et al., 2017), we propose a hierarchical attention mechanism (mirroring the ICD ontology) which yields a multi-level, label-dependent ensemble of attention vectors for predicting each code. Our architecture is shown in Figure 1 and described below.

3.1   Embedding

Documents were pre-processed by lower-casing the text and removing punctuation, followed by tokenisation during which purely numeric tokens were discarded. We used a maximum input length of 4500 tokens and truncated any documents longer than this (260 training, 16 devel-

opment, and 22 test). Tokens were then embedded with a 100-dimensional word2vec model. For each document, token embeddings were concatenated to give a $100 \times N$ document embedding matrix D, where N is the document length.

We pre-trained the word2vec model on the training set using continuous bag-of-words (CBOW) (Mikolov et al., 2013). The vocabulary comprises tokens which occur in at least 3 documents (51,847 tokens). The embedding model was fine-tuned (not frozen) during subsequent supervised training of the complete model.

3.2   Convolutional module

The first part of the network proper consists of a multi-view convolutional module, as introduced by Sadoughi et al. (2018). Multiple one-dimensional convolutional kernels of varying size with stride = 1 and weights W are applied in parallel to the document embedding matrix D along the N dimension. The outputs of these kernels are padded at each end to match the input length N. This yields outputs of size $C \times M \times N$ where C is the number of kernel sizes ("views"), M is the number of filter maps per view, and N is the length of the document. The outputs are max-pooled in the C dimension i.e., across each set of views, to yield a matrix E of dimensions $M \times N$:

$$E = \tanh(\max_{C=[0,3]} W_C * D) \tag{1}$$

Optimal values were $C = 4$ filters of lengths $\{6, 8, 10, 12\}$ with $M = 256$ filter maps each.

3.3   Prediction via label-dependent attention

Label-specific attention vectors are employed to collapse the variable-length E document representations down to fixed-length representations. For each label l, given the matrix E as input, a token-wise linear layer u_l is trained to generate a vector of length N. This is normalised with a softmax operation, resulting in an attention vector a_l:

$$a_l = \text{softmax}(E^T u_l) \tag{2}$$

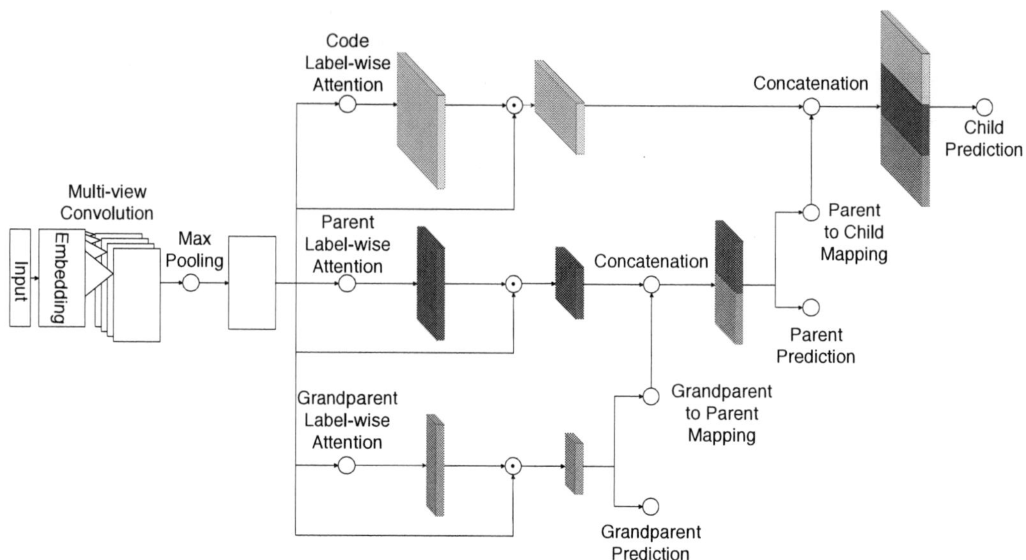

Figure 1: Network architecture. The output of the convolutional module is fed into the ensemble of ancestral attention heads for multi-task learning. Circles with dots represent matrix product operations. Ancestors are mapped to descendants by multiplication with a mapping connectivity matrix based on the ontology structure.

The attention vector is then multiplied with the matrix E which yields a vector v_l of length M, a document representation specific to a label:

$$v_l = a_l E \qquad (3)$$

If multiple linear layers $u_{l,0}, u_{l,1}, \ldots$ are trained for each label at this stage, multiple attention vectors (or "heads") will be generated. Thus, multiple document representations v_l could be made available, each of length M, and concatenated together to form a longer label-specific representation for the document. We experimented with multiple attention vectors and found two vectors per label to be optimal. To make a prediction of the probability of each label, $P(l)$, there is a final dense binary classification layer with sigmoid activation. This is shown for two attention vectors:

$$P(l) = \sigma(W_l[v_{l,0}; v_{l,1}] + \beta_l) \qquad (4)$$

3.4 Prediction via label-dependent ontological attention ensembles

The ICD-9 codes are defined as an ontology, from more general categories down to more specific descriptions of diagnosis and procedure. Rather than simply training two attention heads per code as shown in Section 3.3, we propose to exploit the ontological structure to train shared attention heads between codes on the same branch of the tree, thus pooling information across labels which share ancestry. In this work, we use two levels of ancestry, where the first level corresponds to the pre-floating-point portion of the code. For instance, for the code *425.11 Hypertrophic obstructive cardiomyopathy*, the first-degree ancestor is *425 Cardiomyopathy* and the second-degree ancestor is *420-429 Other forms of heart disease* (the chapter in which the parent occurs). This is illustrated in Figure 2.

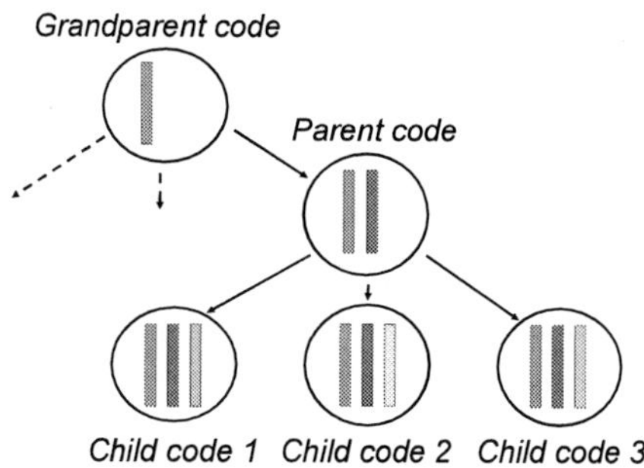

Figure 2: Illustration of inheritance of the linear layers u_l. This yields label-specific ontological attention ensembles of the attention heads a_l and subsequently the document representations v_l.

For the entire set of 8929 labels, we identified 1167 first-degree ancestors and 179 second-degree ancestors. Compared to two attention vectors per code, this reduces the parameter space and memory requirements from 17,858 attention heads (8929 x 2) to 10,275 attention heads (8929 + 1167 + 179) as well as increasing the number of training samples for each attention head.

The label prediction for each code is now derived from the concatenated child (c), parent (p) and grandparent (gp) document representations:

$$P(l_{\text{child}}) = \sigma(W_{l_c}[v_{l,c}; v_{l,p}; v_{l,gp}] + \beta_{l_c}) \quad (5)$$

In order to facilitate learning of multiple attention heads, we employ deep supervision using the ancestral labels, adding auxiliary outputs for predicting the parent and grandparent nodes:

$$P(l_{\text{parent}}) = \sigma(W_{l_p}[v_{l,p}; v_{l,gp}] + \beta_{l_p}) \quad (6)$$

$$P(l_{\text{grandparent}}) = \sigma(W_{l_{gp}}[v_{l,gp}] + \beta_{l_{gp}}) \quad (7)$$

3.5 Training process

We trained our model with weighted binary cross entropy loss using the Adam optimiser (Kingma and Ba, 2014) with learning rate 0.0005.

Stratified shuffling: The network accepts input of any length but all instances within a single batch need to be padded to the same length. To minimise the amount of padding, we used length-stratified shuffling between epochs. For this, documents were grouped by length and shuffled only within these groups; groups were themselves then shuffled before batch selection started.

Dampened class weighting: We employed the standard practice of loss weighting to prevent the imbalanced dataset from affecting performance on rare classes. We used a softer alternative to empirical class re-weighting, by taking the inverse frequencies of positive (label= 1) and negative (label= 0) examples for each code c, and adding a damping factor α. In the equations below, $n_{\text{label}_c=1}$ stands for the number of positive examples for the ICD code c, and n stands for the total number of documents in the dataset.

$$\omega_{(c,1)} = \left(\frac{n}{n_{\text{label}_c=1}}\right)^{\alpha}$$
$$\quad (8)$$
$$\omega_{(c,0)} = \left(\frac{n}{n_{\text{label}_c=0}}\right)^{\alpha}$$

Upweighting for codes with 5 examples or fewer, where we do not expect to perform well in any case, was removed altogether as follows:

$$\omega_{(c,1)} = \begin{cases} \left(\frac{n}{n_{\text{label}_c=1}}\right)^{\alpha} & , n_{\text{label}_c=1} > 5 \\ 1 & , \text{otherwise} \end{cases} \quad (9)$$

Deep supervision: The loss function was weighted in favour of child codes, with progressively less weight given to the codes at higher levels in the ICD ontology. A weighting of 1 was used for the child code loss, a weighting w_h for the parent code auxiliary loss, and w_h^2 for the grandparent code auxiliary loss, i.e.,

$$\text{Loss} = L_{\text{c}} + w_h L_{\text{p}} + w_h^2 L_{\text{gp}} \quad (10)$$

Optimal values were $\alpha = 0.25$ and $w_h = 0.1$.

3.6 Implementation and hyperparameters

The word2vec embedding was implemented with Gensim (Řehůřek and Sojka, 2010) and the ICD coding model was implemented with PyTorch (Paszke et al., 2017). Experiments were run on Nvidia V100 16GB GPUs. Hyperparameter values were selected by maximising the development set macro-F1 score for codes with more than 5 training examples.

4 Experiments

4.1 Results

In our evaluation, we focus on performance across all codes and hence we prioritise macro-averaged metrics, in particular macro-averaged precision, recall, and F1 score. Micro-averaged F1 score and Precision at k ($P@K$) are also reported in order to directly benchmark performance against previously reported metrics. All reported numbers are the average of 5 runs, starting from different random network initialisations.

We compare our model to two previous state-of-the-art models: Mullenbach et al. (2018), and Sadoughi et al. (2018) (published only on arXiv). We trained these models with the hyperparameter values quoted in the respective publications, and used the same early stopping criteria as for our model. Both Mullenbach *et al.* and Sadoughi *et al.* use label regularisation modules, at the output and at the attention layer respectively. In line with their published results, we found that only the method of Sadoughi *et al.* gave an improvement and thus it

Method	R_{macro}	P_{macro}	$F1_{macro}$	$F1_{micro}$	$P@8$
Mullenbach et al. (2018)	0.218	0.195	0.206	0.499	0.651
Sadoughi et al. (2018)	0.261	0.186	0.218	0.498	0.662
Ontological Attention	0.341	0.192	**0.245**	0.497	0.681

Table 2: Benchmark results for the models trained with $F1_{macro}$ stopping criterion.

Method	R_{micro}	P_{micro}	$F1_{macro}$	$F1_{micro}$	$P@8$
Mullenbach et al. (2018)	0.469	0.593	0.172	0.523	0.685
Sadoughi et al. (2018)	0.516	0.560	0.173	0.537	0.695
Ontological Attention	0.514	0.617	0.206	**0.560**	0.727

Table 3: Benchmark results for the models trained with $F1_{micro}$ stopping criterion.

Method	$F1_{macro}$	Relative $F1_{macro}$ change (%)
Ontological Attention	**0.245**	0
Efficacy of ontological attention ensemble		
1. No deep supervision	0.243	-0.82
2. No ontology: One attention head for each label	0.234	-4.5
3. No ontology: Two attention heads for each label	0.242	-1.2
4. Partial ontology: Randomised ontological connections	0.231	-5.7
Efficacy of additional modifications		
5. No class weighting	0.232	-5.3
6. Reduced convolutional filters (70, as in Sadoughi et al. (2018))	0.236	-3.7

Table 4: Ablation study of individual components of the final method. All models are trained with the $F1_{macro}$ stopping criterion. Experiments 2 and 3 do not use the ontological attention mechanism, and instead have one or two attention heads respectively per code-level label. For experiment 4, child-parent and parent-grandparent connections were randomised, removing shared semantics between codes across the full 3 levels.

is included in the model reported here. However, this regularisation is not used in our own model where we observed no benefit.

Overall results are shown in Table 2. Our method significantly outperforms the benchmarks on macro-F1 and $P@8$.

Previous models have optimised for F1 micro-average. Different target metrics require different design choices: after removal of the class weighting in the loss function and when using $F1_{micro}$ as our stopping criterion, we are also able to surpass previous state-of-the-art results on micro-F1. The results are presented in Table 3; our method achieves the highest $F1_{micro}$ score, as well as the highest $P@8$ score. We note that $P@8$ score is consistently higher for models stopped using the $F1_{micro}$ criterion.

In Table 4 we present an ablation study. It can be seen that the improvement in performance of the ontological attention model is not simply due to increased capacity of the network, since even with 73% greater capacity (17,858 compared to 10,275 attention vectors), the two-vector multi-headed model has a 1.2% drop in performance. Experiments with deep supervision and randomisation of the ontology graph connections show the benefit of each component of the ontological architecture. We also measure the effect of additional changes made during optimisation of the architecture and training.

Levels of the ontology: Three levels of the ontology (including the code itself) were found to be optimal for the Ontological Attention model (see Figure 3). Adding parent and grandparent levels provide incremental gains in accuracy. Adding a level beyond the grandparent node (i.e., the great-grandparent level) does not provide further improvement. Since we identified only 22 ancestral nodes at the level directly above the grandparent, we hypothesise that the grouping becomes too coarse to be beneficial. In fact, all procedure codes share the same ancestor at this level; the remaining

21 nodes are split between diagnostic codes.

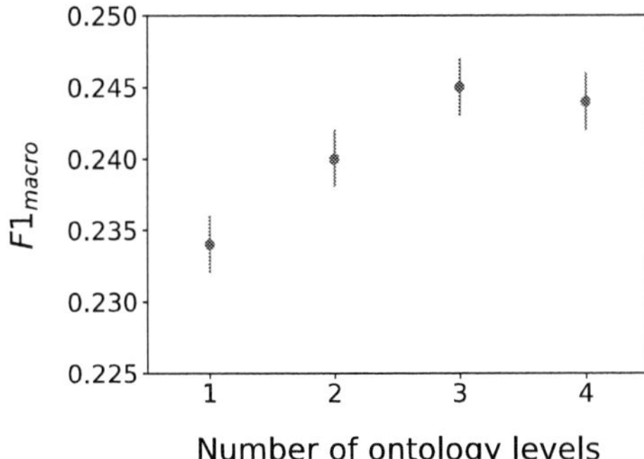

Figure 3: $F1_{macro}$ for models using attention ensembles across different levels of the ontological tree. Error bars represent the standard deviation across 5 different random weight initialisations. The model with 1 level has only the code-level attention head, the model with 2 levels also includes the shared parent attention heads; the model with 3 levels adds the shared grandparent attention heads (this is our reported Ontological Attention model), and finally, the model with 4 levels adds shared great-grandparent attention heads.

4.2 Analysis of the attention weights

In Figure 4 we show how the weights of code-level u_l vectors (which give rise to the attention heads) change when the ontological attention ensemble mechanism is introduced. As expected, we observe that in the case of a single attention head, the weights for different codes largely cluster together based on their position in the ontology graph. Once the parent and grandparent attention heads are trained, the ontological similarity structure on the code level mostly disappears. This suggests that the common features of all codes within a parent group are already extracted by the parent attention. thus, the capacity of the code-level attention is spent on the representation of the differences between the descendants of a single parent.

4.3 Interpretability of the attention heads

In Section 4.2, we showed the links between the ontology and the attention heads within the space of the u_l vector weights. We can widen this analysis to links between the predictions and the input, by examining which words in the input documents are attended by the three levels of attention heads for a given label. A qualitative visual example is shown in Figure 5. We performed quantitative frequency analysis of high-attention terms

(keywords) in the training set. A term was considered a keyword if its attention weight in a document surpassed the threshold t_{kw}:

$$t_{kw}(N, \gamma_{kw}) = \gamma_{kw} \frac{1}{N}, \qquad (11)$$

where N is the length of a document and γ_{kw} is a scalar parameter controlling the strictness of the threshold. With $\gamma_{kw} = 1$, a term is considered a keyword if its attention weight surpasses the uniformly distributed attention. In our analysis we chose $\gamma_{kw} = 17$ for all documents.

We aggregated these keywords across all predicted labels in the training set, counting how many times a term is considered a keyword for a label. The results of this analysis are in line with our qualitative analysis of attention maps. The most frequent keywords for the labels presented in the example in Figure 5 include "cancer", "ca", "tumor", at the grandparent level (focusing on the concept of cancer); "metastatic", "metastases" and "metastasis" at the parent level (focusing on the concept of metastasis); and "brain", "craniotomy", "frontal" at the code-level (focusing on terms relating to specific anatomy). A sibling code (*198.5 Secondary malignant neoplasm of bone and bone marrow*) displays similar behaviour in focusing on anatomy, with "bone", "spine", and "back" being among the most frequent keywords.

Not all codes display such structured behaviour. For instance, the grandparent *401-405 Hypertensive disease* attended to the term "hypertension" most frequently. The parent code *401 Essential hypertension*, does not attend to "hypertension", but neither does it attend to any useful keywords — this may be due to the code being simple compared to its sibling codes, which are more specific (e.g., *402 Hypertensive heart disease*). Interestingly, the children of *401 Essential hypertension* attend to the word "hypertension" again, while also focusing on terms that set them apart from each other — e.g., *401.0 Malignant essential hypertension* focuses on terms implying malignancy, such as "urgency", "emergency", and "hemorrhage".

5 Limitations due to labelling variability

Since performance on this task appears to be much lower than might be acceptable for real-world use, we investigated further. Figure 6 shows the per-label F1 scores; it can be seen that there is high

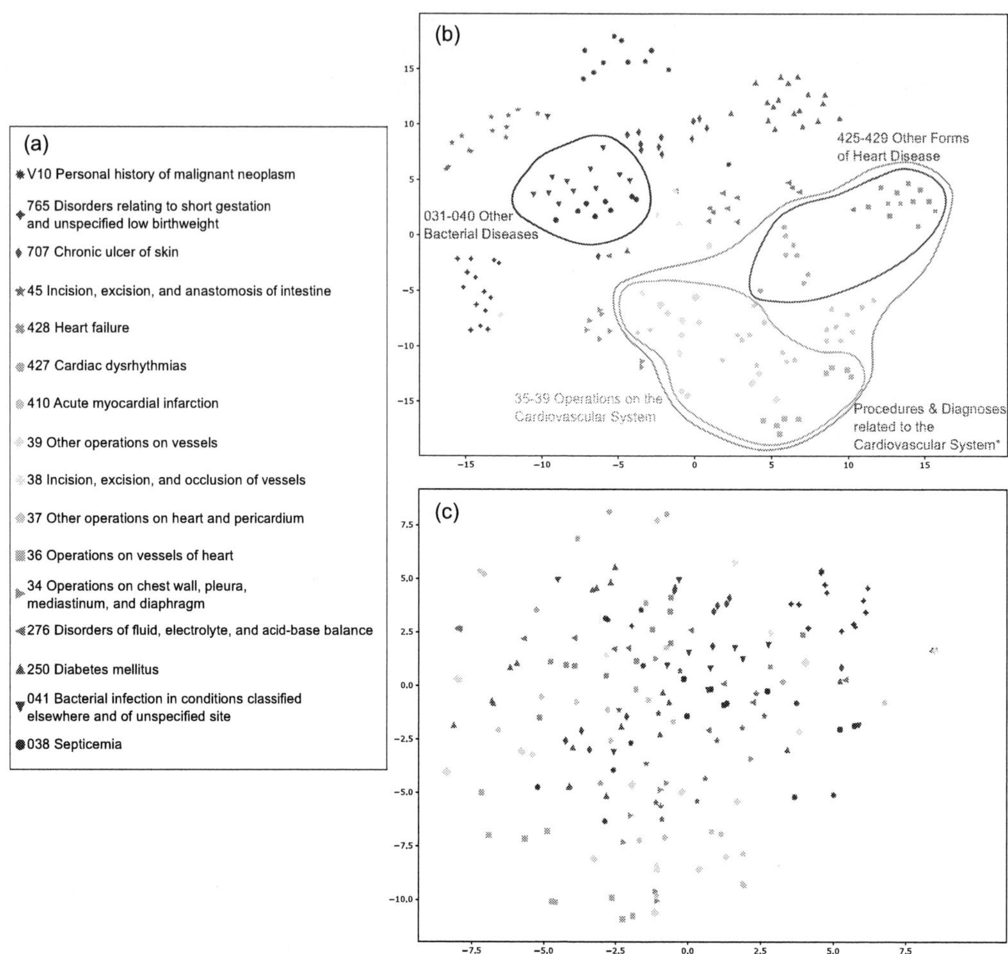

Figure 4: Two-dimensional t-SNE (Maaten and Hinton, 2008) representation of the u_l vectors (which give rise to the attention heads) for a subset of 182 codes with at least 100 occurrences each, in the data belonging to 16 different parent nodes. (a) Legend for the annotation of data points according to their parent node. (b) u_l vectors from the model with only a single attention head for each code (i.e., no ontology). It can be seen that codes naturally cluster by their parent node. Selected higher-level alignments are indicated by additional contours — for grandparent nodes (3 nodes) and for diagnoses/procedure alignment (in the case of cardiovascular disease). (c) u_l vectors in the ontological attention ensemble model for the same set of codes (and the same t-SNE hyperparameters). In most cases the clustering disappears, indicating that the attention weights for the ancestral codes have extracted the similarities from descendants' clusters.

Method	R_{macro}	P_{macro}	$F1_{macro}$	$F1_{micro}$	$P@8$
Mullenbach et al. (2018)	0.226	0.200	0.212	0.500	0.651
Sadoughi et al. (2018)	0.272	0.187	0.222	0.497	0.662
Ontological Attention	0.347	0.199	**0.252**	0.507	0.686

Table 5: Benchmark results for the models trained with $F1_{macro}$ stopping criterion.

variability in accuracy, that is only partially correlated with the number of training examples.

Inspection of examples for some of the poorly performing codes revealed some variability in coding policy, described further below.

5.1 Misreporting of codes

The phenomenon of human coding errors is reported in the literature; for instance, Kokotailo

(a) Brief Hospital Course:
Mr. John Doe is a 68-year-old male with metastatic ████
and brain metastases. He presented with 2/7 of palpitations
and feeling generally unwell. Diagnosed with a saddle PE.

(b) Brief Hospital Course:
Mr. John Doe is a 68-year-old male with metastatic NSCLC
and brain ████████. He presented with 2/7 of palpitations
and feeling generally unwell. Diagnosed with a saddle PE.

(c) Brief Hospital Course:
Mr. John Doe is a 68-year-old male with metastatic NSCLC
and ████ metastases. He presented with 2/7 of palpitations
and feeling generally unwell. Diagnosed with a saddle PE.

Figure 5: Discharge summary snippet with highlights generated from attention heads for (a) the grandparent code (*190-199 Malignant neoplasm of other and unspecified sites*), (b) the parent code (*198 Secondary malignant neoplasm of other specified sites*), and (c) the specific code (*198.3 Secondary malignant neoplasm of brain and spinal cord*). Different words and phrases are attended at each level.

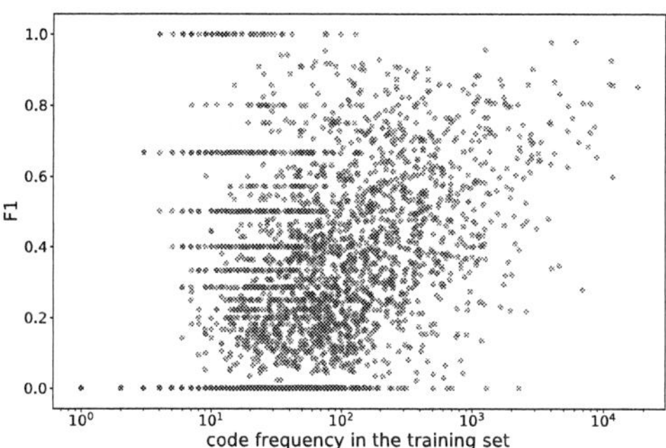

Figure 6: Per-code frequency of training examples v.s. $F1_{macro}$ score from the ontological attention model

and Hill estimated sensitivity and specificity to be 80% and 100% respectively for ICD codes relating to stroke and its risk factors (Kokotailo and Hill, 2005). In the MIMIC-III dataset, we inspected the assignment of smoking codes (current smoker *305.1*, past smoker *V15.82*, or never smoked i.e., no code at all), using regular expression matching to identify examples of possible miscoding, followed by manual inspection of 60 examples (10 relating to each possible miscoding category) to verify our estimates. We estimated that 10% of patients had been wrongly assigned codes, and 30% of patients who had a mention of smoking in their record had not been coded at all. We also observed that often the "correct" code is not clear-cut. For instance, many patients had smoked in the distant past or only smoke occasionally, or had only re-

cently quit; in these cases, where the narrator reliability may be questionable, the decision of how to code is a matter of subjective clinical judgement.

5.2 Revisions to the coding standards

Another limitation of working with the MIMIC-III dataset is that during the deidentification process, information about absolute dates was discarded. This is problematic when we consider that the MIMIC-III dataset contains data that was collected between 2001 and 2012, and the ICD-9 coding standard was reviewed and updated annually between 2006 and 2013 (Centers for Medicare & Medicaid Services) i.e., each year some codes were added, removed or updated in their meaning.

To investigate this issue, we took the 2008 standard and mapped codes created post-2008 back to this year. In total, we identified 380 codes that are present in the dataset but were not defined in the 2008 standard. An example can be seen in Figure 7. We report our results on the 2008 codeset in Table 5. It can be seen that there is an improvement to the metrics on this dataset, which we expect would increase further if all codes were mapped back to the earliest date of 2001. Without time data, it is an unfair task to predict codes which are fundamentally time-dependent. This is an interesting example of conflicting interests between (de)identifiability and task authenticity.

During real-world deployment, codes should be assigned according to current standards. In order to use older data, codes should be mapped forwards rather than backwards. The backwards operation was possible by automated re-mapping of the codes, however the forwards operation is more arduous. Newly introduced codes may require annotation of fresh labels or one-to-many conversion — both operations requiring manual inspection of the original text. A pragmatic approach would be to mask out codes for older documents where they cannot be automatically assigned.

6 Conclusions

We have presented a neural architecture for automated clinical coding which is driven by the ontological graph of relationships between codes. This model establishes a new state-of-the-art result for the task of automated clinical coding with MIMIC-III dataset. Compared to simply doubling the number of attention heads, our ontological attention ensemble mechanism provides improve-

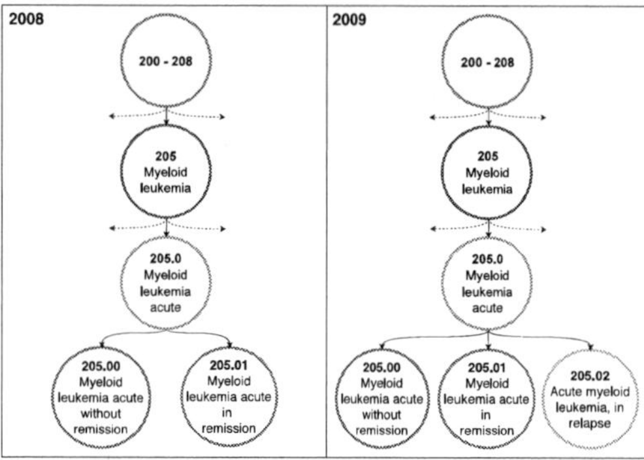

Figure 7: Example code added to the ICD-9 standard.

ments in *accuracy*, in *memory efficiency*, and in *interpretability*. Our method is not specific to an ontology, and in fact could be used for a graph of any formation. If we were to exploit further connections within the ICD ontology e.g., between related diagnoses and procedures, and between child codes which share modifier digits, we would expect to obtain a further performance boost.

We have illustrated that labels may not be reliably present or correct. Thus, even where plenty of training examples are available, the performance may (appear to) be low. In practice, the most successful approach may be to leverage a combination of automated techniques and manual input. An active learning setup would facilitate adoption of new codes by the model as well as allowing endorsement of suggested codes which might otherwise have been missed by manual assignment, and we propose this route for future research.

References

Diane L Adams, Helen Norman, and Valentine J Burroughs. 2002. Addressing medical coding and billing part ii: a strategy for achieving compliance. a risk management approach for reducing coding and billing errors. *Journal of the National Medical Association*, 94(6):430.

Tal Baumel, Jumana Nassour-Kassis, Michael Elhadad, and Noemie Elhadad. 2017. Multi-label classification of patient notes: Case study on ICD code assignment. *ArXiv*.

Centers for Medicare & Medicaid Services. New, deleted, and revised codes - summary tables. `https://www.cms.gov/Medicare/Coding/ICD9ProviderDiagnosticCodes/summarytables.html`.

Yunpeng Chen, Yannis Kalantidis, Jianshu Li, Shuicheng Yan, and Jiashi Feng. 2018. A^ 2-nets: Double attention networks. In *Advances in Neural Information Processing Systems*, pages 352–361.

Edward Choi, Mohammad Taha Bahadori, Le Song, Walter F Stewart, and Jimeng Sun. 2017. Gram: graph-based attention model for healthcare representation learning. In *Proceedings of the 23rd ACM SIGKDD International Conference on Knowledge Discovery and Data Mining*, pages 787–795. ACM.

Jacob Devlin, Ming-Wei Chang, Kenton Lee, and Kristina Toutanova. 2018. Bert: Pre-training of deep bidirectional transformers for language understanding. *arXiv preprint arXiv:1810.04805*.

Francisco Duarte, Bruno Martins, Càtia Sousa Pinto, and Màrio J. Silva. 2018. Deep neural models for icd-10 coding of death certificates and autopsy reports in free-text. *Journal of Biomedical Informatics*, 80:64–77.

Richárd Farkas and György Szarvas. 2008. Automatic construction of rule-based icd-9-cm coding systems. In *BMC bioinformatics*, volume 9, page S10. BioMed Central.

AEW Johnson, TJ Pollard, L Shen, L Lehman, M Feng, M Ghassemi, B Moody, P Szolovits, LA Celi, and RG Mark. 2016. Mimic-iii, a freely accessible critical care database. *Scientific Data*, 3.

Sarvnaz Karimi, Xiang Dai, Hamed Hassanzadeh, and Anthony Nguyen. 2017. Automatic diagnosis coding of radiology reports: a comparison of deep learning and conventional classification methods. *In Proc. of the BioNLP 2017 Workshop*, pages 328–332.

Diederik P Kingma and Jimmy Ba. 2014. Adam: A method for stochastic optimization. *arXiv preprint arXiv:1412.6980*.

Rae A Kokotailo and Michael D Hill. 2005. Coding of stroke and stroke risk factors using international classification of diseases, revisions 9 and 10. *Stroke*, 36(8):1776–1781.

Laurens van der Maaten and Geoffrey Hinton. 2008. Visualizing data using t-sne. *Journal of machine learning research*, 9(Nov):2579–2605.

Tomas Mikolov, Kai Chen, Greg Corrado, and Jeffrey Dean. 2013. Efficient estimation of word representations in vector space. *arXiv preprint arXiv:1301.3781*.

James Mullenbach, Sarah Wiegreffe, Jon Duke, Jimeng Sun, and Jacob Eisenstein. 2018. Explainable Prediction of Medical Codes from Clinical Text. *Proceedings ofNAACL-HLT 2018*, pages 1101–1111.

Adam Paszke, Sam Gross, Soumith Chintala, Gregory Chanan, Edward Yang, Zachary DeVito, Zeming Lin, Alban Desmaison, Luca Antiga, and Adam

Lerer. 2017. Automatic differentiation in PyTorch. In *NIPS Autodiff Workshop*.

Alec Radford, Jeffrey Wu, Rewon Child, David Luan, Dario Amodei, and Ilya Sutskever. 2019. Language models are unsupervised multitask learners. *OpenAI Blog*, 1(8).

Radim Řehůřek and Petr Sojka. 2010. Software Framework for Topic Modelling with Large Corpora. In *Proceedings of the LREC 2010 Workshop on New Challenges for NLP Frameworks*, pages 45–50, Valletta, Malta. ELRA. http://is.muni.cz/publication/884893/en.

Anthony Rios and Ramakanth Kavuluru. 2018. Few-shot and zero-shot multi-label learning for structured label spaces. In *Proceedings of the 2018 Conference on Empirical Methods in Natural Language Processing*, pages 3132–3142.

Najmeh Sadoughi, Greg P. Finley, James Fone, Vignesh Murali, Maxim Korenevski, Slava Baryshnikov, Nico Axtmann, Mark Miller, and David Suendermann-Oeft. 2018. Medical code prediction with multi-view convolution and description-regularized label-dependent attention. *arXiv*.

Haoran Shi, Pengtao Xie, Zhiting Hu, Ming Zhang, and Eric P. Xing. 2018. Towards automated icd coding using deep learning. *ArXiv*.

Hanna Suominen, Filip Ginter, Sampo Pyysalo, Antti Airola, Tapio Pahikkala, Sanna Salanterä, and Tapio Salakoski. 2008. Machine learning to automate the assignment of diagnosis codes to free-text radiology reports: a method description. In *Proceedings of the ICML/UAI/COLT Workshop on Machine Learning for Health-Care Applications*.

Ashish Vaswani, Noam Shazeer, Niki Parmar, Jakob Uszkoreit, Llion Jones, Aidan N Gomez, Łukasz Kaiser, and Illia Polosukhin. 2017. Attention is all you need. In *Advances in Neural Information Processing Systems*, pages 5998–6008.

Neural Token Representations and Negation and Speculation Scope Detection in Biomedical and General Domain Text

Elena Sergeeva
MIT EECS
elenaser@mit.edu
Amir Tahmasebi
CodaMetrix
amir.tahmasebi@gmail.com

Henghui Zhu
Boston University
henghuiz@bu.edu
Peter Szolovits
MIT EECS
psz@mit.edu

Abstract

Since the introduction of context-aware token representation techniques such as Embeddings from Language Models (ELMo) and Bidirectional Encoder Representations from Transformers (BERT), there have been numerous reports on improved performance on a variety of natural language tasks. Nevertheless, the degree to which the resulting context-aware representations can encode information about morpho-syntactic properties of the tokens in a sentence remains unclear.

In this paper, we investigate the application and impact of state-of-the-art neural token representations for automatic cue-conditional speculation and negation scope detection coupled with the independently computed morpho-syntactic information. Through this work, We establish a new state-of-the-art for the BioScope and NegPar corpora.

Furthermore, we provide a thorough analysis of neural representations and additional features interactions, cue-representation for conditioning, discussing model behavior on different datasets and, finally, address the annotation-induced biases in the learned representations.

1 Introduction

In 2018, a set of new state-of-the-art results were established for a variety of Natural Language Processing tasks, the majority of which can be attributed to the introduction of context aware token representations, learned from large amounts of data with Language-modeling like tasks as a training goal (Devlin et al., 2018; Peters et al., 2018). It is, however, unclear to what degree the computed representations capture and encode high-level morphological/syntactic knowledge about the usage of a given token in a sentence. One way of exploring the potential of the learned represen-

tation would be through investigating the performance on a task that would require the representation to acquire some notion of syntactic units such as phrases and clauses, as well as the relationship between the syntactic units and other tokens in the model. An example of such a task is *Speculation* or *Negation Scope Detection*.

The main contributions of this work can be summarized as follows:

- We achieve and report a new state-of-the-art for the negation and speculation scope detection on several biomedical and general domain datasets, which were created using different definitions of what constitutes a scope of a given negation/speculation.[1]

- We investigate different ways of incorporating additional automatically-generated syntactic features into the model and explore the potential improvements resulting from the addition of such features.

- Following Fancellu et al. (2017), we provide a thorough comparison of our proposed model with other state-of-the-art models and analyze their behaviour in the absence of potential "linear clues", the presence of which might result in highly accurate predictions even for syntax-unaware token representations.

2 The Task

In general, speculation or negation scope detection can be constructed as the following conditional token classification task: given a negation or speculation **cue** (*i.e.*, a word or phrase that expresses negation or speculation such as 'No' and 'May'),

---

[1] An implementation of our model together with the pre-trained models for scope detection will be available later.

Proceedings of the 10th International Workshop on Health Text Mining and Information Analysis (LOUHI 2019), pages 178–187
Hong Kong, November 3, 2019. ©2019 Association for Computational Linguistics
https://doi.org/10.18653/v1/D19-62

identify which tokens are affected by the negation or represent an event that is speculative in nature (referred to as **the scope of the negation or speculation**). Consider the following example:

(1) These findings that (*may be from an acute pneumonia*) include minimal bronchiectasis as well.

In this case, the speculation cue is **"may"** and the string of tokens that contains the speculative information is **"may be from an acute pneumonia"**.

Each data point, as such, is a string of tokens paired with the corresponding negation or speculation cue. Note that nested negations in the same sentence would be distinguished only by the associated cue.

From the syntactic structure point of view, it is clear that in most cases, the boundaries of a given scope strongly correlate with the clausal structure of the sentence (Morante and Sporleder, 2012) There is also a strong connection between the fine-grained part-of-speech (POS) of the cue and the scope boundaries.

Consider the following examples where the type of possible adjectives (either attributive or predicative) results in different scope boundaries (scope highlighted as italic):

(2) This is a patient who had ***possible** pyelonephritis* with elevated fever.

(3) *Atelectasis in the right mid zone is, however,* ***possible***.

Such a property of the task requires a well-performing model to be able to determine cue-types and the corresponding syntactic scope structure from a learned representation of cue-sentence pairs. As such, it can be used as an (albeit imperfect) proxy for assessing the knowledge about the structure of the syntax that a sentence aware token representation potentially learns during training.

2.1 Datasets

There are no universal guidelines on what constitutes a scope of a given negation or speculation; different definitions might affect a given model's performance. To take this ambiguity into account, we report our results on two different datasets: BioScope (Vincze et al., 2008) and NegPar (Liu et al., 2018).

- The BioScope corpus (Vincze et al., 2008) consists of three different types of text: Biological publication abstracts from Genia Corpus (1,273 abstracts), Radiology reports from Cincinnati Children's Hospital Medical Center (1,954 reports), and full scientific articles in the bioinformatics domain (nine articles in total). In this work, we focus on two of the sub-corpora: Abstracts and Clinical reports. One should note that BioScope corpus does not allow discontinuous scopes.

- NegPar (Liu et al., 2018) is a corpus of Conan Doyle stories annotated with negation cues and the corresponding scopes. The corpus is available both in English and Chinese. In this work, we only use the English part of the corpus. Unlike BioScope, NegPar provides a canonical split as training (981 negation instances), development (174 instances) and test sets (263 negation instances). NegPar annotation guidelines allows for discontinuous scopes.

3 Previous Work

Negation scope detection algorithms can be classified into two categories: (1) rule-based approaches that rely on pre-defined rules and grammar; and (2) statistical machine learning approaches that utilize surface level features of the input strings to detect the scope of the negation.

Rule-based approaches Due to the somewhat restricted nature of clinical texts syntax, a pre-defined rule-based key-word triggered negation scope detection system achieves competitive performance on a variety of clinical-notes derived data-sets (Chapman et al., 2001; Harkema et al., 2009; Elkin et al., 2005).

Machine learning approaches While rule-based approaches might achieve high performance on medical institution specific datasets, they do not generalize well for other dataset types and they may require customization of the rules to adapt to the new corpus and/or domain. By contrast, machine learning-based systems do not require active human expert participation to adapt to a new dataset/domain. Earlier works utilizing the statistical approaches for negation scope detection include Support Vector Machines (SVM), Conditional Random Fields based models (CRF) (Agarwal and Yu, 2010; Councill et al., 2010) as well as hybrid CRF-SVM ensemble models (Zhu et al., 2010) (Morante and Daelemans, 2009)

Recently, Neural Network-based approaches have been proposed for such tasks, including Convolutional Neural Network (CNN)-based (Qian et al., 2016) and Long Short Term Memory (LSTM)-based (Fancellu et al., 2017; Sergeeva et al., 2019) models.

The work on specifically speculation scope detection is less varied and mainly confined to CONLL-2010 Shared-Task2 submissions (Farkas et al., 2010). It is, however, important to note that due to the similarity in the formulation of the task, the majority of the negation-specific machine learning approaches can be directly applied to the speculation scope detection problem provided the speculation annotated data is available for training.

We also draw inspiration from a large body of work (Linzen et al., 2016; Gulordava et al., 2018; Marvin and Linzen, 2018) examining the nature of modern context aware representations from a linguistic perspective.

4   Model Training and Evaluation

4.1   Neural Token Representation

The use of pre-trained continuous word representations has been ubiquitous in modern statistical natural language processing. The importance of an appropriate word-level representation is especially noticeable in per-token prediction tasks: in such a set-up the model goal is to fine-tune or modify the existing input token representation in such a way that it contains the necessary information to make a correct classification decision at prediction time.

In this work, we consider the following approaches for generating the input token representation:

- **Global Vectors (GloVe)** (Pennington et al., 2014): A pre-trained token representation that relies on the direct matching of tokens and the corresponding ratios of token co-occurrences with their neighbours. Note that the definition of the neighbour in this setup is static (that is, the ultimate representation would incorporate an averaged notion of context) and relies on the bag-of-words representation of the context.

- **Embeddings from Language Models (ELMo)** (Peters et al., 2018): A bidirectional LSTM model-based token representation, pre-trained on the language modeling task. Instead of modeling the bag-of-words neighborhood co-occurrence probabilities directly, this model approximates the conditional probability of a token given the ordered linear context of the token usage.

- **Bidirectional Encoder Representations from Transformers (BERT)** (Devlin et al., 2018): A transformer-based token representation trained on the modified language modeling task together with a broader context next sentence prediction task. In this model, the context of a token is continuously incorporated into the representation of the token itself as a weighted sum of the neighboring token representations through the use of the multi-head-attention mechanism. The linear order of the token information is provided at input time as an additional positional embedding, since the unmodified transformer architecture does not encode any notion of the linear order.

Despite the performance gains achieved by the widespread use of contextual word embeddings like ELMo and BERT, the questions about the nature of the learned representation remain unanswered. Both ELMo and BERT were introduced to incorporate the wider structure of the given input into individual token representation at the time of training; however, both models only have access to the linear order of the context.

The question then arises: To what degree does the word embedding trained on a language modeling like task and computed using the whole linear context of a sentence encode the broader syntax-related characteristics of a token used within a context?

In order to gain insight into the nature of the learned representations and their potential use for negation and speculation scope detection, we introduce the following syntax-informed features to be used together with the token embedding:

POS : Part-Of-Speech of a given token as defined by the Penn Treebank tagging scheme (Marcus et al., 1993).

DEP : Type of dependency between the token and its parent, representing limited dependency tree information of a given token.

PATH : A string of Google Universal POS tags (Petrov et al., 2012) of the three direct ancestors of the token in the dependency tree; this feature captures local constituent-like information of a given token.

LPATH : Depth of the token in the syntactic dependency tree.

CP : The distance between a given token and the negation cue in the constituency parse tree generated using (Kitaev and Klein, 2018). If a negation cue has multiple tokens, the minimum of the distances is used.

Note that all features were automatically generated, and as a result, represent a "noisy" source of information about the syntactic characteristic of a token. If adding syntactic features as additional inputs would not affect or would significantly degrade the model's performance, it is reasonable to assume that the information represented by such features is already present in the token representation in some way.

4.2 Modes of Evaluation

To provide a fair comparison of different types of embeddings, we introduce two different modes of evaluation. The first mode (referred to as **Feature-based embeddings** later in the paper) is designed to test the embeddings in the same setup as previously used to get the state-of-the-art performance on the dataset. The second mode (referred to as **BERT fine-tuning** later in the paper) is designed to test BERT embeddings in their native direct fine-tuning setting.

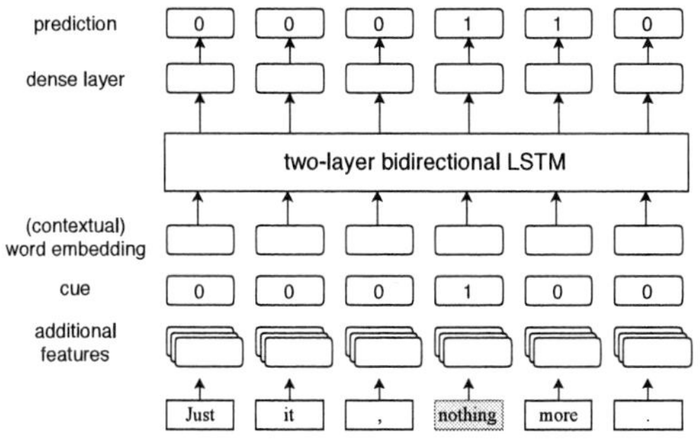

Figure 1: A diagram of the proposed bi-directional LSTM model for negation and speculation detection with additional features.

Feature-based Embeddings using Bi-directional LSTM: Figure 1 demonstrates the proposed framework for the desired task. One should note that the factor that differentiates the two experiments from one another is the embeddings. The task specific layers (two-layer Bi-directional LSTM) remains the same across all experiments. To properly condition each scope on a given cue, we concatenate a specific cue embedding to the input embedding, before computing the final representation for each token. Additional syntactic information is also provided by concatenating the input embedding with all of the syntactic feature embeddings.

BERT Fine-tuning: The original setup for the use of BERT embedding does not require an elaborate task-specific layer; the task specific model is a copy of the original transformer-based BERT architecture with the corresponding pre-trained model-parameters, and the top prediction layer swapped for a new task specific layer that predicts the probability of a given label for a token representation. Crucially, the token representation is allowed to change during the fine-tuning. For this particular setup, it is unclear how to account for the conditional nature of the scope prediction task. In other words, a sentence can potentially contain more than one negation/piece of speculative information.

We consider two different testing scenarios to evaluate the different ways of providing the cue information to the model:

1. Providing the embedded cue at the top layer of the model by concatenating it to the learned token embedding.

2. Providing the embedded cue at the bottom as a part of the input to the transformer layer before the fine-tuning by adding the cue embeddings (initialized randomly at the fine-tuning stage) to the initial token representation.

To test if the additional syntactic information provides any additional benefit to our framework, we also add the mean of all of the syntactic feature embeddings to the initial pre-transformer representation of the input.

4.3 Hyperparameter Settings

Feature-based Embeddings For the aforementioned set of experiments, the following architecture parameters have been considered:

Table 1: Performance of the negation scope detection task on BioScope and NegPar corpora using different approaches. Results are reported as the percentage of number of predicted scopes that exactly match the golden scope (PCS)

Model	BioS_Abstracts	BioS_Clinical	NegPar	NegPar(CV)
Fancellu et al. (2017)	81.38%	94.21%	68.93 % [a]	N/A
Fancellu et al. (2018)	N/A	N/A	61.98% [b]	N/A
Bi-LSTM$_{GloVe}$	63.24%(1.80%)	90.46%(3.64%)	51.48%(4.45%)	49.18%(4.97%)
Bi-LSTM$_{ELMo}$	81.62%(1.87%)	93.10%(2.18%)	71.52%(1.98%)	75.29%(3.35%)
Bi-LSTM$_{BERT}$	79.29%(3.06%)	91.26%(2.82%)	66.78%(3.50%)	69.45%(3.55%)
Bi-LSTM$_{GloVe}$ + AF	79.00%(2.07%)	94.02%(1.98%)	69.70%(2.81%)	73.11%(3.19%)
Bi-LSTM$_{ELMo}$ + AF	83.30%(3.16%)	**94.25%**(2.86%)	69.96%(2.12%)	75.43%(4.82%)
Bi-LSTM$_{BERT}$ + AF	80.68%(3.23%)	93.10%(2.77%)	67.42%(2.10%)	73.39%(4.12%)
BERT (c-top)	74.63%(3.23%)	92.87%(2.04%)	63.14% (2.08%)	—
BERT (c-bottom)	86.97%(2.24%)	93.68%(2.37%)	76.78%(2.04%)	81.91%(3.04%)
BERT (c-bottom) + AF	**87.03%**(2.38%)	93.45%(1.63%)	**79.00%**(1.37%)	80.64%(2.57%)

The number in the parenthesis indicates the standard deviation of the score.

[a] These results are generated using an older version of the corpus annotation.

[b] Since this work is aimed at cross-lingual negation detection, the reported results are based on using cross-language word embeddings, which are likely to degrade a single-language model performance.

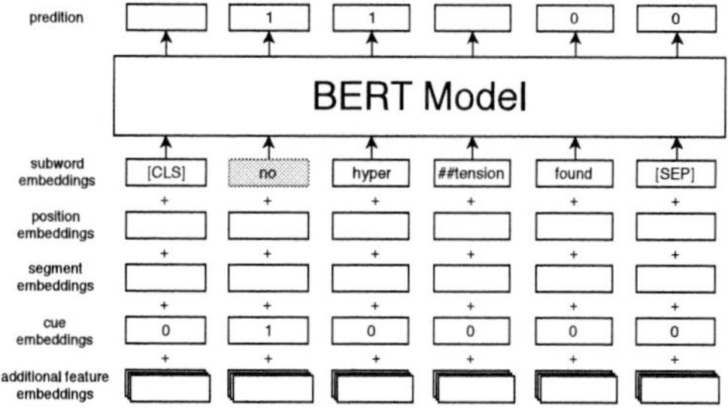

Figure 2: A diagram of the proposed BERT-based architecture for negation and speculation scope detection with inclusion of additional features.

- Word embedding dimension: GloVe: 300; ELMo, BERT: 1024

- Syntactic feature embedding dimension: 10 per feature

- Task-specific LSTM embedding dimension: 400

During training, a dropout rate of 0.5 (Gal and Ghahramani, 2016) was used to prevent overfitting. The Adam optimizer (Kingma and Ba, 2014) was used with step size of 10^{-3} and batch size of 32 for 50 epochs for BioScope and 200 epochs for NegPar. The reason why we use different epochs is that there are fewer training examples for NegPar than BioScope. Therefore, it takes more epochs for the NegPar models to converge.

BERT Fine-tuning The BERT models have the following architecture parameters:

- Word embedding dimension: 1024

- BERT$_{LARGE}$ layer transformer configuration (Devlin et al., 2018)

- Syntactic features embedding dimension: 1024 for each feature

- Cue embedding dimensions: 1024

We perform fine tuning on the negation/speculation task for 20 epochs. The Adam optimizer was used with learning rate of 10^{-5} and batch size of 2 for 10 epochs for the BioScope corpus and 50 epochs for the NegPar corpus.

4.4 Evaluation Procedure

We report our results in terms of the percentage of number of predicted scopes that exactly match the golden scopes (PCS). Since pre-trained BERT models use their own tokenization algorithm, it results in inconsistent final number of tokens in the dataset across evaluation modes. As a result, other traditional evaluation metrics such as precision, recall and F1 are inappropriate to be used in this study as they depend on the number of tokens.

Since the BioScope dataset does not have a canonical training/development/test set split, we report 10-fold cross-validation results together with the standard deviation of the resulting scores.

For the NegPar dataset, we report the result on the test set as well as 10-fold cross-validation results. To overcome the possible random initialization influences on the results, we report the average score for 10 random seeds on the test set together with the associated standard deviation.

5 Results

The performance of different approaches on Bio-Scope and NegPar corpora for the negation scope detection and the speculating scope detection are shown in Table 1 and Table 2, respectively. Bi-LSTM-marked entries of the table correspond to Feature-based and BERT-marked entries correspond to BERT fine-tuning approaches.

5.1 Feature-based Approach

Effect of embedding on performance: Except for the negation scope detection task on Bio-Scope clinical notes, ELMo embeddings significantly outperformed GloVe embeddings as well as the feature-based use of BERT embeddings, but not the fined-tuned version of BERT. While the former is expected, the latter is noteworthy: for NER task (Devlin et al., 2018), for example, the difference in performance between the fine-tuning and feature-based approach results is 1.5% of the F1 score. For negation scope detection the difference is a striking 7.68% on BioScope-abstracts and 10% on a test set of the NegPar dataset. For speculation scope detection the difference remains as large (7.93%). We theorize that this differences comes from the different syntactic nature of the target strings of tokens: NER systems are concerned with finding named entities in text, where the majority of the named entities are represented by relatively short (token-wise) noun phrases, negation/speculation scope detection requires recognition of a much more diverse set of syntactic phenomena. This suggests an important difference between the featurized and fine-tuned approaches for highly syntax-dependent token classification tasks.

Syntactic features induced gains: In general, we observe consistent small gains in performance for all types of embedding on BioScope (both speculation and negation detection modalities) but not on the NegPar dataset. The only exception to this pattern is in non-context aware GloVe embeddings. Adding syntactic features embeddings has inconsistent effects on standard deviations over modalities and datasets.

5.2 BERT fine-tuning approach

Cue-conditioning influence on the results The way to condition a given instance on a particular cue greatly influences the model performance: providing cue information at the top layer of the model results in poor performance of the model for all datasets and both negation and speculation modalities.

Syntactic features induced gains and the importance of Cross Validation evaluation: Adding features to the best performing BERT fine-tuned models does not result in any significant differences on the BioScope dataset. We observe a significant gain in performance on NegPar: note that in this case the gain is purely train/test set split induced and disappears entirely in a cross-validation mode of evaluation.

Artificial noise and the model performance: Even though the experimental results suggest no to minimal contribution of the additional features to the best model performance, natural questions to ask are: "Does the feature enriched model rely on the provided features during the prediction phase?" and "Do the final learned representations differs significantly for feature-enriched and featureless inputs?" We introduce noise into the trained model inputs to check if artificial noise undermines its performance. In particular, we consider the model BERT(cue-bottom) + AF, as it provides the best performance out of all feature-enriched models.

With a given probability, which we call the noise level, we replace a given feature value with a random value: for categorical features (POS, DEP, PATH), we replace it with a random category, and for numerical features (LPATH,CP), we replace it with a random integer drawn from a uniform distribution bounded by the feature's possible minimum and maximum values. We observe a consistent and significant decrease in performance as the probability of seeing the incorrect features increases (see Figure 3). This suggests that the additional features introduced in this paper play an important role in decision making. This is supported by the fact that the performance on clini-

Table 2: Performance of speculation scope detection task on BioScope corpus using different approaches. Results are reported as the percentage of number of the predicted scopes that exactly match the golden scope (PCS).

Model	BioScope_Abstracts	BioScope_Clinical
Qian et al. (2016)	85.75%	73.92%
Bi-LSTM$_{\text{GloVe}}$	47.99%(4.07%)	46.90%(2.87%)
Bi-LSTM$_{\text{ELMo}}$	84.62%(2.33%)	81.82%(2.74%)
Bi-LSTM$_{\text{BERT}}$	81.35%(1.95%)	78.75%(3.24%)
Bi-LSTM$_{\text{GloVe}}$ + AF	85.07%(2.66%)	80.73%(3.01%)
Bi-LSTM$_{\text{ELMo}}$ + AF	86.57%(2.65%)	81.55%(2.74%)
Bi-LSTM$_{\text{BERT}}$ + AF	84.43%(1.08%)	81.37%(4.32%)
BERT (cue-top)	57.32%(2.14%)	60.49%(4.77%)
BERT (cue-bottom)	**89.28%(1.65%)**	**83.71%(2.77%)**
BERT (cue-bottom) + AF	88.91%(1.65%)	82.36%(4.27%)

The number within the parentheses indicates the standard deviation of the score.

cal reports negation detection remains almost unaffected by the change, since the majority of the negation scopes in this dataset can be captured by structure-independent heuristics.

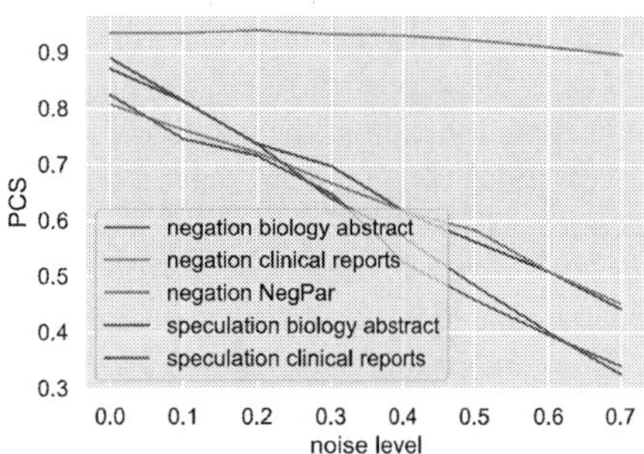

Figure 3: Performance of the BERT + AF models with respect to the noise level, averaged for 10 fold CV

5.3 Linear punctuation cues and model performance

Even though the scope boundaries correlate to the syntactic structures of the sentence, a good performance on a given dataset does not necessarily prove the model acquired any kind of a structural knowledge: as was noted in Fancellu et al. (2017), the majority of scopes in the BioScope corpus consist of cases where the punctuation boundaries match the scope boundaries directly. For those cases, the model does not have to learn any kinds of underlying syntactic phenomena: learning a simple heuristic to mark everything between a cue and the next punctuation mark as a scope would produce an illusion of a more complex syntax-informed performance.

To see if our model's performance is significantly affected by the punctuation clues, we remove all the punctuation from the training corpus, re-train all the models on the modified dataset and evaluate the learned models on the test set. We also report the performance on "hard" (non-punctuation bound) instances of scopes separately.

As can be seen in Table 3, removing punctuation affects all models' behaviour similarly: model performance degrades by losing 2-3 percent of PCS on average. Interestingly, the performance on the non-punctuation boundaries scopes declines similarly, which suggest that punctuation plays an important role in computing a given token representation, and not only as a direct linear cue that signifies the scope's start and end.

5.4 Error overlap

Given the difference in the model architectures, a natural question to ask is: "Is the best performing model strictly better than the others, or do they make different types of errors?" We compute the error overlap between BERT and ELMo on the negation detection task as shown in Figure 4. About half of ELMo and slightly more than a quarter of BERT errors appear to be model specific, suggesting the potential for ensemble-induced improvements.

We also compute the error overlap for the Neg-Par test set performance for the top 3 performing models: almost half of the ELMo errors and about 3/4 of BERT fine-tuned and BERT fine-tuned with

Table 3: Performance on percentage of correct span on BioScope Abstracts sub-corpus trained under different schemes.

	Trained w/ punctuation		Trained w/o punctuation	
	all	hard cases	all	hard cases
Bi-LSTM$_{\text{GloVe}}$	63.24%(1.80%)	51.83%(3.47%)	57.19%(2.48%)	48.82%(3.36%)
Bi-LSTM$_{\text{ELMo}}$	81.62%(1.87%)	73.07%(3.60%)	76.90%(2.72%)	69.88%(3.78%)
Bi-LSTM$_{\text{BERT}}$	79.29%(3.06%)	70.37%(6.60%)	77.54%(3.18%)	70.28%(5.41%)
Bi-LSTM$_{\text{GloVe}}$ + AF	79.00%(2.07%)	68.79%(4.06%)	76.21%(1.76%)	67.96%(2.64%)
Bi-LSTM$_{\text{ELMo}}$ + AF	83.30%(3.16%)	75.56%(4.46%)	82.31%(2.95%)	76.58%(3.67%)
Bi-LSTM$_{\text{BERT}}$ + AF	80.68%(3.23%)	72.68%(6.71%)	80.45%(3.24%)	73.19%(5.52%)
BERT (c-bottom)	86.97%(2.24%)	**82.51%(3.78%)**	83.48%(3.22%)	**79.42%(4.46%)**
BERT (c-bottom +AF)	**87.03%(2.38%)**	82.38%(4.48%)	**84.58%(3.58%)**	79.27%(5.82%)

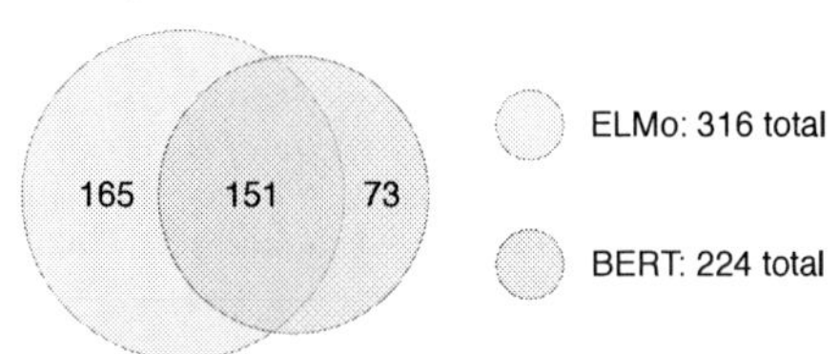

Figure 4: Distribution of error overlaps: BERT vs. ELMo on BioScope Abstracts dataset.

features are common for all of the models. It is interesting to note that the the errors of BERT without the features are not a subset of BERT with the features, suggesting the possibility of a performance trade-off instead of a straight feature-induced performance improvement.

Qualitatively, on average ELMo tends to prefer longer scopes, sometimes extending the scope for an additional clause. Both models have trouble with common words that can be encountered in a variety of different contexts, such as certain prepositions and personal pronouns.

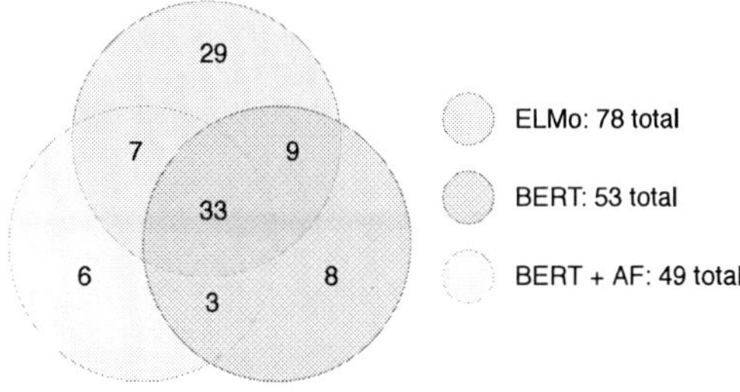

Figure 5: Distribution of error overlaps: BERT vs. BERT with features (BERT + AF) vs. ELMo on NegPar test set.

6 Conclusions and Future Work

This work presents a comparison among different context-aware neural token representations and the corresponding performance on the negation and speculation scope detection tasks. Furthermore, we introduce a new state-of-the-art BERT-based cue-conditioned feature-enriched framework for negation/speculation scope detection. Based on the empirical results, we are inclined to recommend BERT fine-tuning over using a feature-based approach with BERT for syntax-dependent tasks.

We used two commonly used publicly available datasets, BioScope and NegPar for our evaluation. Despite the observed gains on the test set of the NegPar corpus, the effect of the syntactic features on BERT (fine-tuned) performance remains largely inconclusive.

It is also important to note that the syntactic information we have been trying to incorporate into the model was generated automatically; one of the possible avenues of research would be comparing the possible golden annotation induced gains with the imperfect information gain we observe when incorporating silver syntactic features.

We were unable to find any consistent grammatical explanation for the errors context-aware models result in on the test data; however, this does not conclusively mean that such an explanation does not exist. An appropriate next step would be annotating a smaller set of sentences, grouped by the corresponding syntactic construction and see if a given token representation yields improved performance on such a construction.

Acknowledgments

Research partially supported by the Office of Naval Research under MURI grant N00014-16-1-2832

References

Shashank Agarwal and Hong Yu. 2010. Biomedical negation scope detection with conditional random fields. *Journal of the American Medical Informatics Association*.

Wendy W. Chapman, Will Bridewell, Paul Hanbury, Gregory F. Cooper, and Bruce G. Buchanan. 2001. A Simple Algorithm for Identifying Negated Findings and Diseases in Discharge Summaries. *Journal of Biomedical Informatics*, 34(5):301–310.

Isaac Councill, Ryan McDonald, and Leonid Velikovich. 2010. What's great and what's not: learning to classify the scope of negation for improved sentiment analysis. In *Proceedings of the Workshop on Negation and Speculation in Natural Language Processing*, pages 51–59. University of Antwerp.

Jacob Devlin, Ming-Wei Chang, Kenton Lee, and Kristina Toutanova. 2018. BERT: pre-training of deep bidirectional transformers for language understanding. *CoRR*, abs/1810.04805.

Peter L. Elkin, Steven H. Brown, Brent A. Bauer, Casey S. Husser, William Carruth, Larry R. Bergstrom, and Dietlind L. Wahner-Roedler. 2005. A controlled trial of automated classification of negation from clinical notes. *BMC Medical Informatics and Decision Making*, 5(1):13.

Federico Fancellu, Adam Lopez, Bonnie Webber, and Hangfeng He. 2017. Detecting negation scope is easy, except when it isn't. In *Proceedings of the 15th Conference of the European Chapter of the Association for Computational Linguistics: Volume 2, Short Papers*, pages 58–63, Stroudsburg, PA, USA. Association for Computational Linguistics.

Federico Fancellu, Adam Lopez, and Bonnie L. Webber. 2018. Neural networks for cross-lingual negation scope detection. *CoRR*, abs/1810.02156.

Richárd Farkas, Veronika Vincze, György Móra, János Csirik, and György Szarvas. 2010. The CoNLL-2010 Shared Task: Learning to Detect Hedges and their Scope in Natural Language Tex. In *Proceedings of the Fourteenth Conference on Computational Natural Language Learning – Shared Task*, pages 1–12. Association for Computational Linguistics.

Yarin Gal and Zoubin Ghahramani. 2016. A theoretically grounded application of dropout in recurrent neural networks. In *Advances in Neural Information Processing Systems 29*, pages 1019–1027. Curran Associates, Inc.

Kristina Gulordava, Piotr Bojanowski, Edouard Grave, Tal Linzen, and Marco Baroni. 2018. Colorless green recurrent networks dream hierarchically. In *Proceedings of the 2018 Conference of the North American Chapter of the Association for Computational Linguistics: Human Language Technologies, Volume 1 (Long Papers)*, pages 1195–1205. Association for Computational Linguistics.

Henk Harkema, John N. Dowling, Tyler Thornblade, and Wendy W. Chapman. 2009. ConText: An algorithm for determining negation, experiencer, and temporal status from clinical reports. *Journal of Biomedical Informatics*, 42(5):839–851.

Diederik P. Kingma and Jimmy Ba. 2014. Adam: A method for stochastic optimization. *CoRR*, abs/1412.6980.

Nikita Kitaev and Dan Klein. 2018. Constituency parsing with a self-attentive encoder. In *Proceedings of the 56th Annual Meeting of the Association for Computational Linguistics (Volume 1: Long Papers)*, pages 2676–2686. Association for Computational Linguistics.

Tal Linzen, Emmanuel Dupoux, and Yoav Goldberg. 2016. Assessing the ability of lstms to learn syntax-sensitive dependencies. *Transactions of the Association for Computational Linguistics*, 4:521–535.

Qianchu Liu, Federico Fancellu, and Bonnie Webber. 2018. NegPar: a parallel corpus annotated for negation. In *Proceedings of the Eleventh International Conference on Language Resources and Evaluation (LREC-2018)*.

Mitchell P Marcus, Mary Ann Marcinkiewicz, and Beatrice Santorini. 1993. Building a large annotated corpus of english: The penn treebank. *Computational linguistics*, 19(2):313–330.

Rebecca Marvin and Tal Linzen. 2018. Targeted syntactic evaluation of language models. In *Proceedings of the 2018 Conference on Empirical Methods in Natural Language Processing*, pages 1192–1202. Association for Computational Linguistics.

Roser Morante and Walter Daelemans. 2009. A metalearning approach to processing the scope of negation. In *Proceedings of the Thirteenth Conference on Computational Natural Language Learning (CoNLL-2009)*, pages 21–29. Association for Computational Linguistics.

Roser Morante and Caroline Sporleder. 2012. Modality and negation: An introduction to the special issue. *Computational linguistics*, 38(2):223–260.

Jeffrey Pennington, Richard Socher, and Christopher Manning. 2014. Glove: Global Vectors for Word Representation. In *Proceedings of the 2014 Conference on Empirical Methods in Natural Language Processing (EMNLP)*, pages 1532–1543, Stroudsburg, PA, USA. Association for Computational Linguistics.

Matthew Peters, Mark Neumann, Mohit Iyyer, Matt Gardner, Christopher Clark, Kenton Lee, and Luke Zettlemoyer. 2018. Deep contextualized word representations. In *Proceedings of the 2018 Conference of the North American Chapter of the Association for Computational Linguistics: Human Language Technologies, Volume 1 (Long Papers)*, pages 2227–2237. Association for Computational Linguistics.

Slav Petrov, Dipanjan Das, and Ryan McDonald. 2012. A universal part-of-speech tagset. In *Proceedings of the Eight International Conference on Language Resources and Evaluation (LREC'12)*, Istanbul, Turkey. European Language Resources Association (ELRA).

Zhong Qian, Peifeng Li, Qiaoming Zhu, Guodong Zhou, Zhunchen Luo, and Wei Luo. 2016. Speculation and Negation Scope Detection via Convolutional Neural Networks. In *Proceedings of the 2016 Conference on Empirical Methods in Natural Language Processing*, pages 815–825, Stroudsburg, PA, USA. Association for Computational Linguistics.

Elena Sergeeva, Henghui Zhu, Peter Prinsen, and Tahmasebi Amir. 2019. Negation scope detection in clinical notes and scientific abstracts: A feature-enriched lstm-based approach. *AMIA Jt Summits Transl Sci Proc. 2019*, pages 212–221.

Veronika Vincze, György Szarvas, Richárd Farkas, György Móra, and János Csirik. 2008. The BioScope corpus: biomedical texts annotated for uncertainty, negation and their scopes. *BMC Bioinformatics*, 9(S11):S9.

Qiaoming Zhu, Junhui Li, Hongling Wang, and Guodong Zhou. 2010. A unified framework for scope learning via simplified shallow semantic parsing. In *Proceedings of the 2010 Conference on Empirical Methods in Natural Language Processing*, pages 714–724. Association for Computational Linguistics.

Author Index